MULTIRATE DIGITAL
SIGNAL PROCESSING

PRENTICE-HALL SIGNAL PROCESSING SERIES

Alan V. Oppenheim, Editor

MULTIRATE DIGITAL
SIGNAL PROCESSING

RONALD E. CROCHIERE

LAWRENCE R. RABINER

Acoustics Research Department
Bell Laboratories
Murray Hill, New Jersey

Prentice-Hall, Inc., Upper Saddle River, New Jersey 07458

Library of Congress Cataloging in Publication Data

Crochiere, Ronald E.
    Multirate digital signal processing.

    (Prentice-Hall signal processing series)
    Includes index.
    1. Signal processing—Digital techniques.
I. Rabiner, Lawrence R.   II. Title.
III. Series.
TK5102.5.C76   1983   621.38′043   82-23113
ISBN 0-13-605162-6

Text processing: *Donna Manganelli*
Editorial/production and supervision by *Barbara Cassel* and *Mary Carnis*
Cover design: *Mario Piazza*
Manufacturing buyer: *Anthony Caruso*

Printed in the United States of America

10 9 8 7

ISBN 0-13-605162-6

Prentice-Hall International (UK) Limited,London
Prentice-Hall of Australia Pty. Limited, Sydney
Prentice-Hall Canada Inc., Toronto
Prentice-Hall Hispanoamericana, S.A., Mexico
Prentice-Hall of India Private Limited, New Delhi
Prentice-Hall of Japan, Inc., Tokyo
Pearson Education Asia Pte. Ltd., Singapore
Editora Prentice-Hall do Brasil, Ltda., Rio de Janeiro

To our families,
Gail and *Ryan*
and
Suzanne, Sheri, Wendi, and *Joni*
for their love, encouragement, and support.

Contents

Preface

The idea for a monograph on multirate digital signal processing arose out of an IEEE Workshop held at Arden House in 1975. At that time it became clear that the field of multirate signal processing was a burgeoning area of digital signal processing with a specialized set of problems. Several schools of thought had arisen as to how to solve the problems of multirate systems, and a wide variety of applications were being discussed. At that time, however, the theoretical basis for multirate processing was just emerging. Thus no serious thought was given to writing a monograph in this area at that time.

A steady stream of theoretical and practical papers on multirate signal processing followed the Arden House Workshop and by 1979 it became clear to the authors that the time was ripe for a careful and thorough exposition of the theory and implementation of multirate digital systems. Our initial estimate was that a moderate size monograph would be adequate to explain the basic theory and to illustrate principles of implementation and application of the theory. However once the writing started it became clear that a full and rich field had developed, and that a full sized text was required to do justice to all aspects of the field. Thus the current text has emerged.

The area of multirate digital signal processing is basically concerned with problems in which more than one sampling rate is required in a digital system. It is an especially important part of modern (digital) telecommunications theory in which digital transmission systems are required to handle data at several rates (e.g. teletype, Fascimile, low bit-rate speech, video etc.). The main problem is the design of an efficient system for either raising or lowering the sampling rate of a signal by an arbitrary factor. The process of lowering the sampling rate of a signal has been called decimation; similarly the process of raising the sampling rate of a signal has been called interpolation. Both processes have been studied for a long time, especially by numerical analysts who were interested in efficient methods for tabulating functions and providing accurate procedures for interpolating the table entries. Although the numerical analysts learned a great deal about temporal aspects of multirate systems it was not until recently that more modern techniques were applied that enabled people to understand both temporal and spectral aspects of these systems.

The goal of this book is to provide a theoretical exposition of all aspects of multirate digital signal processing. The basis for the theory is the Nyquist sampling theorem and the general theory of lowpass and bandpass sampling. Thus in Chapter 2 we present a general discussion of the sampling theorem and show how it leads to a straightforward digital system for changing the sampling rate of a signal. The canonic form of the system is a realization of a linear, periodically time-varying digital system. However for most cases of interest considerable simplifications in the form of the digital system are realized. Thus in the special cases of integer reductions in sampling rate, integer increases in sampling rate, and rational fraction changes in sampling rate, significant reductions in complexity of the digital system are achieved. In such cases the digital system basically involves filtering and combinations of subsampling (taking 1 of M samples and throwing away the rest) and sampling rate expansion (inserting $L-1$ zero valued samples between each current sample).

It is also shown in Chapter 2 how to efficiently change the sampling rate of bandpass (rather than lowpass) signals. Several modulation techniques are discussed including bandpass translation (via quadrature or single-sideband techniques) and integer-band sampling.

In Chapter 3 we show how to apply standard digital network theory concepts to the structures for multirate signal processing. After a brief review of some simple signal-flow graph principles, we show in what ways these same general principles can (or often cannot) be used in multirate structures. We then review the canonic structures for linear time-invariant filters, such as the FIR direct form and IIR cascade form, and show how efficient multirate structures can be realized by combining the signal-flow graph principles with the filtering implementations. We also introduce, in this chapter, the ideas of polyphase structures in which subsampled versions of the filter impulse response are combined in a parallel structure with a commutator switch at either the input or output. We conclude this chapter with a discussion of advanced network concepts which provide additional insights and interpretations into the properties of multirate digital systems.

In Chapter 4 we discuss the basic filter design techniques used for linear digital filters in multirate systems. We first describe the canonic lowpass filter characteristics of decimators and interpolators and discuss the characteristics of ideal (but nonrealizable) filters. We then review several practical lowpass filter design procedures including the window method and the equiripple design method. It is shown that in some special cases the lowpass filter becomes a multistopband filter with "don't care" bands between each stopband. In such cases improved efficiency in filter designs can be achieved.

Two special classes of digital filters for multirate systems are also introduced in Chapter 4, namely half-band designs, and minimum mean-square error designs. Such filters provide improved efficiency in some fairly general applications. A brief discussion of IIR filter design techniques is also given in this chapter and the design methods are compared with respect to their efficiency in multirate systems.

In Chapter 5 we show that by breaking up a multirate system so that the processing occurs in a series of stages, increased efficiency results when large changes in sampling rate occur within the system. This increased efficiency results primarily from the relaxed filter design requirements in all stages. It is shown that there are three possible ways of realizing a multistage structure, namely by using a small number of stages (typically 2 or 3) and optimizing the efficiency for each stage; by using a series of power of 2 stages with a generalized final stage; or by using one of several predesigned comb filter stages (systems) and choosing which one is most appropriate at each stage of the processing. The advantages and disadvantages of each of these methods is discussed in this chapter.

The last two chapters of the book illustrate how the techniques discussed in the previous chapters can be applied to basic signal processing operations (Chapter 6) and filter bank structures (Chapter 7). In particular, in Chapter 6 we show how to efficiently implement multirate, multistage structures for lowpass and bandpass filters, for fractional sample phase shifters and Hilbert transformers, and for narrow band, high resolution spectrum analyzers. In Chapter 7 we first define the basic properties of uniformly spaced filter banks and show how combinations of polyphase and DFT structures can be used to provide efficient implementations of these filter banks. An alternative structure, the weighted overlap-add method, is introduced next and it is shown to yield efficiencies comparable to the polyphase implementations. An extensive discussion of design methods for the filter bank is given in this chapter, along with a derivation of how aliasing and imaging can occur in both the time and frequency domains. Specific design rules are given which enable the designer to minimize (and often eliminate) aliasing and imaging by careful selection of filter bank parameters. An important result shown in this chapter is how the well-known methods of fast digital convolution (namely overlap-add and overlap-save) can be derived from the general filter bank analysis-synthesis methodology. The remainder of Chapter 7 gives a thorough discussion of several generalizations of the DFT filter bank structure which allow arbitrary frequency stacking of the filters. Finally the chapter concludes with a brief discussion of some nonuniform filter bank structures. In particular we discuss filter banks based on tree like structures including the well-known quadrature mirror filter banks.

The material in this book is intended as a one-semester advanced graduate course in digital signal processing or as a reference for practicing engineers and researchers. It is anticipated that all students taking such a course would have completed one or two semesters of courses in basic digital signal processing. The material presented here is, as much as possible, self-contained. Each chapter builds up basic concepts so the student can follow the ideas from basic theory to applied designs. Although the chapters are closely tied to each other, the material was written so that the reader can study, almost independently, any of Chapters 3-7 once he has successfully learned the material in Chapters 1 and 2. In this manner the book can also serve as a worthwhile reference text for graduate engineers.

Our goal in writing this book was to make the vast area of multirate digital signal processing theory and application available in a complete, self-contained text. We hope we have succeeded in reaching this goal.

Ronald E. Crochiere
Lawrence R. Rabiner

Acknowledgments

As with any book of this size, a number of individuals have had a significant impact on the material presented. We first would like to acknowledge the contributions of Dr. James L. Flanagan who has served as our supervisor, our colleague, and our friend. His appreciation of the importance of the area of this book along with his encouragement and support while working on the writing have lessened the difficulty of writing this book.

Although a great deal of the material presented in this book came out of original research done at Bell Laboratories by the authors and their colleagues, a number of people have contributed significantly to our understanding of the theory and application of multirate digital signal processing. Perhaps the biggest contribution has come from Dr. Maurice Bellanger who first proposed the polyphase filter structure and showed how it could be used in a wide variety of applications related to telecommunications. Major contributions to the field are also due to Professor Ronald Schafer, Dr. Wolfgang Mecklenbrauker, Dr. Theo Claasen, Professor Hans Schuessler, Dr. Geerd Oetken, Dr. Peter Vary, Dr. David Goodman, Dr. Michael Portnoff, Dr. Jont Allen, Dr. Daniel Esteban, Dr. Claude Galand, Professor David Malah, Professor Vijay Jain, and Professor Thomas Parks. To each of these people and many others we owe a debt of thanks for what they have taught us about this field.

We would like to thank Professor Alan Oppenheim, as series editor for Prentice-Hall books on Digital Signal Processing, for inviting us to write this book and for his encouragement and counsel throughout the period in which the book was written.

Finally, we would like to acknowledge the contributions of Mrs. Donna (Dos Santos) Manganelli who has worked with us since the inception of the book and typed the many revisions through which this manuscript has undergone. Donna's pleasant personality and warmth have made working with her a most rewarding experience.

1

Introduction

1.0 BASIC CONSIDERATIONS

One of the most fundamental concepts of digital signal processing is the idea of sampling a continuous process to provide a set of numbers which, in some sense, is representative of the characteristics of the process being sampled. If we denote a continuous function from the process being sampled as $x_C(t)$, $-\infty < t < \infty$, where x_C is a continuous function of the continuous variable t (t may represent time, space, or any other continuous physical variable), then we can define the set of samples as $x_D(n)$, $-\infty < n < \infty$, where the correspondence between t and n is specified by the sampling process, that is,

$$n = q(t) \qquad (1.1)$$

Many types of sampling have been discussed in the literature [1.1-1.3], including nonuniform sampling, uniform sampling, random sampling, and multiple-function (e.g. sampling a signal and its derivatives) uniform sampling. The most common form of sampling, and the one that we will refer to throughout this book, is uniform (periodic) sampling in which

$$q(t) = \frac{t}{T} = n, \quad n \text{ an integer} \qquad (1.2)$$

That is, the samples $x_D(n)$ are uniformly spaced in the dimension t, occurring T apart. Figure 1.1 illustrates uniform sampling (with period T) of a signal. For uniform sampling we define the sampling period as T and the sampling rate as

1

$$F = \frac{1}{T} \tag{1.3}$$

It should be clear from the discussion above that $x_C(t)$ can be sampled with any sampling period T. However, for a unique correspondence between the continuous function $x_C(t)$ and the discrete sequence $x_D(n)$, it is necessary that the sampling period T be chosen to satisfy the requirements of the Nyquist sampling theorem (discussed more thoroughly in Chapter 2). This concept of a unique analog waveform corresponding to a digital sequence will often be used in the course of our discussion to provide greater intuitive insights into the nature of the processing algorithms that we will be considering.

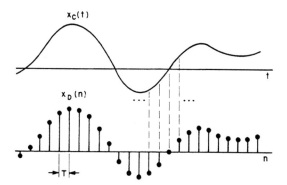

Figure 1.1 Uniform sampling of a signal.

The sampling period T is a fundamental consideration in many signal processing techniques and applications. It often determines the convenience, efficiency, and/or accuracy with which the signal processing can be performed. In some cases an input signal may already be sampled at some predetermined sampling period T and the goal is to convert this sampled signal to a new sampled signal with a different sampling period T' such that the resulting signal corresponds to the same analog function. In other cases it may be more efficient or convenient to perform different parts of a processing algorithm at different sampling rates, in which case it may be necessary to convert the sampling rates of the signals in the system from one rate to another. We refer to such systems as *multirate systems* [1.4].

A question that naturally arises is what is the most effective way to convert a sampled signal to a new sampled signal with a different sampling period. The answer to this fundamental question is the purpose of this book. In the course of this book we will see that this answer involves a variety of concepts in digital signal processing, including the sampling theorem, digital filter design and implementation methods, spectral analysis ideas, and digital hardware methods. We will also show

how sampling rate conversion concepts can be used to form efficient implementations of a variety of digital signal processing operations, such as digital filtering, phase shifting, modulation, and filter banks.

1.1 SAMPLING RATE CONVERSION

The process of digitally converting the sampling rate of a signal from a given rate $F = 1/T$ to a different rate $F' = 1/T'$ is called *sampling rate conversion*. When the new sampling rate is higher than the original sampling rate, that is,

$$F' > F \qquad\qquad (1.4a)$$

or

$$T' < T \qquad\qquad (1.4b)$$

the process is generally called *interpolation* since we are creating samples of the original physical process from a reduced set of samples. Historically, the mathematical process of interpolation, or "reading between the lines," received widespread attention from mathematicians who were interested in the problem of tabulating useful mathematical functions. The question was how often a given function had to be tabulated (sampled) so that someone could use some simple interpolation rule to obtain accurate values of the function at any higher sampling rate [1.5]. Not only did this work lead to an early appreciation of the sampling process, but it also led to several interesting classes of "interpolation functions" which could provide almost arbitrarily high accuracy in the interpolated values, provided that sufficient tabulated values of the function were available.

The process of digitally converting the sampling rate of a signal from a given rate F to a lower rate F', that is,

$$F' < F \qquad\qquad (1.5a)$$

or

$$T' > T \qquad\qquad (1.5b)$$

is called *decimation*.[1] It will be shown in Chapter 3 that decimation and interpolation of signals are dual processes - e.g. a digital system that implements a decimator can be transformed into a dual digital system that implements an interpolator using straightforward transformation techniques.

---

[1] Strictly speaking, decimation means a reduction by 10%. In signal processing, decimation has come to mean a reduction in sampling rate by any factor.

1.2 EXAMPLES OF MULTIRATE DIGITAL SYSTEMS

The techniques described in this book have been applied in a wide variety of areas, of signal processing including:

1. Communications systems [1.6-1.8]
2. Speech and audio processing systems [1.9-1.12]
3. Antenna systems [1.13]
4. Radar systems [1.14, 1.15]

This list contains only a few representative areas where applications of multirate digital systems can be found. In this section we present several specific examples of practical multirate digital systems from some of these areas and briefly outline the principles upon which they work.

1.2.1 Sampling Rate Conversion in Digital Audio Systems

An important practical application of multirate digital signal processing and sampling rate conversion is in the field of professional digital audio [1.12]. A variety of different types of digital processing systems have emerged for storage, transmission, and processing of audio program material. For a number of reasons such systems may have different sampling rates depending on whether they are used for broadcasting, digital storage, consumer products, or other professional applications. Also in digital processing of audio material, signals may be submitted to different types of digital rate control for varying the speed of the program material. This process can inherently vary the sampling frequency of the digital signal [1.12].

In practice it is often desired to convert audio program material from one digital format to another. One way to achieve this format conversion is to convert the audio signal back to analog form and digitize it in the new format. This process inherently introduces noise at each stage of conversion due to the limited dynamic range of the analog circuitry associated with the D/A (digital-to-analog) and A/D (analog-to-digital) conversion process. Furthermore this noise accumulates at each new interface.

An alternative and more attractive approach is to convert directly between the two digital formats by a process of waveform interpolation. This process is depicted in Fig. 1.2 and it is seen to be basically a sampling rate conversion problem. Since the accuracy of this sampling rate conversion can be maintained with any desired degree of precision (by controlling the wordlengths and the interpolator designs), essentially a noise free interconnection between the two systems can be achieved. The techniques discussed in this book apply directly to this application.

1.2.2 Conversion Between Delta Modulation and PCM Signal Coding Formats

A second application of sampling rate conversion is in the area of digital communications. In communication networks a variety of different coding formats may be used in different parts of the network to achieve flexibility and efficiency [1.11]. Conversion between these coding formats often involves a conversion of the basic sampling rate.

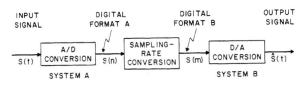

Figure 1.2 Example of a digital-to-digital translation between two audio signal formats.

By way of example, delta modulation (DM) is sometimes used in A/D conversion or in voice terminals because of its simplicity and low cost. DM is basically a one bit per sample coding technique in which only the sign of the sample-to-sample difference of a highly oversampled signal is encoded. This approach eliminates the need for expensive anti-aliasing filters and allows the signal to be manipulated in a simple unframed serial bit stream format.

Alternatively, in long distance transmission or in signal processing operations, such as digital filtering, it is generally desired to have the signal in a PCM (pulse code modulation) format. Thus it is necessary to convert between the high sampling rate, single bit format, of DM and the low sampling rate, multiple bit format, of PCM. Fig. 1.3a shows an example of this process for the DM to PCM conversion and Fig. 1.3b shows the reverse process of converting from a PCM to DM format. When used as a technique for A/D conversion, this approach can combine the advantages of both the DM and PCM signal formats. Again, the techniques involved in this conversion process are based on the multirate signal processing concepts discussed in this book.

1.2.3 Digital Time Division Multiplexing (TDM) to Frequency Division Multiplexing (FDM) Translation

A third example of multirate digital systems is the translation of signals in a telephone system between time division multiplexed (TDM) and frequency division multiplexed (FDM) formats [1.6-1.8]. The FDM format is often used for long distance transmission, whereas the TDM format is more convenient for digital switching.

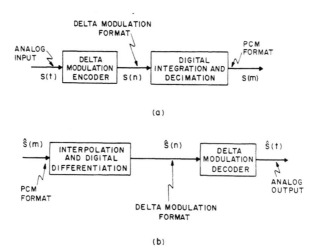

Figure 1.3 Illustration of (a) a delta modulation-to-PCM conversion; (b) a PCM-to-delta modulation conversion

Fig. 1.4a illustrates the basic process of translating a series of 12 TDM digital speech signals, $s_1(n)$, $s_2(n)$,..., $s_{12}(n)$, to a single FDM signal $r(m)$ and Fig. 1.4b illustrates the reverse (FDM-to-TDM) translation process. The sampling rate of the TDM speech signals is 8 kHz whereas the sampling rate of the FDM signal is much higher to accommodate the increased total bandwidth. In each channel of the TDM-to-FDM translator the sampling rate is effectively increased (by interpolation) to the higher FDM sampling rate. The signal is then modulated by single-sideband modulation techniques to its appropriate frequency band location in the range 56 to 112 kHz as illustrated in Fig. 1.5. The interpolated and modulated channel signals are then digitally summed to give the desired FDM signal. In the FDM-to-TDM translator the reverse process takes place.

As seen from Fig. 1.4 the process of translation between TDM and FDM formats involves sampling rate conversion and therefore these systems are inherently multirate systems. In practice it is often convenient to integrate the signal processing operations of modulation and interpolation (or decimation) and share the processing that is common to all of the channels in a way that reduces the computation, in the overall translator, substantially. Thus the block diagrams in Fig. 1.4 only serve as conceptual models of the system whereas the actual computational structures for efficient implementation of these systems may take on considerably different forms. The methods for design of efficient multirate signal processing structures are also covered in this book.

1.2.4 Sub-Band Coding of Speech Signals

A fourth example of a practical multirate digital system is that of sub-band coding [1.10]. Sub-band coding is a technique that is used to efficiently encode speech

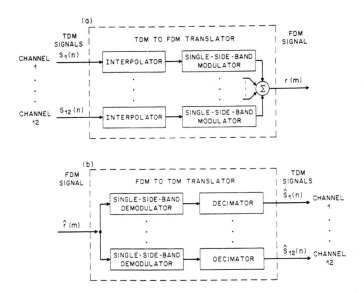

Figure 1.4 Illustration of (a) a TDM-to-FDM translator; and (b) an FDM-to-TDM translator.

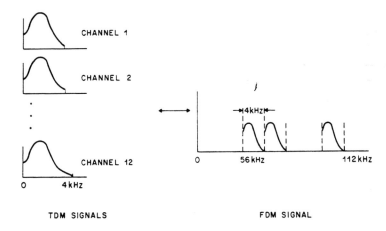

Figure 1.5 Spectral interpretation of TDM-to-FDM signal translation.

signals at low bit-rates by taking advantage of the time varying properties of the speech spectrum as well as some well known properties of speech perception. It is based on the principle of decomposing the speech signal into a set of sub-band signals and separately encoding each of these signals with an adaptive PCM quantizer. By carefully selecting the number of bits/sample used to quantize each sub-band, according to perceptual criteria, an efficient encoding of the speech can be achieved.

Fig. 1.6a shows a block diagram of a sub-band coder-decoder, and Fig. 1.6b shows a typical 5-band filter bank arrangement for sub-band coding. A key element in this design is the implementation of the filter bank analysis and synthesis systems. Each sub-band in the filter bank analyzer is effectively obtained by a process of bandpass filtering, modulation to zero frequency, and decimation in a manner similar to that of the FDM-to-TDM translator described in Section 1.2.3.

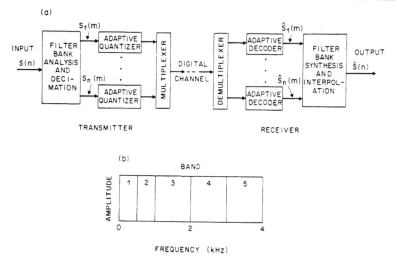

Figure 1.6 (a) Illustration of an N-band sub-band coder; and (b) an example of the frequency bands for a 5-band sub-band coder design.

In the receiver the reverse process takes place with the filter bank synthesizer reconstructing an output signal from the decoded sub-band signals. This process is similar to that of the TDM-to-FDM translator. Again, the concepts and the computational techniques for efficiently designing and implementing filter bank analyzers and synthesizers of this type are covered in this book.

1.2.5 Short-Time Spectral Analysis and Synthesis

A final example of multirate signal processing systems is a short-time spectral analysis and synthesis system [1.16]. Such systems are widely used in the areas of speech processing, antenna systems, and radar systems. Fig. 1.7 gives a simplified explanation of a spectrum analysis-synthesis system. The signal $s(n)$ is periodically windowed with a sliding window (shown in Fig. 1.7b) to form a short-time, finite duration, piece of the signal at each time slot. This short-time signal is then transformed with a fast DFT (discrete Fourier transform) algorithm to form a short-time spectral estimate of the signal for that time slot. Based on the spectral information, the signal may be modified in a variety of ways (or the short-time spectrum may be an end result in itself). The modified short-time spectrum is then inverse transformed to form a modified short-time segment of the signal. Finally, a reconstructed, modified signal $\hat{s}(n)$ is obtained by appropriately overlapping and summing the modified signal segments from each time slot.

In Chapter 7, the above short-time spectral analysis/synthesis framework is discussed in considerable detail in terms of the multirate concepts established earlier in the book. Efficient computation structures are discussed and the relationship of this framework to that of the filter bank framework is carefully presented.

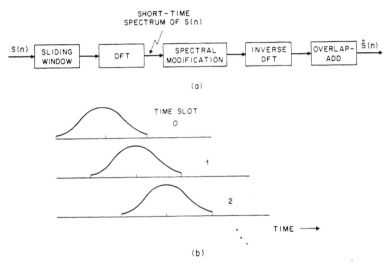

Figure 1.7 (a) Example of a short-time spectral analysis/synthesis system with spectral modification; and (b) an illustration of the sliding window framework of this process.

1.3 SCOPE OF THE BOOK

From a digital signal processing point of view, both the process of interpolation and decimation can be well formulated in terms of linear filtering operations. This is the basic point of view we have taken in this book. We begin in Chapter 2 with the mathematical (and signal processing) framework of sampling, interpolation, and decimation. We give several alternative interpretations of systems for decimation and interpolation of both lowpass and bandpass signals. The concepts of quadrature and single-sideband modulation of bandpass signals are discussed.

In Chapter 3 we discuss digital networks (structures) which can be used to implement the conversion from one sampling rate to another. Included in this chapter is a brief review of signal-flow graph representations of digital systems, and of structures for implementing the digital filters required for all sampling rate conversion systems. It is then shown how efficient implementations of sampling rate conversion systems can be obtained by simple manipulations of the proposed canonic structures. In this chapter the polyphase structure is introduced.

In Chapter 4 we discuss the question of how to design the digital filters used in the systems presented in Chapters 2 and 3. Following a brief review of the types and properties of digital filters that have been proposed for general purpose signal processing, it is shown that two general structures can be used to aid in the design of the special filters required in sampling rate conversion systems. Based on these structures, a number of special-purpose design algorithms are described. The advantages, disadvantages, and special properties of the resulting digital filters are discussed in this chapter.

Chapter 5 addresses the question of structures for handling two special cases of sampling rate conversion:

1. Large changes in sampling rates within the system.
2. Changes in sampling rate requiring large sampling rate changes internally in the structure: for example, sampling rate conversion by a factor of 97/151.

Each of these cases can be handled most efficiently in a multistage structure in which the sampling rate conversion occurs in a series of two or more distinct stages. Questions of computational, storage, and control efficiency are of paramount concern in the discussions in this chapter.

Chapter 6 is devoted to a discussion of how some basic signal processing algorithms can be efficiently implemented using multirate structures. Included in this chapter are algorithms for lowpass and bandpass filtering, phase shifting, Hilbert transformation, and narrow-band, high-resolution spectrum analysis.

Finally, in Chapter 7, we provide a comprehensive discussion of the theory and practical implementation of filter banks and spectrum analyzers and synthesizers based on multirate techniques. Included in this chapter are treatments of both uniform and nonuniform filter banks. It is shown that efficient implementations can often be realized using fast transform methods (e.g., the fast Fourier transform) combined with polyphase structures. An interesting result, shown in this chapter, is that the standard fast convolution methods of overlap-add and overlap-save are specific cases of a generalized structure for implementing a spectrum analysis and synthesis system.

REFERENCES

1.1   C. E. Shannon, "Communications in the Presence of Noise," *Proc. IRE*, Vol. 37, No. 1, pp. 10-21, January 1949.

1.2   D. A. Linden, "A Discussion of Sampling Theorems," *Proc. IRE*, Vol. 47, No. 7, pp. 1219-1226, July 1959.

1.3   H. Nyquist, "Certain Topics in Telegraph Transmission Theory," *Trans. AIEE*, Vol. 47, pp. 617-664, February 1928.

1.4   R. E. Crochiere and L. R. Rabiner, "Interpolation and Decimation of Digital Signals - A Tutorial Review," *Proc. IEEE*, Vol. 69, No. 3, pp. 300-331, March 1981.

1.5   R. W. Hamming, *Numerical Methods for Scientists and Engineers*. New York: McGraw-Hill, 1962.

1.6   S. L. Freeny, R. B. Kieburtz, K. V. Mina, and S. K. Tewksbury, "Design of Digital Filters for an All Digital Frequency Division Multiplex-Time Division Multiplex Translator," *IEEE Trans. Circuit Theory*, Vol. CT-18, No. 6, pp. 702-711, November 1971.

1.7   Special Issue on TDM-FDM Conversion, *IEEE Trans. on Comm.*, Vol. COM-26, No. 5, May 1978.

1.8   S. L. Freeny, J. F. Kaiser, and H. S. McDonald, "Some Applications of Digital Signal Processing in Telecommunications," in *Applications of Digital Signal Processing* (A. V. Oppenheim, Ed.). Englewood Cliffs, N.J.: Prentice-Hall, 1978, pp. 1-28.

1.9   L. R. Rabiner and R. W. Schafer, *Digital Processing of Speech Signals*. Englewood Cliffs, N.J.: Prentice-Hall, 1978.

1.10   R. E. Crochiere, S. A. Webber, and J. L. Flanagan, "Digital Coding of Speech in Subbands," *Bell Sys. Tech. J.*, Vol. 55, No. 8, pp. 1069-1085, October 1976.

1.11   D. J. Goodman and J. L. Flanagan, "Direct Digital Conversion between Linear and Adaptive Delta Modulation Formats," *Proc. IEEE Int. Comm. Conf.*, Montreal, June 1971.

1.12   R. Lagadec, D. Pelloni, and D. Weiss, "A 2-Channel, 16-Bit Sampling Frequency Converter for Professional Digital Audio," *Proc. of the 1982 IEEE Int. Conf. on Acoust. Speech Signal Process.*, pp. 93-96, Paris, May 1982.

1.13   R. G. Pridhman and R. A. Mucci, "Digital Interpolation Beam Forming for Lowpass and Bandpass Signals," *Proc. IEEE* Vol. 67, pp. 904-919, June 1979.

1.14   L. R. Rabiner and B. Gold, *Theory and Application of Digital Signal Processing*. Englewood Cliffs, N.J.: Prentice-Hall, 1975, Chap. 13.

1.15   J. H. McClellan and R. J. Purdy, "Applications of Digital Signal Processing to Radar," in *Applications of Digital Signal Processing* (A. V. Oppenheim, Ed.). Englewood Cliffs, N.J.: Prentice-Hall, 1978, pp. 239-330.

1.16   M. R. Portnoff, "Time-Frequency Representation of Digital Signals and Systems Based on Short-Time Fourier Analysis," *IEEE Trans. Acoust. Speech Signal Process.*, Vol. ASSP-28, No. 1, pp. 55-69, February 1980.

2

Basic Principles of Sampling and Sampling Rate Conversion

2.0 INTRODUCTION

The purpose of this chapter is to provide the basic theoretical framework for uniform sampling and for the signal processing operations involved in sampling rate conversion. As such we begin with a discussion of the sampling theorem and consider its interpretations in both the time and frequency domains. We then consider sampling rate conversion systems (for decimation and interpolation) in terms of both analog and digital operations on the signals for integer changes in the sampling rate. By combining concepts of integer decimation and interpolation, we generalize the results to the case of rational fraction changes of sampling rates for which a general input-output relationship can be obtained. These operations are also interpreted in terms of concepts of periodically time-varying digital systems.

Next we consider more complicated sampling techniques and modulation techniques for dealing with bandpass signals instead of lowpass signals (which are assumed in the first part of the chapter). We show that sampling rate conversion techniques can be extended to bandpass signals as well as lowpass signals and can be used for purposes of modulation as well as sampling rate conversion.

2.1 UNIFORM SAMPLING AND THE SAMPLING THEOREM

2.1.1 Uniform Sampling Viewed as a Modulation Process

Let $x_C(t)$ be a continuous function of the continuous variable t. We are interested in sampling $x_C(t)$ at the uniform rate

$$t = nT, \quad -\infty < n < \infty \tag{2.1}$$

that is, once every interval of duration T.[1] We denote the sampled signal as $x(n)$. Figure 2.1 shows an example of a signal $x_C(t)$ and the associated sampled signal $x(n)$ for two different values of T.

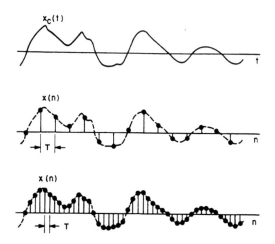

Figure 2.1 Continuous signal and two sampled versions of it.

One convenient way of interpreting this sampling process is as a modulation or multiplication process, as shown in Fig. 2.2(a). The continuous signal $x_C(t)$ is multiplied (modulated) by the periodic impulse train (sampling function) $s(t)$ to give the pulse amplitude modulated (PAM) signal $x_C(t)s(t)$. This PAM signal is then discretized in time to give $x(n)$, that is,

$$x(n) = \lim_{\epsilon \to 0} \int_{t=nT-\epsilon}^{nT+\epsilon} x_C(t)s(t)\, dt \tag{2.2}$$

where

$$s(t) = \sum_{l=-\infty}^{\infty} u_0(t - lT) \tag{2.3}$$

and where $u_0(t)$ denotes an ideal unit impulse function. In the context of this interpretation, $x(n)$ denotes the area under the impulse at time nT. Since this area is equal to the area under the unit impulse (area $= 1$), at time nT, weighted

---

[1] A more general derivation of the sampling theorem would sample at $t = nT + \delta$, where δ is an arbitrary constant. The results to be presented are independent of δ for stationary signals $x(t)$. Later we consider cases where δ is nonzero.

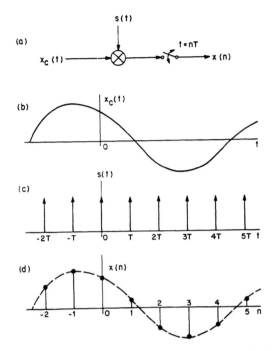

Figure 2.2 Periodic sampling of $x_C(t)$ via modulation to obtain $x(n)$.

by $x_C(nT)$, it is easy to see that

$$x(n) = x_C(nT) \tag{2.4}$$

Figure 2.2 (b), (c), and (d) show $x_C(t)$, $s(t)$, and $x(n)$ for a sampling period of T seconds.

2.1.2 Spectral Interpretations of Sampling

We assume that $x_C(t)$ has a Fourier transform $X_C(j\Omega)$ defined as

$$X_C(j\Omega) = \int_{-\infty}^{\infty} x_C(t)e^{-j\Omega t} \, dt \tag{2.5}$$

where Ω denotes the analog frequency (in radians/sec). Similarly, the Fourier transform of the sampling function $s(t)$ can be defined as

$$S(j\Omega) = \int_{-\infty}^{\infty} s(t)e^{-j\Omega t} \, dt \tag{2.6}$$

and it can be shown that by applying Eq. (2.3) to Eq. (2.6), $S(j\Omega)$ has the form

$$S(j\Omega) = \frac{2\pi}{T} \sum_{l=-\infty}^{\infty} u_0\left[\Omega - \frac{2\pi l}{T}\right] \tag{2.7}$$

By defining

$$F = \frac{1}{T} \tag{2.8}$$

$$\Omega = 2\pi f \tag{2.9a}$$

and

$$\Omega_F = 2\pi F \tag{2.9b}$$

$S(j\Omega)$ also has the form

$$S(j\Omega) = \Omega_F \sum_{l=-\infty}^{\infty} u_0(\Omega - l\Omega_F) \tag{2.10}$$

That is, a uniformly spaced impulse train in time, $s(t)$, transforms to a uniformly spaced impulse train in frequency, $S(j\Omega)$.

Since multiplication in the time domain is equivalent to convolution in the frequency domain, we have the relation

$$X_C(j\Omega) * S(j\Omega) = \int_{-\infty}^{\infty} [x_C(t)s(t)]e^{-j\Omega t}\, dt \tag{2.11}$$

where $*$ denotes a linear convolution of $X_C(j\Omega)$ and $S(j\Omega)$ in frequency. Figure 2.3 shows typical plots of $X_C(j\Omega)$, $S(j\Omega)$, and the convolution $X_C(j\Omega) * S(j\Omega)$, where it is assumed that $X_C(j\Omega)$ is bandlimited and its highest-frequency component $2\pi F_C$ is less than one-half of the sampling frequency, $\Omega_F = 2\pi F$. From this figure it is seen that the process of pulse amplitude modulation periodically repeats the spectrum $X_C(j\Omega)$ at harmonics of the sampling frequency due to the convolution of $X_C(j\Omega)$ and $S(j\Omega)$.

Because of the direct correspondence between the sequence $x(n)$ and the pulse amplitude modulated signal $x_C(t)s(t)$, as seen by Eqs. (2.2) and (2.4) it is clear that the information content and the spectral interpretations of the two signals are synonymous. This correspondence can be shown more formally by considering the (discrete) Fourier transform of the sequence $x(n)$, which is defined as

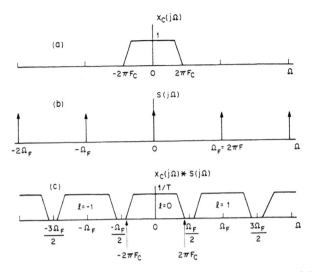

Figure 2.3 Spectra of signals obtained from periodic sampling via modulation.

$$X(e^{j\omega}) = \sum_{n=-\infty}^{\infty} x(n)e^{-j\omega n} \qquad (2.12)$$

where ω denotes the frequency (in radians relative to the sampling rate F), defined as

$$\omega = \Omega T = \frac{\Omega}{F} \qquad (2.13)$$

Since $x_C(t)$ and $x(n)$ are related by Eq. (2.4), a relation can be derived between $X_C(j\Omega)$ and $X(e^{j\omega})$ with the aid of Eqs. (2.5) and (2.12) as follows. The inverse Fourier transform of $X_C(j\Omega)$ gives $x_C(t)$ as

$$x_C(t) = \frac{1}{2\pi} \int_{-\infty}^{\infty} X_C(j\Omega)e^{j\Omega t} \, d\Omega \qquad (2.14)$$

Evaluating Eq. (2.14) for $t = nT$, we get

$$x(n) = x_C(nT) = \frac{1}{2\pi} \int_{-\infty}^{\infty} X_C(j\Omega)e^{j\Omega nT} \, d\Omega \qquad (2.15)$$

The sequence $x(n)$ may also be obtained as the (discrete) inverse Fourier transform of $X(e^{j\omega})$,

$$x(n) = \frac{1}{2\pi} \int_{-\pi}^{\pi} X(e^{j\omega}) e^{j\omega n} \, d\omega \qquad (2.16)$$

Combining Eqs. (2.15) and (2.16), we get

$$\frac{1}{2\pi} \int_{-\pi}^{\pi} X(e^{j\omega}) e^{j\omega n} \, d\omega = \frac{1}{2\pi} \int_{-\infty}^{\infty} X_C(j\Omega) e^{j\Omega n T} \, d\Omega \qquad (2.17)$$

By expressing the right-hand side of Eq. (2.17) as a sum of integrals (each of width $2\pi/T$), we get

$$\frac{1}{2\pi} \int_{-\infty}^{\infty} X_C(j\Omega) e^{j\Omega n T} \, d\Omega = \frac{1}{2\pi} \sum_{l=-\infty}^{\infty} \int_{(2l-1)\pi/T}^{(2l+1)\pi/T} X_C(j\Omega) e^{j\Omega n T} \, d\Omega$$

$$= \frac{1}{2\pi} \sum_{l=-\infty}^{\infty} \int_{-\pi/T}^{\pi/T} \left[X_C\left(j\Omega + j\frac{2\pi l}{T}\right) \right] e^{j\Omega n T} e^{j2\pi l n} \, d\Omega$$

$$= \frac{1}{2\pi} \int_{-\pi/T}^{\pi/T} \left[\sum_{l=-\infty}^{\infty} X_C\left(j\Omega + j\frac{2\pi l}{T}\right) \right] e^{j\Omega n T} \, d\Omega \qquad (2.18)$$

since $e^{j2\pi l n} = 1$ for all integer values of l and n. Combining Eqs. (2.17) and (2.18), setting $\Omega = \omega/T$ and $\Omega_F = 2\pi/T$, gives

$$\frac{1}{2\pi} \int_{-\pi}^{\pi} [X(e^{j\omega})] e^{j\omega n} \, d\omega = \frac{1}{2\pi} \int_{-\pi}^{\pi} \left[\frac{1}{T} \sum_{l=-\infty}^{\infty} X_C(j\Omega + jl\Omega_F) \right] e^{j\omega n} \, d\omega \qquad (2.19)$$

Finally, by equating terms within the brackets, we get

$$X(e^{j\omega}) = \frac{1}{T} \sum_{l=-\infty}^{\infty} X_C(j(\Omega + l\Omega_F)) = \frac{1}{T} \sum_{l=-\infty}^{\infty} X_C\left[\frac{j}{T}(\omega + 2\pi l) \right] \qquad (2.20)$$

Equation (2.20) provides the fundamental link between continuous and digital systems. The correspondence between these relations and the spectral interpretation of the PAM signal $X_C(j\Omega) * S(j\Omega)$ in Figure 2.3 is also apparent; that is, the spectrum of the digital signal corresponds to harmonically translated and amplitude scaled repetitions of the analog spectrum.

2.1.3 The Sampling Theorem

Given the analog signal $x_C(t)$ it is always possible to obtain the digital signal $x(n)$. However, the reverse process is not always true; that is, $x_C(t)$ uniquely specifies

$x(n)$; but $x(n)$ does not necessarily uniquely specify $x_C(t)$. In practice it is generally desired to have a unique correspondence between $x(n)$ and $x_C(t)$ and the conditions under which this uniqueness holds is given by the well-known *sampling theorem*:

> If a continuous signal $x_C(t)$ has a bandlimited Fourier transform $X_C(j\Omega)$, that is, $|X_C(j\Omega)| = 0$ for $|\Omega| \geqslant 2\pi F_C$, then $x_C(t)$ can be uniquely reconstructed without error from equally spaced samples $x_C(nT)$, $-\infty < n < \infty$, if $F \geqslant 2F_C$, where $F = 1/T$ is the sampling frequency.

The sampling theorem can be conveniently understood in terms of the spectral interpretations of the sampling process and Eq. (2.20). Figure 2.4 shows an example of the spectrum of a bandlimited signal [part (a)] and the resulting spectrum of the digital signal for a sampling period which is shorter than required by the sampling theorem [part (b)], a sampling period equal to that required by the sampling theorem [part (c)], and a sampling period longer than required by the sampling theorem [part d)]. From Fig. 2.4 we readily see that for parts (b) and (c) (when the conditions of the sampling theorem are met) the higher-order spectral components (the terms in Eq. (2.20) for $|l| \geqslant 1$) do not overlap the baseband and distort the digital spectrum. Thus one basic interpretation of the sampling theorem is that the spectrum of the sampled signal must be the same as (to within a constant multiplier) the spectrum of the continuous signal for the baseband of frequencies $(-2\pi F_C < \Omega < 2\pi F_C)$.

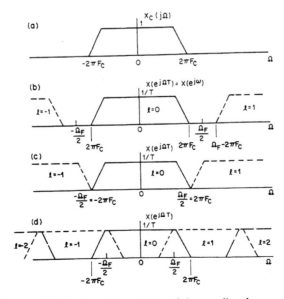

Figure 2.4 Spectral interpretations of the sampling theorem.

2.1.4 Reconstruction of an Analog Signal from Its Samples

The major consequence of the sampling theorem is that the original sequence $x_C(t)$ can be uniquely and without error reconstructed from its samples $x(n)$ if the samples are obtained at a sufficiently high rate. To see how this reconstruction is accomplished, we consider the spectrum of the continuous-time modulated signal $x_C(t)s(t)$ as shown in Figure 2.3(c). This spectrum is identical to that of the sampled signal $x(n)$. To recover $X_C(j\Omega)$ from the convolution $X_C(j\Omega) * S(j\Omega)$, we merely have to filter the signal $x_C(t)s(t)$ by an ideal lowpass filter whose cutoff frequency is between $2\pi F_C$ and $\Omega_F - 2\pi F_C$. This processing is illustrated in Figure 2.5. To implement this process, an ideal digital-to-analog converter is required to get $x_C(t)s(t)$ from $x(n)$. Assuming that we do not worry about the realizability of such an ideal converter, the reconstruction formula from Figure 2.5 is

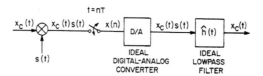

Figure 2.5 Sampling and reconstruction of a continuous signal.

$$x_C(t) = \int_{\tau=-\infty}^{\infty} x_C(\tau)s(\tau)\hat{h}(t - \tau)d\tau \qquad (2.21)$$

and applying Eqs. (2.2) to (2.4) gives

$$x_C(t) = \sum_{n=-\infty}^{\infty} x(n)\hat{h}(t - nT) \qquad (2.22)$$

For an ideal lowpass filter with cutoff frequency F_{LP}, the ideal impulse response $\hat{h}_I(t)$ is of the form

$$\hat{h}_I(t) = \frac{\sin(2\pi F_{LP}t)}{2\pi F_{LP}t} \qquad (2.23)$$

Generally, F_{LP} is chosen as

$$F_{LP} = \frac{F}{2} = \frac{1}{2T} \qquad (2.24)$$

leading to the well-known reconstruction formula

$$x_C(t) = \sum_{n=-\infty}^{\infty} x(n) \left[\frac{\sin \left[\pi (t - nT)/T \right]}{\pi(t - nT)/T} \right] \qquad (2.25)$$

Figure 2.6 illustrates the application of Eq. (2.25) to a typical signal. It is seen that the ideal lowpass filter acts like an interpolator for the bandlimited signal $x_C(t)$, allowing the determination of *any* value of $x_C(t)$ from the *infinite* set of its samples taken at a sufficiently high rate.

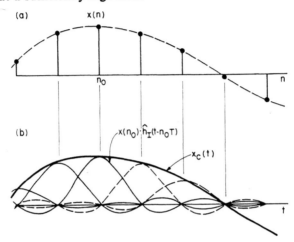

Figure 2.6 Illustration of a bandlimited reconstruction from shifted and scaled lowpass filter responses.

In practice the "ideal" filter is unrealizable because it requires values of $x(n)$ for $-\infty < n < \infty$ in order to evaluate a single value of $x_C(t)$. Therefore, some realizable approximation to $\hat{h}_I(t)$ must be used. Figure 2.7 illustrates an example of an impulse response for a realizable reconstruction or interpolating lowpass filter, $\hat{h}(t)$, that extends over a finite number of samples of $x(n)$. In this figure we show plots of $x_C(t)$ (bottom figure) and $x(n)$, and the range of $\hat{h}(t-nT)$ evaluated in the region of the n_oth sample [i.e., at $t = n_o T$ (top figure)]. To the extent that the frequency response of the actual lowpass filter approximates the ideal lowpass filter, the reconstruction error of $x_C(t)$ can be kept small.

2.1.5 Summary of the Implications of the Sampling Theorem

The main result of the sampling theorem is that there is a minimum rate (related directly to the bandwidth of the signal) at which a signal can be sampled and for which theoretically exact reconstruction of the signal is possible from its samples. If the signal is sampled below this minimum rate, then distortions, in the form of spectral foldover or *aliasing* [e.g., see Fig. 2.4(d)], occur from which no recovery is

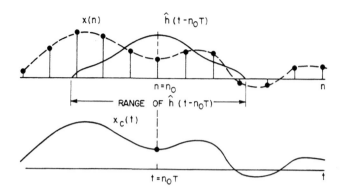

Figure 2.7 Illustration of reconstruction of a bandlimited signal from its samples using a nonideal finite duration impulse response lowpass filter.

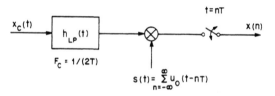

Figure 2.8 Representation of a practical sampling system with prefiltering to avoid aliasing.

generally possible. Thus to ensure that the conditions of the sampling theorem are met for a given application, the signal to be sampled is generally first filtered by a lowpass filter whose cutoff frequency is less than (or equal to) half the sampling frequency. Such a filter is often called an *anti-aliasing prefilter* because its purpose is to guarantee that no aliasing occurs due to sampling. Thus the standard representation of a system for sampling a signal (analog-to-digital conversion) is as shown in Fig. 2.8. We will see in the following sections that a lowpass filter of the type shown in Fig. 2.8 is required for almost all sampling rate conversion systems.

2.2 SAMPLING RATE CONVERSION — AN ANALOG INTERPRETATION

The process of sampling rate conversion is one of converting the sequence $x(n)$ obtained from sampling $x_C(t)$ with a period T, to another sequence $y(m)$ obtained from sampling $x_C(t)$ with a period T'. The most straightforward way to perform this conversion is to reconstruct $x_C(t)$ (or the lowpass filtered version of it) from the samples of $x(n)$ and then resample $x_C(t)$ (assuming that it is sufficiently bandlimited for the new sampling rate) with period T' to give $y(m)$. The processing involved in this procedure is illustrated in Fig. 2.9. Figure 2.10 shows typical waveforms which illustrate the signal processing involved in implementing the system of Fig. 2.9. Because $\hat{h}(t)$, the impulse response of the analog lowpass filter, is assumed to be of finite duration, the value of $x_C(t)$ at $t = m_o T'$ is determined *only* from the finite set of samples of $x(n)$ shown in part (a) of this figure. Thus for any m, the value of $y(m)$ can be obtained as

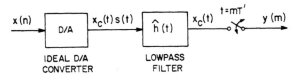

Figure 2.9 Conversion of a sequence $x(n)$ to another sequence $y(m)$ by analog reconstruction and resampling.

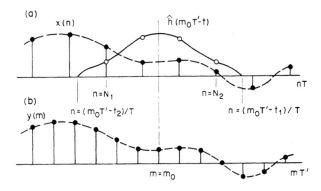

Figure 2.10 Typical waveforms for sampling rate conversion by analog reconstruction and resampling.

$$y(m) = x_C(t)\big|_{t=mT'} = \sum_{n=N_1}^{N_2} x(n)\hat{h}(mT' - nT) \qquad (2.26)$$

where N_1 and N_2 denote minimum and maximum of the range of values of n involved in the computation of $y(m)$. From Eq. (2.26) and Fig. 2.10 we see that only specific values of n and specific values of $\hat{h}(t)$ are used to generate $y(m)$, that is,

$$y(m) = x(N_1)\hat{h}(mT' - N_1 T) + \cdots + x(N_2)\hat{h}(mT' - N_2 T) \qquad (2.27)$$

The values of $\hat{h}(t)$ that are used to give $y(m)$ are spaced T apart in time. In effect the signal $x(n)$ samples (and weights) the impulse response $\hat{h}(t)$ to give $y(m)$. It is interesting to note that when $T' = T$, the form of Eq. (2.26) reduces to that of the familiar discrete convolution

$$y(m) = \sum_n x(n)\hat{h}((m-n)T) \tag{2.28}$$

The limits on the summation of Eq. (2.26) are determined from the range of values for which $\hat{h}(t)$ is nonzero. If we assume that $\hat{h}(t)$ is zero for $t < t_1$ and $t > t_2$, that is,

$$\hat{h}(t) = 0, \quad t > t_2, t < t_1 \tag{2.29}$$

this leads to the result

$$\hat{h}(mT' - nT) = 0, \quad mT' - nT > t_2, mT' - nT < t_1 \tag{2.30}$$

or (see Figure 2.10)

$$n < \frac{mT' - t_2}{T} \tag{2.31a}$$

$$n > \frac{mT' - t_1}{T} \tag{2.31b}$$

Thus by integerizing Eqs. (2.31) we get

$$N_1 = \lceil \frac{mT' - t_2}{T} \rceil \tag{2.32a}$$

$$N_2 = \lfloor \frac{mT' - t_1}{T} \rfloor \tag{2.32b}$$

where $\lceil u \rceil$ denotes the smallest integer greater than or equal to u and $\lfloor u \rfloor$ denotes the largest integer less than or equal to u. It can be seen from Eqs. (2.32) that the set of samples $x(n)$ involved in the determination of $y(m)$ is a complicated function of the sampling periods T and T', the endpoints of the filter t_1 and t_2, and the sample m being determined.

Figure 2.11 illustrates this effect for the case $T' = T/2$, and for two impulse response durations, $t_2 = -t_1 = 2.3T$ and $t_2 = -t_1 = 2.8T$. As shown in parts (a) and (b), the determination of $y(m)$ for m even for both $t_2 = -t_1 = 2.3T$ [part (a)] and $t_2 = -t_1 = 2.8T$ [part (b)] involves the identical set of samples of $x(n)$. However, the determination of $y(m)$ for m odd involves different sets of samples of $x(n)$ for $t_2 = -t_1 = 2.3T$ [part (c)] than for $t_2 = -t_1 = 2.8T$ [part (d)].

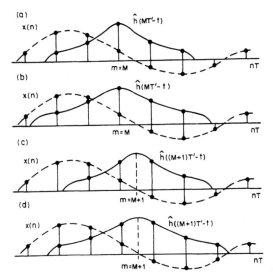

Figure 2.11 Examples showing the samples of $x(n)$ involved in the computation of $y(m)$ for two different impulse response durations and for even and odd samples of $y(m)$ for a 2-to-1 increase in the sampling rate.

A second important issue in the implementation described above involves the set of samples of $\hat{h}(t)$ used in the determination of $y(m)$. For each value of m, a *distinct set* of samples of $\hat{h}(t)$ are used to give $y(m)$. Figure 2.12 illustrates this point for the case $T' = T/2$ (i.e., a 2-to-1 increase in the sampling rate). Figure 2.12(a) shows $x(n)$, $\hat{h}(m_e T' - t)$, and $y(m)$ for the computation of $y(m_e)$, where m_e is an even integer, and Fig. 2.12(b) shows the same waveforms for $m_o = m_e + 1$ (i.e., an odd value of m). It can be seen that two distinctly different sets of values of $\hat{h}(t)$ are involved in the computation of $y(m)$ for even and odd m. For the case $T' = 2T$ (i.e., a 2-to-1 decrease in the sampling rate), the same set of samples of $\hat{h}(t)$ are used to determine *all* output samples $y(m)$.

By introducing the change of variables

$$k = \left\lfloor \frac{mT'}{T} \right\rfloor - n \qquad (2.33)$$

the form of Eq. (2.26) can be modified to another form that more explicitly reveals the nature of the indexing problem associated with the evaluation of $y(m)$ in the sampling rate conversion process described above. This form will be used extensively in later sections. Applying Eq. (2.33) to Eq. (2.26) gives the expression

$$y(m) = \sum_{k=K_1}^{K_2} x\left[\left\lfloor \frac{mT'}{T} \right\rfloor - k \right] \hat{h}\left[mT' - \left\lfloor \frac{mT'}{T} \right\rfloor T + kT \right] \qquad (2.34)$$

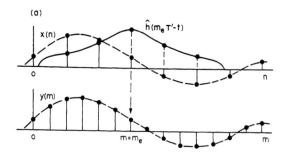

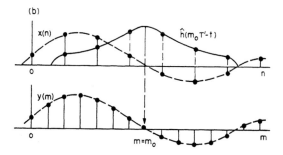

Figure 2.12 Examples showing the samples of $\hat{h}(t)$ involved in the computation of; (a) even values of $y(m)$, and; (b) odd values of $y(m)$ for a 2-to-1 increase in the sampling rate.

and rearranging terms gives the desired form,

$$y(m) = \sum_{k=K_1}^{K_2} \hat{h}((k + \delta_m)T)x\left[\lfloor \frac{mT'}{T} \rfloor - k\right] \qquad (2.35)$$

where δ_m is defined as

$$\delta_m = \frac{mT'}{T} - \lfloor \frac{mT'}{T} \rfloor \qquad (2.36)$$

It is clear that since δ_m corresponds to the difference of a number mT'/T and its next lowest integer,

$$0 \leqslant \delta_m < 1 \qquad (2.37)$$

Thus from Eq. (2.35) it can be seen that the determination of a sample value $y(m)$ involves samples of $\hat{h}(t)$ spaced T apart and offset by the fractional sample time

$\delta_m T$, where δ_m varies as a function of m. It is also interesting to note that when $T' = T$, Eq. (2.35) again reduces to a familiar convolutional form

$$y(m) = \sum_k \hat{h}(kT)x(m-k) \tag{2.38}$$

Figure 2.13 depicts the samples of $\hat{h}(t)$ and $x(n)$ involved in determining $y(m)$ based on Eq. (2.35). As in the earlier interpretation of Eq. (2.26), it is seen that the finite range of $\hat{h}(t)$ restricts the number of samples $x(n)$ that are actually used in determining $y(m)$. By again applying the conditions of Eq. (2.29) it can be shown that the limits on the summation, K_1 and K_2, can be determined from the condition

$$\hat{h}((k+\delta_m)T) = 0, \quad (k+\delta_m)T > t_2, \ (k+\delta_m)T < t_1 \tag{2.39}$$

or

$$k > \frac{t_2}{T} - \delta_m \tag{2.40a}$$

$$k < \frac{t_1}{T} - \delta_m \tag{2.40b}$$

and integerizing Eqs. (2.40) gives

$$K_1 = \left\lceil \frac{t_1}{T} - \delta_m \right\rceil = \left\lceil \frac{t_1}{T} - \frac{mT'}{T} \right\rceil + \left\lfloor \frac{mT'}{T} \right\rfloor \tag{2.41a}$$

$$K_2 = \left\lfloor \frac{t_2}{T} - \delta_m \right\rfloor = \left\lfloor \frac{t_2}{T} - \frac{mT'}{T} \right\rfloor + \left\lfloor \frac{mT'}{T} \right\rfloor \tag{2.41b}$$

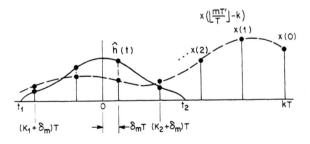

Figure 2.13 Alternative form for sampling rate conversion process showing samples of $\hat{h}(t)$ and $x(n)$ involved in determining $y(m)$.

By comparing Eqs. (2.26) and (2.35), it is seen that the first sample, $\hat{h}((K_1 + \delta_m)T)$, is multiplied by $x(N_2)$ and the last sample, $\hat{h}((K_2 + \delta_m)T)$, is multiplied by $x(N_1)$ in the determination of $y(m)$.

Thus far we have considered the conversion of a discrete signal $x(n)$ with sampling period T to the discrete signal $y(m)$ with sampling period T', where T and T' are *arbitrary* values subject to the constraints of the sampling theorem. In most practical systems it is often useful to restrict this generality slightly by requiring that the two sampling periods T and T' be related by some common (perhaps higher rate) clock. In this case we can specify that the ratio T'/T can be expressed as

$$\frac{T'}{T} = \frac{M}{L} \qquad (2.42)$$

where M and L are integers. This is particularly true in the case where the digital system involving signals $x(n)$ and $y(m)$ is, in fact, an all-digital system, as will be discussed in the next sections. We assume throughout the remainder of this book that Eq. (2.42) is valid.

One consequence of the restriction imposed by Eq. (2.42) can be seen in the offset δ_m in the samples of $\hat{h}(t)$ used in Eq. (2.35) and defined in Eq. (2.36). Applying Eq. (2.42) to Eq. (2.36) gives

$$\delta_m = \frac{mM}{L} - \lfloor \frac{mM}{L} \rfloor$$

$$= \frac{1}{L}\left[mM - \lfloor \frac{mM}{L} \rfloor L \right] \qquad (2.43)$$

By noting that

$$mM - \lfloor \frac{mM}{L} \rfloor L = (mM) \oplus L \qquad (2.44)$$

where $(i) \oplus L$ denotes the value of i modulo L, Eq. (2.43) becomes

$$\delta_m = \frac{1}{L}[(mM) \oplus L] \qquad (2.45)$$

From this relation it is seen that δ_m can take on only L unique values, $0, 1/L, 2/L, ..., (L-1)/L$, for all values of m. Thus there are only L unique *sets* of samples of $\hat{h}(t)$ that are used in computing $y(m)$ from $x(n)$ in Eqs. (2.26) or (2.35). As shown previously, if $L = 1$, there is one unique set of filter coefficients used in computing $y(m)$ from $x(n)$. If $M = 1$, however, the number of unique sets of filter samples is still L. In general, if T'/T cannot be expressed as in Eq.

(2.42) (i.e., the quantity is irrational), then an infinite number of *sets* of samples of $\hat{h}(t)$ are used to compute $y(m)$ for $-\infty < m < \infty$.

2.3 SAMPLING RATE CONVERSION — A DIRECT DIGITAL APPROACH

2.3.1 Relationship to Time-Varying Systems

In the preceding section we have shown how sampling rate conversion can be interpreted in terms of an analog signal processing formulation. We have shown that, in general, a distinct set of samples of the original sequence $x(n)$, and a distinct set of samples of an analog lowpass filter response $\hat{h}(t)$, are involved in the conversion process to give the desired new sampled sequence $y(m)$.

In many signal processing applications it is desirable to be able to perform sampling rate conversion directly by digital computation without an intermediate analog stage as illustrated in Fig. 2.14(a). More explicitly, we seek to derive a linear digital system $g_m(n)$ such that $x(n)$ can be processed by $g_m(n)$ to give $y(m)$ directly as shown in Fig. 2.14(b). From the discussion of the preceding section it should be clear that in cases when T'/T, the ratio of sampling periods of $y(m)$ to $x(n)$, can be expressed as a rational fraction M/L, then *only* L distinct sets of *samples* of the filter $\hat{h}(t)$ are used in the conversion. For such cases it is clear that a system of the type given in Fig. 2.14(b) can easily be obtained. One simple way of achieving such an implementation, as shown in this section, is to design the analog filter $\hat{h}(t)$ and uniformly sample its impulse response at a sampling rate that is L times higher than the sampling rate of $x(n)$. This set of samples forms the unit sample response of a *digital* filter, $h(k)$, which can be broken down into the L distinct sets of coefficients of $g_m(n)$ (one set for each value of δ_m). Alternative methods of designing the required lowpass digital filter can be used and a discussion of such methods is the topic of Chapter 4.

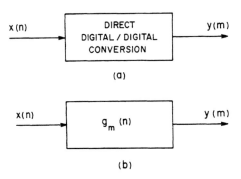

Figure 2.14 (a) Direct digital/digital conversion of $x(n)$ to $y(m)$; (b) a more explicit interpretation.

A close examination of the proposed structure of Fig. 2.14(b) shows that the systems we are dealing with for digital-to-digital sampling rate conversion are inherently time-varying systems; that is, $g_m(n)$ is the response of the system at the output sample time m to an input at the input sample time $\lfloor mM/L \rfloor - n$. From the point of view of the analog interpretation it is seen that sampling rate conversion is a combination of a linear time-invariant filtering process [i.e., the convolution by $\hat{h}(t)$] and two well-defined, linear, time-varying operations: the conversion of $x(n)$ to a PAM representation, $x_C(t)$, and sampling of $x_C(t)$ at the new rate to give $y(m)$. Thus the overall system $g_m(n)$ is linear but time varying.

Since the system is linear, each output sample $y(m)$ can be expressed as a linear combination of input samples $x(n)$. From Eqs. (2.35) to (2.45) it is seen that a general form for this expression is

$$y(m) = \sum_{n=-\infty}^{\infty} g_m(n)x\left[\lfloor \frac{mM}{L} \rfloor - n\right] \qquad (2.46a)$$

where $g_m(n)$ can be expressed in the form

$$g_m(n) = \hat{h}((n + \delta_m)T) \qquad (2.46b)$$

Since $g_m(n)$ can take on L distinct sets of values, it is periodic in m; that is,

$$g_m(n) = g_{m+rL}(n), \quad r = 0, \pm 1, \pm 2, \cdots \qquad (2.47)$$

Thus the system $g_m(n)$ is a linear, digital, *periodically* time-varying system. Such systems have been extensively studied for a wide range of applications [2.1, 2.2].

In the trivial case when $T' = T$, or $L = M = 1$, Eq. (2.46) reduces to the simple time-invariant digital convolution equation

$$y(m) = \sum_{n=-\infty}^{\infty} g(n)x(m - n) \qquad (2.48)$$

since the period of $g_m(n)$ is 1, in this case, and the integer part of $m - n$ is the same as $m - n$.

The above formulation of a digital periodically time-varying system is particularly suitable for defining models and deriving practical structures for decimators and interpolators. As such, it is used extensively in this chapter and in the derivation of the structures in Chapter 3. Later, in Chapter 3, Section 3.5, we show an alternative formulation of linear, periodically time-varying, digital systems which is more appropriate for the development of advanced theoretical network concepts of such systems.

In the next few sections we study in some detail the structure and properties of systems that perform two special cases of sampling rate conversion: decimation by integer factors, and interpolation by integer factors [2.3]. We then consider the general case of a sampling rate change by a factor of L/M.

2.3.2 Sampling Rate Reduction — Decimation by an Integer Factor M

Consider the process of reducing the sampling rate (decimation) of $x(n)$ by an integer factor M, that is,

$$\frac{T'}{T} = \frac{M}{1} \tag{2.49}$$

Then the new sampling rate is

$$F' = \frac{1}{T'} = \frac{1}{MT} = \frac{F}{M} \tag{2.50}$$

Assume that $x(n)$ represents a full band signal; that is, its spectrum is nonzero for all frequencies in the range $-F/2 \leqslant f \leqslant F/2$, with $\omega = 2\pi f T$:

$$|X(e^{j\omega})| \neq 0, \ |\omega| = |2\pi f T| \leqslant \frac{2\pi F T}{2} = \pi \tag{2.51}$$

except possibly at an isolated set of points. Based on the analog interpretation of sampling we see that in order to lower the sampling rate and to avoid aliasing at this lower rate, it is necessary to filter the signal $x(n)$ with a *digital* lowpass filter that approximates the ideal characteristic

$$\widetilde{H}(e^{j\omega}) = \begin{cases} 1, & |\omega| \leqslant \dfrac{2\pi F' T}{2} = \dfrac{\pi}{M} \\ 0, & \text{otherwise} \end{cases} \tag{2.52}$$

The sampling rate reduction is then achieved by forming the sequence $y(m)$ by saving only every Mth sample of the filtered output. This process is illustrated in Fig. 2.15(a). If we denote the actual lowpass filter unit sample response as $h(n)$, then we have

$$w(n) = \sum_{k=-\infty}^{\infty} h(k) x(n-k) \tag{2.53}$$

where $w(n)$ is the filtered output as seen in Fig. 2.15(a). The final output $y(m)$ is then obtained as

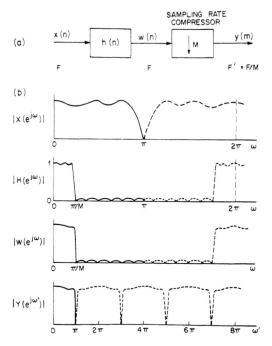

Figure 2.15 Block diagram and typical spectra for decimation by an integer factor M.

$$y(m) = w(Mm) \qquad (2.54)$$

as denoted by the operation of the second box in Fig. 2.15(a). This block diagram symbol, which will be referred to as a *sampling rate compressor*, will be used consistently throughout this book, and it corresponds to the resampling operation given by Eq. (2.54). It is the digital equivalent of the analog sampler; i.e., the block diagram of Fig. 2.15(a) is the digital equivalent of the analog sampling system of Fig. 2.8.

Figure 2.15(b) shows typical spectra (magnitude of the discrete Fourier transforms) of the signals $x(n)$, $h(n)$, $w(n)$, and $y(m)$ for an M-to-1 reduction in sampling rate. Note that in this figure the frequencies ω and ω' are normalized respectively to the sampling frequencies F and F'.

By combining Eqs. (2.53) and (2.54) the relation between $y(m)$ and $x(n)$ is of the form

$$y(m) = \sum_{k=-\infty}^{\infty} h(k)x(Mm - k) \qquad (2.55a)$$

or, by a change of variables, it becomes

$$y(m) = \sum_{n=-\infty}^{\infty} h(Mm - n)x(n) \qquad (2.55b)$$

Eq. (2.55a) is seen to be a special case of Eq. (2.46). Thus for decimation by integer factors of M, we have

$$g_m(n) = g(n) = h(n), \text{ for all } m \text{ and all } n \qquad (2.56)$$

Although $g_m(n)$ is *not* a function of m for this case, it can readily be seen, by considering the output signal obtained when $x(n)$ is shifted by an integer number of samples, that the overall system of Eq. (2.55) and Fig. 2.15(a) is not time invariant. For this case, unless the shift is a multiple of M, the output is *not* a shifted version of the output for 0 shift; that is,

$$x(n) \rightarrow y(m) \qquad (2.57a)$$

but

$$x(n - \delta) \not\rightarrow y(m - \frac{\delta}{M}) \text{ unless } \delta = rM \qquad (2.57b)$$

It is of value to derive the relationship between the z-transforms of $y(m)$ and $x(n)$ so as to be able to study the nature of the errors in $y(m)$ caused by a practical (non-ideal) lowpass filter. To obtain this relationship, we define the signal

$$w'(n) = \begin{cases} w(n), & n = 0, \pm M, \pm 2M, \ldots \\ 0, & \text{otherwise} \end{cases} \qquad (2.58)$$

That is, $w'(n) = w(n)$ at the sampling instants of $y(m)$, but is zero otherwise. A convenient and useful representation of $w'(n)$ is then

$$w'(n) = w(n) \left[\frac{1}{M} \sum_{l=0}^{M-1} e^{j2\pi ln/M} \right], \quad -\infty < n < \infty \qquad (2.59)$$

where the term in brackets corresponds to a discrete Fourier series representation of a periodic impulse train with a period of M samples. (This pulse train is the digital equivalent of the analog PAM sampling function described in Section 2.1.) Thus we have

$$y(m) = w'(Mm) = w(Mm) \qquad (2.60)$$

We now write the z-transform of $y(m)$ as

$$Y(z) = \sum_{m=-\infty}^{\infty} y(m)z^{-m}$$

$$= \sum_{m=-\infty}^{\infty} w'(Mm)z^{-m} \tag{2.61}$$

and since $w'(m)$ is zero except at integer multiples of M, Eq. (2.61) becomes

$$Y(z) = \sum_{m=-\infty}^{\infty} w'(m)z^{-m/M}$$

$$= \sum_{m=-\infty}^{\infty} w(m)\left[\frac{1}{M}\sum_{l=0}^{M-1} e^{j2\pi lm/M}\right]z^{-m/M}$$

$$= \frac{1}{M}\sum_{l=0}^{M-1}\left[\sum_{m=-\infty}^{\infty} w(m)e^{j2\pi lm/M}z^{-m/M}\right]$$

$$= \frac{1}{M}\sum_{l=0}^{M-1} W(e^{-j2\pi l/M}z^{1/M}) \tag{2.62}$$

Since

$$W(z) = H(z)X(z) \tag{2.63}$$

we can express $Y(z)$ as

$$Y(z) = \frac{1}{m}\sum_{l=0}^{M-1} H(e^{-j2\pi l/M}z^{1/M})X(e^{-j2\pi l/M}z^{1/M}) \tag{2.64}$$

Evaluating $Y(z)$ on the unit circle, $z = e^{j\omega'}$, leads to the result

$$Y(e^{j\omega'}) = \frac{1}{M}\sum_{l=0}^{M-1} H(e^{j(\omega'-2\pi l)/M})X(e^{j(\omega'-2\pi l)/M}) \tag{2.65a}$$

where

$$\omega' = 2\pi fT' \text{ (in radians relative to sampling period } T') \tag{2.65b}$$

Equation (2.65a) expresses the Fourier transform of the output signal $y(m)$ in terms of the transforms of the aliased components of the filtered input signal $x(n)$. The similarity between this equation and Eq. (2.20) should be clear. By writing the individual components of Eq. (2.65) directly, we see that

$$Y(e^{j\omega'}) = \frac{1}{M}\left[H(e^{j\omega'/M})X(e^{j\omega'/M}) + H(e^{j(\omega'-2\pi)/M})X(e^{j(\omega'-2\pi)/M}) + ... \right] \quad (2.66)$$

The purpose of the lowpass filter $H(e^{j\omega})$ is to filter $x(n)$ sufficiently so that its spectral components above the frequency $\omega = \pi/M$ are negligible (see Fig. 2.15). Thus it serves as an *anti-aliasing* filter. In terms of Eq. (2.65a) this implies that all terms for $l \neq 0$ are removed and if the filter $H(e^{j\omega})$ closely approximates the ideal response of Eq. (2.52), then Eq. (2.65a) becomes

$$Y(e^{j\omega'}) \simeq \frac{1}{M}X(e^{j\omega'/M}), \text{ for } |\omega'| \leqslant \pi \quad (2.67)$$

2.3.3 Sampling Rate Increase — Interpolation by an Integer Factor L

If the sampling rate is increased by an integer factor L, then the new sampling period, T', is

$$\frac{T'}{T} = \frac{1}{L} \quad (2.68)$$

and the new sampling rate F' is

$$F' = LF \quad (2.69)$$

This process of increasing the sampling rate (interpolation) of a signal $x(n)$ by L implies that we must interpolate $L - 1$ new sample values between each pair of sample values of $x(n)$. The process is similar to that of digital-to-analog conversion, discussed in Section 2.1.4, in which all continuous-time values of a signal $x_C(t)$ must be interpolated from its sequence $x(n)$. (In this case only specific values must be determined.)

Figure 2.16 illustrates an example of interpolation by a factor $L = 3$. The input signal $x(n)$ is "filled in" with $L - 1$ zero-valued samples between each pair of samples of $x(n)$, giving the signal

$$w(m) = \begin{cases} x\left[\dfrac{m}{L}\right], & m = 0, \pm L, \pm 2L, ... \\ 0, & \text{otherwise} \end{cases} \quad (2.70)$$

This process is the digital equivalent to the digital-to-PAM conversion process

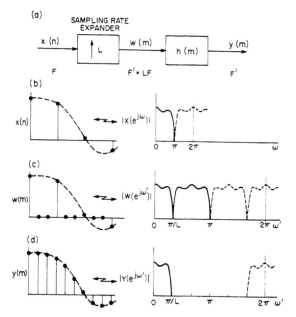

Figure 2.16 Block diagram and typical waveforms and spectra for interpolation by an integer factor L.

discussed in Section 2.1.4 and it is illustrated by the first box in the block diagram of Fig. 2.16(a). As with the resampling operation, the block diagram symbol of an up-arrow with an integer corresponds to increasing the sampling rate as given by Eq. (2.70) and it will be referred to as a *sampling rate expander*. The resulting signal $w(n)$ has the z-transform

$$W(z) = \sum_{m=-\infty}^{\infty} w(m)z^{-m} \tag{2.71a}$$

$$= \sum_{m=-\infty}^{\infty} x(m)z^{-mL} \tag{2.71b}$$

$$= X(z^L) \tag{2.72}$$

Evaluating $W(z)$ on the unit circle, $z = e^{j\omega'}$, gives the result

$$W(e^{j\omega'}) = X(e^{j\omega'L}) \tag{2.73}$$

which is the Fourier transform of the signal $w(m)$ expressed in terms of the spectrum of the input signal $x(n)$ (where $\omega' = 2\pi fT'$ and $\omega = 2\pi fT$).

As illustrated by the spectral interpretation in Fig. 2.16(c), the spectrum of

$w(m)$ contains not only the baseband frequencies of interest (i.e., $-\pi/L$ to π/L) but also *images* of the baseband centered at harmonics of the original sampling frequency $\pm 2\pi/L, \pm 4\pi/L, \ldots$. To recover the baseband signal of interest and eliminate the unwanted image components, it is necessary to filter the signal $w(m)$ with a digital lowpass (anti-imaging) filter which approximates the ideal characteristic

$$\tilde{H}(e^{j\omega'}) = \begin{cases} G, & |\omega'| \leqslant \dfrac{2\pi FT'}{2} = \dfrac{\pi}{L} \\ 0, & \text{otherwise} \end{cases} \tag{2.74}$$

It will be shown that in order to ensure that the amplitude of $y(m)$ is correct, the gain of the filter, G, must be L in the passband.

Letting $H(e^{j\omega'})$ denote the frequency response of an actual filter that approximates the characteristic in Eq. (2.74), it is seen that

$$Y(e^{j\omega'}) = H(e^{j\omega'})X(e^{j\omega'L}) \tag{2.75}$$

and within the approximation of Eq. (2.74),

$$Y(e^{j\omega'}) \cong \begin{cases} GX(e^{j\omega'L}), & |\omega'| \leqslant \dfrac{\pi}{L} \\ 0, & \text{otherwise} \end{cases} \tag{2.76}$$

It is easy to see why we need a gain of G in $\tilde{H}(e^{j\omega})$, whereas for the decimation filter a gain of 1 is adequate. For the "ideal" sampling system (with no aliasing error) we have seen from Eq. (2.20) that we desire

$$X(e^{j\omega}) = \frac{1}{T} X_C\left(\frac{j\omega}{T}\right)$$

For the "ideal" decimator we have shown [Eq. (2.67)] that

$$Y(e^{j\omega'}) = \frac{1}{M} X(e^{j\omega'/M})$$

$$= \frac{1}{MT} X_C\left(\frac{j\omega'}{MT}\right)$$

$$= \frac{1}{T'} X_C\left(\frac{j\omega'}{T'}\right)$$

Thus the necessary scaling is taken care of directly in the decimation process and a filter gain of 1 is suitable. For the "ideal" interpolator, however, we have

$$Y(e^{j\omega'}) = GX(e^{j\omega'L})$$

$$= \frac{G}{T} X_c \left[\frac{j\omega'L}{T} \right]$$

$$= \frac{G}{L} \left[\frac{1}{T'} \right] X_c \left[\frac{j\omega'}{T'} \right]$$

Clearly, a gain $G = L$ is required to meet the conditions of Eq. (2.20).

If $h(m)$ denotes the unit sample response of $H(e^{j\omega'})$, then from Fig. 2.16 $y(m)$ can be expressed as

$$y(m) = \sum_{k=-\infty}^{\infty} h(m-k)w(k) \tag{2.77}$$

Combining Eqs. (2.70) and (2.77) leads to the time domain input-to-output relation of the interpolator

$$y(m) = \sum_{k=-\infty}^{\infty} h(m-k)x\left[\frac{k}{L}\right], \quad k/L \text{ an integer}$$

$$= \sum_{r=-\infty}^{\infty} h(m-rL)x(r) \tag{2.78}$$

An alternative formulation of this equation can be obtained by introducing the change of variables

$$r = \lfloor \frac{m}{L} \rfloor - n \tag{2.79}$$

where $\lfloor u \rfloor$ again denotes the integer less than or equal to u. Then applying Eq. (2.44) for $M = 1$ gives

$$y(m) = \sum_{n=-\infty}^{\infty} h\left[m - \lfloor \frac{m}{L} \rfloor L + nL\right] x\left[\lfloor \frac{m}{L} \rfloor - n\right]$$

$$= \sum_{n=-\infty}^{\infty} h(nL + m \oplus L)x\left[\lfloor \frac{m}{L} \rfloor - n\right] \tag{2.80}$$

Equation (2.80) expresses the output $y(m)$ in terms of the input $x(n)$ and the filter coefficients $h(m)$ and it is seen to be a special case of Eq. (2.46). Thus for interpolation by integer factors of L, we have

$$g_m(n) = h(nL + m \oplus L), \quad \text{for all } m \text{ and all } n \tag{2.81}$$

and it is seen that $g_m(n)$ is periodic in m with period L, as indicated by Eq. (2.47).

2.3.4 Sampling Rate Conversion by a Rational Factor M/L

In the preceding two sections we have considered the cases of decimation by an integer factor M and interpolation by an integer factor L. In this section we consider the general case of conversion by the ratio

$$\frac{T'}{T} = \frac{M}{L} \tag{2.82}$$

or

$$F' = \frac{L}{M} F \tag{2.83}$$

This conversion can be achieved by a cascade of the two processes of integer conversions above by first increasing the sampling rate by L and then decreasing it by M. Figure 2.17(a) illustrates this process. It is important to recognize that the interpolation by L *must* precede the decimation process by M so that the width of the baseband of the intermediate signal $s(k)$ is greater than or equal to the width of the basebands of $x(n)$ or $y(m)$.

It can be seen from Fig. 2.17(a) that the two filters $h_1(k)$ and $h_2(k)$ are operating in cascade at the same sampling rate LF. Thus a more efficient implementation of the overall process can be achieved if the filters are combined into one composite lowpass filter, as shown in Fig. 2.17(b). Since this digital filter, $h(k)$, must serve the purposes of both the decimation and interpolation operations described in the preceding two sections, it is clear from Eqs. (2.52) and (2.74) that it must approximate the ideal digital lowpass characteristic

$$\widetilde{H}(e^{j\omega''}) = \begin{cases} L, & |\omega''| \leqslant \min|\frac{\pi}{L}, \frac{\pi}{M}| \\ 0, & \text{otherwise} \end{cases} \tag{2.84}$$

where

$$\omega'' = 2\pi f T'' = 2\pi f \frac{T}{L} \tag{2.85}$$

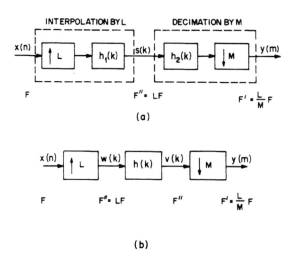

Figure 2.17 (a) Cascade of an integer interpolator and an integer decimator for achieving sampling rate changes by rational fractions; (b) a more efficient implementation of this process.

That is, the ideal cutoff frequency must be the minimum of the two cutoff frequency requirements for the decimator and interpolator, and the sampling rate of the filter is $F'' = LF$.

The time domain input-to-output relation for the general conversion circuit of Fig. 2.17(b) can be derived by considering the integer interpolation and decimation relations derived in Sections 2.3.2 and 2.3.3; that is, from Eq. (2.78) it can be seen that $v(k)$ can be expressed as

$$v(k) = \sum_{r=-\infty}^{\infty} h(k - rL)x(r) \tag{2.86}$$

and from Eq. (2.54) $y(m)$ can be expressed in terms of $v(k)$ as

$$y(m) = v(Mm) \tag{2.87}$$

Combining Eqs. (2.86) and (2.87) gives the desired result

$$y(m) = \sum_{r=-\infty}^{\infty} h(Mm - rL)x(r) \tag{2.88}$$

Alternatively, by making the change of variables

$$r = \left\lfloor \frac{mM}{L} \right\rfloor - n \tag{2.89}$$

and applying Eq. (2.44) gives

$$
y(m) = \sum_{n=-\infty}^{\infty} h\left[Mm - \lfloor \frac{mM}{L} \rfloor L + nL\right] x\left[\lfloor \frac{mM}{L} \rfloor - n\right]
$$

$$
= \sum_{n=-\infty}^{\infty} h(nL + mM \oplus L) x\left[\lfloor \frac{mM}{L} \rfloor - n\right] \tag{2.90}
$$

It is seen that Eq. (2.90) corresponds to the general form of the time-varying digital-to-digital conversion system described by Eq. (2.46) and that the time-varying unit sample response $g_m(n)$ can be expressed as

$$
g_m(n) = h(nL + mM \oplus L), \quad \text{for all } m \text{ and all } n \tag{2.91}
$$

where $h(k)$ is the time-invariant unit sample response of the lowpass digital filter at the sampling rate LF [2.4].

Similarly, by considering the transform relationships of the individual integer decimation and interpolation systems, the output spectrum $Y(e^{j\omega'})$ can be determined in terms of the input spectrum $X(e^{j\omega})$ and the frequency response of the filter $H(e^{j\omega''})$. From Eq. (2.75) it is seen that $V(e^{j\omega''})$ can be expressed in terms of $X(e^{j\omega})$ and $H(e^{j\omega''})$ as

$$
V(e^{j\omega''}) = H(e^{j\omega''}) X(e^{j\omega''L}) \tag{2.92}
$$

and from Eq. (2.62) $Y(e^{j\omega'})$ can be expressed in terms of $V(e^{j\omega''})$ as

$$
Y(e^{j\omega'}) = \frac{1}{M} \sum_{l=0}^{M-1} V(e^{j(\omega'-2\pi l)/M})
$$

$$
= \frac{1}{M} \sum_{l=0}^{M-1} H(e^{j(\omega'-2\pi l)/M}) X(e^{j(\omega'L-2\pi l)/M}) \tag{2.93}
$$

When $H(e^{j\omega''})$ closely approximates the ideal characteristic of Eq. (2.84), it is seen that this expression reduces to

$$
Y(e^{j\omega'}) \cong \begin{cases} \dfrac{L}{M} X(e^{j\omega'L/M}), & \text{for } |\omega'| \leqslant \min\left[\pi, \pi\dfrac{M}{L}\right] \\[2ex] 0, & \text{otherwise} \end{cases} \tag{2.94}
$$

Thus far, we have developed the general system for sampling rate conversion of lowpass signals by arbitrary rational factors, L/M. It was shown that the process of sampling rate conversion could be modeled as a linear, periodically time-varying system, and that the unit sample response of this system, $g_m(n)$, could be expressed in terms of the unit sample response, $h(k)$, of a time-invariant digital filter designed for the highest system sampling rate LF. In Chapter 4 we will consider several methods for designing the filter $h(k)$ and in Chapter 3 we will use the foregoing model to define practical methods for implementing sampling rate conversion systems.

2.4 DECIMATION AND INTERPOLATION OF BANDPASS SIGNALS

2.4.1 The Sampling Theorem Applied to Bandpass Signals

In the preceding sections it was assumed that the signals that we are dealing with are lowpass signals and therefore the filters required for decimation and interpolation are lowpass filters which preserve the baseband signals of interest. In many practical systems, however, it is often necessary to deal with bandpass signals, as well as lowpass signals (as will be seen in later chapters). In this section we show how the concepts of decimation and interpolation can be applied to systems in which bandpass signals are present.

Figure 2.18(a) shows an example of the discrete Fourier transform of a digital bandpass signal $S(e^{j2\pi fT})$ which contains spectral components only in the frequency range $f_l < |f| < f_l + f_\Delta$. If we apply directly the concepts of lowpass sampling as discussed in Section 2.1 it is seen that the sampling rate, F_w, necessary to represent this signal must be twice that of the highest-frequency component in $S(e^{j2\pi fT})$, that is, $F_w \geq 2(f_l + f_\Delta)$. Alternatively, let S^+ denote the component of $S(e^{j2\pi fT})$ associated with $f > 0$ and S^- denote the component of $S(e^{j2\pi fT})$ associated with $f < 0$, as seen in Fig. 2.18. Then, by lowpass translating (modulating) S^+ to the band 0 to f_Δ and S^- to the band $-f_\Delta$ to 0, as illustrated by Fig. 2.18(b), it is seen that a new signal $S_\gamma(e^{j2\pi fT})$ can be generated which is "equivalent" to $S(e^{j2\pi fT})$ in the sense that $S(e^{j2\pi fT})$ can uniquely be reconstructed from $S_\gamma(e^{j2\pi fT})$ by the inverse process of bandpass translation. [Actually, it is seen that $S(e^{j2\pi fT})$ is the "single-sideband" modulated version of $S_\gamma(e^{j2\pi fT})$]. By applying concepts of lowpass sampling to $S_\gamma(e^{j2\pi fT})$, however, it can be seen that the sampling frequency necessary to represent this signal is now $F_\Delta \geq 2f_\Delta$, which can be much lower than the value of F_w specified above (if $f_l \gg f_\Delta$). Thus it is seen that by an appropriate combination of modulation followed by lowpass sampling, any real bandpass signal with (positive frequency) bandwidth f_Δ can be uniquely sampled at a rate $F_\Delta \geq 2f_\Delta$ (i.e., such that the original bandpass signal can be uniquely reconstructed from the sampled representation).

In practice, there are many ways in which the combination of modulation and sampling described above can be carried out. In this section we consider three specific methods in detail: integer-band sampling, quadrature modulation, and

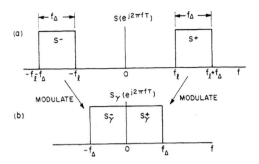

Figure 2.18   Bandpass signal and its lowpass translated representation.

single-sideband modulation (based on a quadrature implementation).

2.4.2 Integer-Band Decimation and Interpolation

Perhaps the simplest and most direct approach to decimating or interpolating digital bandpass signals is to take advantage of the inherent frequency translating (i.e., aliasing or imaging) properties of decimation and interpolation. As discussed in Sections 2.1 and 2.3, sampling and sampling rate conversion can be viewed as a modulation process in which the spectrum of the digital signal contains periodic repetitions of the baseband signal (images) spaced at harmonics of the sampling frequency. This property can be used to advantage when dealing with bandpass signals by associating the bandpass signal with one of these images instead of with the baseband.

Figure 2.19(a) illustrates an example of this process for the case of decimation by the integer factor M. The input signal $x(n)$ is first filtered by the bandpass filter $h_{BP}(n)$ to isolate the frequency band of interest. The resulting bandpass signal, $x_{BP}(n)$, is then directly reduced in sampling rate by an M-sample compressor giving the final output, $y(m)$. It is seen that this system is identical to that of the integer lowpass decimator discussed in Section 2.3.2, with the exception that the filter is a bandpass filter instead of a lowpass filter. Thus the output signal $Y(e^{j\omega'})$ can be expressed as

$$Y(e^{j\omega'}) = \frac{1}{M}\sum_{l=0}^{M-1} H_{BP}(e^{j(\omega'-2\pi l)/M})X(e^{j(\omega'-2\pi l)/M}) \qquad (2.95)$$

From Eq. (2.95) it is seen that $Y(e^{j\omega'})$ is composed of M aliased components of $X(e^{j\omega'})H_{BP}(e^{j\omega'})$ modulated by factors of $2\pi l/M$. The function of the filter $H_{BP}(e^{j\omega})$ is to remove (attenuate) all aliasing components except those associated with the desired band of interest. Since the modulation is restricted to values of

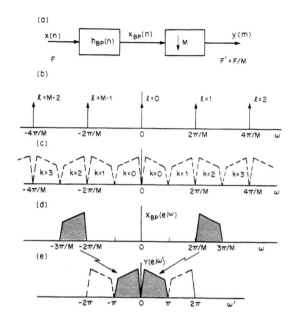

Figure 2.19 Integer-band decimation and a spectral interpretation for the $k = 2$ band.

$2\pi l/M$, it can be seen that only specific frequency bands are allowed by this method. As a consequence the choice of the filter $H_{BP}(e^{j\omega})$ is restricted to approximate one of the M ideal characteristics

$$\widetilde{H}_{BP}(e^{j\omega}) = \begin{cases} 1, & k\dfrac{\pi}{M} < |\omega| < (k+1)\dfrac{\pi}{M} \\ 0, & \text{otherwise} \end{cases} \qquad (2.96)$$

where $k = 0, 1, 2, ..., M - 1$; that is, $H_{BP}(e^{j\omega})$ is restricted to bands $\omega = k\pi/M$ to $\omega = (k+1)\pi/M$, where π/M is the bandwidth.

Figure 2.19(b) to (e) illustrates this approach. Figure 2.19(b) shows the M possible modulating frequencies which are a consequence of the M-to-1 sampling rate reduction; that is, the digital sampling function (a periodic train of unit samples spaced M samples apart) has spectral components spaced $2\pi l/M$ apart. Figure 2.19(c) shows the "sidebands" that are associated with these spectral components, which correspond to the M choices of bands as defined by Eq. (2.96). They correspond to the bands that are aliased into the baseband of the output signal $Y(e^{j\omega'})$ according to Eq. (2.95). [As seen by Eqs. (2.95) and (2.96) and Figs. 2.19(b) and 2.19(c), the relationship between k and l is nontrivial.]

Figure 2.19(d) illustrates an example in which the $k = 2$ band is used, such that $X_{BP}(e^{j\omega})$ is bandlimited to the range $2\pi/M < |\omega| < 3\pi/M$. Since the process of sampling rate compression by M to 1 corresponds to a convolution of the spectra of $X_{BP}(e^{j\omega})$ [Fig. 2.19(d)] and the sampling function [Fig. 2.19(b)], this band is lowpass translated to the baseband of $Y(e^{j\omega'})$ as seen in Fig. 2.19(e). Thus the processes of modulation and sampling rate reduction are achieved simultaneously by the M-to-1 compressor.

Figure 2.20 illustrates a similar example for the $k = 3$ band such that $X_{BP}(e^{j\omega})$ is bandlimited to the band $3\pi/M < |\omega| < 4\pi/M$. In this case it is seen that the spectrum is inverted in the process of lowpass translation. If the noninverted representation of $y(m)$ is desired, it can easily be achieved by modulating $y(m)$ by $(-1)^m$ [i.e., $\hat{y}(m) = (-1)^m y(m)$], which corresponds to inverting the signs of odd samples of $y(m)$. In general, bands associated with even values of k are directly lowpass translated to the baseband of $Y(e^{j\omega'})$, whereas bands associated with odd values of k are translated and inverted [see Fig. 2.19(c)]. This is a consequence of the fact that even numbered bands (k even) correspond to "upper sidebands" of the modulation frequencies $2\pi l/M$, whereas odd-numbered bands (k odd) correspond to "lower sidebands" of the modulation frequencies (e.g., the $k = 2$ band is an upper sideband for $l = 1$ and $l = M - 1$ and the $k = 3$ band is a lower sideband for $l = 2$ and $l = M - 2$, as seen in Fig. 2.19).

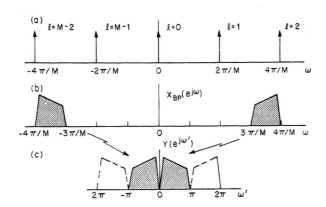

Figure 2.20 Spectral interpretation of integer-band decimation for the band $k = 3$.

Figure 2.21 illustrates an example in which the integer-band constraints of Eq. (2.96) are not satisfied. It is seen that nonrecoverable aliasing occurs in the baseband of $Y(e^{j\omega'})$, and therefore the signal $X_{BP}(e^{j\omega})$, when integer-band constraints are violated, cannot be reconstructed from its decimated version.

The process of integer-band interpolation is the inverse to that of integer-band decimation; that is, it performs the reconstruction (interpolation) of a bandpass

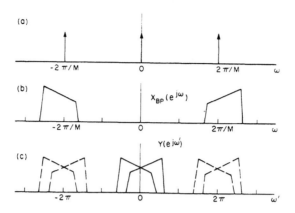

Figure 2.21 Spectral interpretation of integer-band decimation when integer-band constraints are violated.

signal from its integer-band decimated representation. Figure 2.22(a) illustrates this process. The input signal, $x(n)$, is sampling rate expanded by L [by inserting $L - 1$ zero-valued samples between each pair of samples of $x(n)$] to produce the signal $w(m)$. From the discussion of integer interpolation in Section 2.3.3 it is seen that the spectrum of $w(m)$ can be expressed as

$$W(e^{j\omega'}) = X(e^{j\omega'L}) \tag{2.97}$$

and it corresponds to periodically repeated images of the baseband of $X(e^{j\omega})$ centered at the harmonics $\omega' = 2\pi l/L$ [as depicted in Fig. 2.22(b) and (c)]. A bandpass filter $h_{BP}(m)$ is then used to select the appropriate image of this signal. It can be seen that to obtain the kth image, the bandpass filter must approximate the ideal characteristic

$$\widetilde{H}_{BP}(e^{j\omega'}) = \begin{cases} L, & k\dfrac{\pi}{L} < |\omega'| < (k+1)\dfrac{\pi}{L} \\ 0, & \text{otherwise} \end{cases} \tag{2.98}$$

where $k = 0, 1, 2, ..., L - 1$. Figure 2.22(d) shows an example of the output spectrum of the bandpass signal $Y(e^{j\omega'})$ for the $k = 2$ band and Fig. 2.22(e) illustrates an example for the $k = 3$ band. As in the case of integer-band decimation, it is also seen that the spectrum of the resulting bandpass signal is inverted for odd values of k. If this inversion is not desired, the input signal $x(n)$ can first be modulated by $(-1)^n$, which inverts the spectrum of the baseband and consequently the bandpass signal.

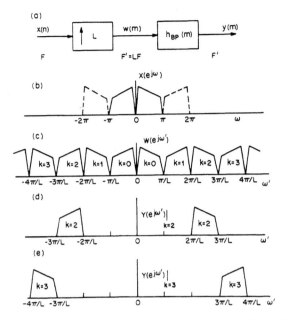

Figure 2.22 Spectral interpretation of integer-band interpolation of bandpass signals.

From the discussion in Section 2.3.3 it is seen that the integer-band interpolator is identical to that of the lowpass interpolator with the exception that the filter may be a bandpass filter. Thus the spectrum of the output signal can be expressed as

$$Y(e^{j\omega'}) = H_{BP}(e^{j\omega'})X(e^{j\omega'L}) \qquad (2.99)$$

or

$$Y(e^{j\omega'}) \simeq \begin{cases} LX(e^{j\omega'L}), & k\dfrac{\pi}{L} < |\omega'| < (k+1)\dfrac{\pi}{L} \\ 0, & \text{otherwise} \end{cases} \qquad (2.100)$$

The processes of integer band decimation and interpolation can also be used for translating or modulating bandpass signals from one integer band to another. For example, a cascade of an integer-band decimator with the bandpass characteristic for $k = 2$ and an integer-band interpolator (with $L = M$) with a bandpass characteristic for $k = 3$ results in a system that translates the band $2\pi/M$ to $3\pi/M$ to a band $3\pi/M$ to $4\pi/M$. Other examples of integer-band interpolators and decimators are presented in later chapters.

2.4.3 Quadrature Modulation of Bandpass Signals

Although the methods of integer-band decimation and interpolation provide a simple and direct way to modulate bandpass signals, they are inherently limited to specific choices of bands. In practice, of course, this is not always convenient, and therefore it is necessary to consider other methods of modulation that do not have these restrictions. In this section we consider methods of quadrature modulation.

Figure 2.23(a) depicts the spectrum of a bandpass signal $X(e^{j\omega})$ that has arbitrary band edges $|\omega_l|$ and $|\omega_l + \omega_\Delta|$ and a center frequency for the band

$$\omega_0 = \omega_l + \frac{\omega_\Delta}{2} \qquad (2.101)$$

It is desired to modulate this bandpass signal to the form of a lowpass signal so that it can be reduced in sampling rate without aliasing. We first consider a number of properties of the signal $X(e^{j\omega})$. The spectrum $X(e^{j\omega})$ is related to the time sequence $x(n)$ via the Fourier transform relation

$$X(e^{j\omega}) = \sum_{n=-\infty}^{\infty} x(n)e^{-j\omega n}$$

$$= \sum_{n=-\infty}^{\infty} x(n)[\cos(\omega n) - j\sin(\omega n)] \qquad (2.102)$$

and by assuming that $x(n)$ is real it can be seen that

$$\text{Re}\,[X(e^{j\omega})] = \text{Re}\,[X(e^{-j\omega})] = \sum_{n=-\infty}^{\infty} x(n)\cos(\omega n) \qquad (2.103a)$$

and

$$\text{Im}\,[X(e^{j\omega})] = -\text{Im}\,[X(e^{-j\omega})] = -\sum_{n=-\infty}^{\infty} x(n)\sin(\omega n) \qquad (2.103b)$$

where $\text{Re}[X(e^{j\omega})]$ and $\text{Im}\,[X(e^{j\omega})]$ denote the real and imaginary components of $X(e^{j\omega})$, respectively. From Eqs. (2.103) it is clear that $X(e^{j\omega})$ is conjugate symmetric. Also, by denoting X^+ as the component of $X(e^{j\omega})$ for $\omega \geq 0$ and X^- as the component of $X(e^{j\omega})$ for $\omega \leq 0$, as denoted in Fig. 2.23(a), it is clear that X^+ and X^- are conjugate symmetric and therefore each component can be uniquely reconstructed from the other. Therefore, we only need to focus on one component.

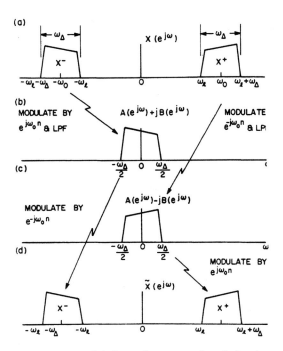

Figure 2.23 Quadrature modulation and reconstruction of a bandpass signal.

By modulating the real signal $x(n)$ by the complex signal $e^{j\omega_0 n}$, it is seen that the resulting complex signal corresponds to a frequency translation

$$X(e^{j(\omega-\omega_0)}) = \sum_{n=-\infty}^{\infty} [x(n)e^{j\omega_0 n}]e^{-j\omega n} \qquad (2.104)$$

This modulation moves the center of the X^- band to $\omega = 0$ and the center of the X^+ band to $\omega = 2\omega_0$. Since we only need to consider one component for a unique reconstruction of $X(e^{j\omega})$, it is apparent that the modulated signal $x(n)e^{j\omega_0 n}$ can be lowpass filtered to the band $-\omega_\Delta/2 < \omega < \omega_\Delta/2$. Consequently, we can define the filtered signal as

$$a(n) + jb(n) = \sqrt{2}[x(n)e^{j\omega_0 n}] * h(n) \qquad (2.105)$$

$$= [\sqrt{2}x(n)\cos(\omega_0 n)] * h(n) + j[\sqrt{2}x(n)\sin(\omega_0 n)] * h(n)$$

or its transform as

$$A(e^{j\omega}) + jB(e^{j\omega}) = \sqrt{2}X(e^{j(\omega-\omega_0)})H(e^{j\omega}) \qquad (2.106)$$

where * in Eq. (2.105) denotes a discrete convolution and the filter $H(e^{j\omega})$ [or $h(n)$] approximates the ideal lowpass characteristic

$$|\widetilde{H}(e^{j\omega})| = \begin{cases} 1, & 0 \leqslant |\omega| < \dfrac{\omega_\Delta}{2} \\ 0, & \text{otherwise} \end{cases} \tag{2.107}$$

The resulting lowpass signal $A(e^{j\omega}) + jB(e^{j\omega})$ is depicted in Fig. 2.23(b) and the components $A(e^{j\omega})$ and $B(e^{j\omega})$ [or their inverse transforms $a(n)$ and $b(n)$] are often referred to as *quadrature* components of the bandpass signal $X(e^{j\omega})$. The factor of $\sqrt{2}$ in Eqs. (2.105) and (2.106) is included for purposes of normalization. Alternatively, it is possible to modulate $x(n)$ by $e^{-j\omega_0 n}$ and lowpass filter the resulting signal by $h(n)$. This leads to the conjugate signal $A(e^{j\omega}) - jB(e^{j\omega})$, as illustrated in Fig. 2.23(c).

To reconstruct the signal $x(n)$ from its quadrature components $a(n)$ and $b(n)$, it is a simple matter to modulate the signal $a(n) + jb(n)$ by $e^{-j\omega_0 n}$ to form the signal corresponding to X^- and similarly to modulate the signal $a(n) - jb(n)$ by $e^{j\omega_0 n}$ to form the signal corresponding to X^+. Summing these two signals X^- and X^+ then gives the reconstructed version of $X(e^{j\omega})$ denoted as $\widetilde{X}(e^{j\omega})$, as illustrated in Fig. 2.23(d). This sequence of operations can be expressed as follows:

$$\widetilde{x}(n) = \frac{[a(n) + jb(n)]e^{-j\omega_0 n} + [a(n) - jb(n)]e^{j\omega_0 n}}{\sqrt{2}}$$

$$= \sqrt{2}a(n) \cos(\omega_0 n) + \sqrt{2}b(n) \sin(\omega_0 n) \tag{2.108}$$

The process above assumes that the quadrature signals $a(n)$ and $b(n)$ are bandlimited to the band $-\omega_\Delta/2$ to $\omega_\Delta/2$. In general, this may not be true and they may first have to be lowpass filtered by a filter $h(n)$ which approximates the ideal characteristic in Eq. (2.107).

Figure 2.24 depicts a block diagram of the sequence of operations generally involved in quadrature modulating a real signal $x(n)$ to produce the quadrature signals $a(n)$ and $b(n)$ according to Eq. (2.105). The second part of the figure illustrates the sequence of operations involved in reconstructing a signal $\widetilde{x}(n)$ from an arbitrary set of quadrature signals $a(n)$ and $b(n)$ (assuming that they are not necessarily bandlimited). This follows from Eq. (2.108).

Since the complex quadrature signal $a(n) + jb(n)$ obtained from the quadrature modulator is bandlimited to a bandwidth of $\omega_\Delta/2 < \pi/2$, its sampling rate can potentially be reduced (by the process of decimation), and if $\omega_\Delta/2$ is much less than $\pi/2$, it can be reduced substantially. In the process of reconstructing the original bandpass signal, the sampling rate of the quadrature signal can first be increased (by interpolation) to its original sampling rate prior to demodulation.

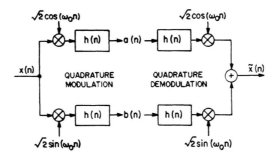

Figure 2.24 Block diagram of quadrature modulation and demodulation of a bandpass signal.

For an integer decimation by M and a subsequent interpolation by $L = M$, it can be seen that the relationship between M and ω_Δ is

$$\frac{\pi}{M} \geqslant \frac{\omega_\Delta}{2}$$

or

$$M = L \leqslant \frac{2\pi}{\omega_\Delta} \tag{2.109}$$

Figure 2.25 illustrates a block diagram of this process, where it is seen that the lowpass filter $h(n)$ must approximate the ideal characteristic in Eq. (2.107) and $\hat{a}(m)$ and $\hat{b}(m)$ denote the quadrature signals at the reduced sampling rate F/M (F is assumed to be the initial sampling rate). Note that the lowpass filter $h(n)$ in the decimator and the lowpass filter for the quadrature modulator are in fact one and the same filter. Similarly, the lowpass filter and the interpolation filter in the demodulator are the same.[2]

In the case where it is desired to reduce the sampling rate of the quadrature signal by a noninteger, rational ratio L/M, the block diagram configuration in Fig. 2.26 applies. From the previous discussion, it is seen that the ratio L/M must satisfy the constraint

$$\frac{L\pi}{M} \geqslant \frac{\omega_\Delta}{2}$$

or

[2] Recall also that the filter $h(n)$ in the L-to-1 interpolator should have an additional gain of L, as discussed in Section 2.3.3.

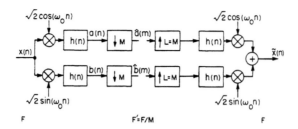

Figure 2.25 Block diagram of a quadrature modulator and sampling of bandpass signals using integer decimation and interpolation.

$$1 > \frac{L}{M} \geqslant \frac{\omega_\Delta}{2\pi} \tag{2.110}$$

As seen in Fig. 2.26, the outputs of the cosine and sine modulators are first increased in sampling rate by a factor L. They are then lowpass filtered by the filters $h(k)$, which must approximate the ideal characteristic (referenced to the sampling rate $F' = LF$)

$$\widetilde{H}(e^{j\omega'}) = \begin{cases} 1, & 0 \leqslant |\omega'| \leqslant \dfrac{\omega_\Delta}{2L} \\ 0, & \text{otherwise} \end{cases} \tag{2.111}$$

and then decimated by M. The resulting quadrature signals are at the sampling rate $F'' = (L/M)F$. In the quadrature demodulator the reverse process takes place, as seen in Fig. 2.26.

2.4.4 Single-Sideband Modulation

The quadrature modulation approach coupled with methods of interpolation and decimation as discussed in the preceding section permits the sampling of bandpass signals, with arbitrary band edges, at their minimum required sampling rate. One consequence of this method, however, is that the resulting sampled signal $\hat{a}(m) + j\hat{b}(m)$ is a complex signal, and in some applications this may not be desired. By a slight modification of the quadrature approach, however, it can be shown that a real signal output can be obtained which corresponds to a "single-sideband" modulated format at the minimal sampling rate [2.5, 2.6].

Figure 2.27 illustrates this approach. The complex signal $a(n) + jb(n)$ is modulated by $e^{-j(\omega_\Delta/2)n}$ and its conjugate is modulated by $e^{j(\omega_\Delta/2)n}$ to produce the respective sideband signals X_γ^- and X_γ^+, as shown in Fig. 2.27(b). Summing these

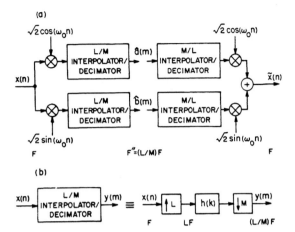

Figure 2.26 Block diagram of a quadrature modulator using decimation of the quadrature signals by rational fractions L/M.

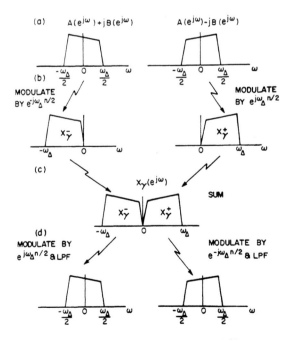

Figure 2.27 Constructing the single-sideband signal $X_\gamma(e^{j\omega})$ from the quadrature signal.

two signals then gives the signal $X_\gamma(e^{j\omega})$, as shown by Fig. 2.27(c). This sequence of operations corresponds to the time domain operations

$$x_\gamma(n) = \frac{[a(n) + jb(n)]e^{-j(\omega_\Delta/2)n} + [a(n) - jb(n)]e^{j(\omega_\Delta/2)n}}{\sqrt{2}}$$

$$= \sqrt{2}a(n) \cos\left[\frac{\omega_\Delta n}{2}\right] + \sqrt{2}b(n) \sin\left[\frac{\omega_\Delta n}{2}\right] \tag{2.112}$$

Similarly, by modulating the signal $x_\gamma(n)$ by $e^{j(\omega_\Delta/2)n}$ and $e^{-j(\omega_\Delta/2)n}$, respectively, and lowpass filtering the resulting signals by the lowpass filter $h(n)$, which approximates the ideal characteristic

$$\tilde{H}(e^{j\omega}) = \begin{cases} 1, & |\omega| \leqslant \dfrac{\omega_\Delta}{2} \\ 0, & \text{otherwise} \end{cases} \tag{2.113}$$

the quadrature signals $A(e^{j\omega}) \pm jB(e^{j\omega})$ can be recovered as shown by Fig. 2.27(d). This corresponds to the sequence of operations

$$a(n) \pm jb(n) = \sqrt{2}\left[x_\gamma(n)e^{\pm j(\omega_\Delta/2)n}\right] * h(n) \tag{2.114}$$

from which it follows that

$$a(n) = \left[\sqrt{2}x_\gamma(n) \cos\left[\frac{\omega_\Delta n}{2}\right]\right] * h(n) \tag{2.115a}$$

$$b(n) = \left[\sqrt{2}x_\gamma(n) \sin\left[\frac{\omega_\Delta n}{2}\right]\right] * h(n) \tag{2.115b}$$

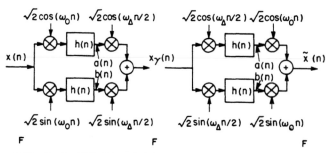

Figure 2.28 Single-sideband modulation based on a modification of quadrature modulation.

Thus from Eqs. (2.112) and (2.115) a block diagram of operations can be obtained for transferring between the complex quadrature signal $a(n) + jb(n)$ and the real signal $x_\gamma(n)$, as shown in Fig. 2.28. This approach is sometimes referred to as the *Weaver modulator* [2.5].

Since the signal $x_\gamma(n)$ is bandlimited to $|\omega| \leqslant \omega_\Delta$, it is clear that it can be reduced in sampling rate to its minimum rate $F' = (\omega_\Delta/\pi)F$. Alternatively, the signals $a(n)$ and $b(n)$ can first be decimated to this sampling rate [i.e., twice the actual minimum rate necessary to represent $a(n) + jb(n)$] and then used to generate the decimated version of $x_\gamma(n)$, in which case the modulating frequency $\omega_\Delta/2$ must also be referenced to this lower frequency F'. Letting $\omega'_\Delta/2$ denote this modulating frequency, it is seen that

$$\frac{\omega'_\Delta}{2} = \frac{\pi}{2} \qquad\qquad (2.116)$$

That is, the modulating frequency used to convert the decimated values of $a(n)$ and $b(n)$ to decimated values of $x_\gamma(n)$ is one-fourth of the sampling rate F'. At this sampling rate, however, it is seen that

$$\cos\left(\frac{\omega'_\Delta}{2}m\right) = \cos\left(\frac{\pi}{2}m\right) = 1, 0, -1, 0, 1, \ldots \qquad (2.117a)$$

and

$$\sin\left(\frac{\omega'_\Delta}{2}m\right) = \sin\left(\frac{\pi}{2}m\right) = 0, 1, 0, -1, \ldots \qquad (2.117b)$$

for $m = 0, 1, 2, 3, \ldots$ respectively. Figure 2.29(a) illustrates this process of generating $\hat{x}_\gamma(m)$ from the decimated quadrature components $\hat{a}(m)$ and $\hat{b}(m)$, respectively, where $\hat{x}_\gamma(m)$, $\hat{a}(m)$, and $\hat{b}(m)$ correspond to the decimated versions of $x_\gamma(n)$, $a(n)$, and $b(n)$, respectively. It is assumed that F/F' is equal to an integer factor M' and that the filter $h(n)$ [see Fig. 2.29(b)] approximates the ideal characteristic

$$\widetilde{H}(e^{j\omega}) = \begin{cases} 1, & 0 \leqslant |\omega| < \dfrac{\omega_\Delta}{2} = \dfrac{\pi}{2M'} \\[2mm] 0, & \text{otherwise} \end{cases} \qquad (2.118)$$

From Eqs. (2.117) and Fig. 2.29(a) it is seen that even numbered samples of $\hat{x}_\gamma(m)$ correspond to even-numbered samples of $\hat{a}(m)$ (with appropriate changes of signs and a scale factor of $\sqrt{2}$). Similarly, odd-numbered samples of $\hat{x}_\gamma(m)$ correspond to odd-numbered samples of $\hat{b}(m)$ (with appropriate sign changes and a

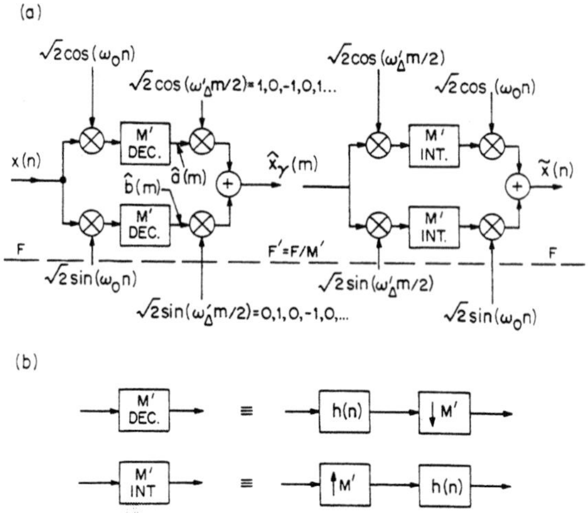

Figure 2.29 Single-sideband modulation with integer decimation to the minimum sampling rate.

scale factor of $\sqrt{2}$). Since only even-numbered samples of $\hat{a}(m)$ and only odd-numbered samples of $\hat{b}(m)$ are needed for constructing $\hat{x}_\gamma(m)$, the signals $a(n)$ and $b(n)$ can in fact be decimated by a total factor of $M = 2M'$ (to their minimum allowed sampling rates), giving the decimated signals $\hat{a}(k)$ and $\hat{b}(k)$, where the samples $\hat{a}(k)$ correspond to the even-numbered samples of $\hat{a}(m)$ and the samples $\hat{b}(k)$ correspond to the odd-numbered samples of $\hat{b}(m)$. The sign changes can be achieved by modulations of $(-1)^k$ (i.e., reversing the signs of alternate samples) on the signals $\hat{a}(k)$ and $\hat{b}(k)$. In addition, since odd samples of $\hat{b}(m)$ are required (instead of even samples) an offset delay of one sample at the rate $F/M' = F/2M$ or equivalently $M/2$ samples at the rate F must be introduced in the channel for $b(n)$. The corresponding sequence of operations is thus illustrated in Fig. 2.30 and it is seen that the sideband signal $\hat{x}_\gamma(m)$, sampled at its minimum sampling rate, consists of interleaved samples of the quadrature signal $\hat{a}(k)$ and the offset quadrature signal $\hat{b}(k)$, also sampled at their minimum rates. Although the delay $z^{-M/2}$ and the advance $z^{M/2}$ appear to add additional complexity to the system, they in fact do not result in any additional complexity in a practical implementation as will be seen in Chapters 3 and 7.

2.4.5 Discussion

In this section we have discussed concepts of sampling and reconstruction of bandpass signals at their minimum required rate by combined methods of modulation, interpolation, and decimation. Although we have discussed three particular methods of modulation a variety of other techniques, such as Hilbert

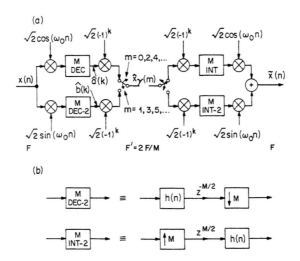

Figure 2.30 More efficient implementation of single-sideband based on sample interleaving of the quadrature signals $\hat{a}(m)$ and (offset) $\hat{b}(m)$.

transforms and second-order sampling, exist and can be conveniently combined with decimation and interpolation techniques for minimum rate sampling of bandpass signals [2.7, 2.8]. A discussion of these methods is beyond the scope of this book.

2.5 SUMMARY

This chapter has been directed at the basic concepts of digital sampling of analog lowpass and bandpass signals and at the fundamentals of converting the sampling rates of sampled signals by a direct digital-to-digital approach. Basic interpretations of the operations of sampling rate conversion have been given in terms of their analog equivalents and in terms of modulation concepts. In the remainder of the book we depend heavily on these fundamental principles and investigate, in more detail, methods of designing the required filters and methods for efficiently implementing the signal processing operations involved in sampling rate conversion. Finally, we will apply these principles to a variety of applications and show how they can be used not only for sampling rate conversion but also for the efficient implementation of a number of other signal processing operations as well.

REFERENCES

2.1   R. A. Meyer and C. S. Burrus, "A Unified Analysis of Multirate and Periodically Time-Varying Digital Filters," *IEEE Trans. Circuits Syst.*, Vol. CAS-22, No. 3, pp. 162-168, March 1975.

2.2   B. Liu and P. A. Franaszek, "A Class of Time-Varying Digital Filters," *IEEE Trans. Circuit Theory,* Vol. CT-16, No. 4, pp. 467-471, November 1969.

2.3   R. W. Schafer and L. R. Rabiner, "A Digital Signal Processing Approach to Interpolation," *Proc. IEEE,* Vol. 61, No. 6, pp. 692-702, June 1973.

2.4   R. E. Crochiere and L. R. Rabiner, "Optimum FIR Digital Filter Implementations for Decimation, Interpolation, and Narrow-Band Filtering," *IEEE Trans. Acoust., Speech Signal Process.,* Vol. ASSP-23, No. 5, pp. 444-456, October 1975.

2.5   S. Darlington, "On Digital Single-Side-Band Modulators," *IEEE Trans. Circuit Theory,* Vol. CT-17, pp. 409-414, August 1970.

2.6   R. E. Crochiere, S. A. Webber, and J. L. Flanagan, "Digital Coding of Speech in Sub-bands," *Bell Syst. Tech. J.,* Vol. 55, No. 8, pp. 1069-1085, October 1976.

2.7   D. A. Linden, "A Discussion of Sampling Theorems," *Proc. IRE,* Vol. 47, No. 7, pp. 1219-1226, July 1959.

2.8   R. G. Pridham and R. A. Mucci, "Digital Interpolation Beam Forming for Lowpass and Bandpass Signals," *Proc. IEEE* Vol. 67, No. 6, pp. 904-919, June 1979.

3

Structures and Network Theory for Multirate Digital Systems

3.0 INTRODUCTION

In Chapter 2 we presented an extensive treatment of the theoretical issues involved in systems for sampling rate conversion for both lowpass and bandpass signals. In the next two chapters we investigate several practical implications of the theory: numerically efficient structures for realizing multirate systems (Chapter 3) and filter design techniques which are suited to the design problems encountered in such systems (Chapter 4).

It is easy to understand the need for studying structures for realizing sampling rate conversion systems by examining the simple block diagram of Fig. 3.1, which can be used to convert the sampling rate of a signal by a factor of L/M. As discussed in Chapter 2, the theoretical model for this system is increasing the signal sampling rate by a factor of L [by filling in $L - 1$ zero-valued samples between each sample of $x(n)$ to give the signal $w(k)$], filtering $w(k)$ to eliminate the images of $X(e^{j\omega})$ by a standard linear time-invariant lowpass filter, $h(k)$, to give $v(k)$, and sampling rate compressing $v(k)$ by a factor M [by retaining 1 of each M samples of $v(k)$]. A direct implementation of the system of Fig. 3.1 is grossly inefficient since the lowpass filter, $h(k)$, is operating at the high sampling rate on a signal for which $L - 1$ out of each L input values are zero, and the values of the filtered output are required only once each M samples. For this example, one can directly apply this knowledge in implementing the system of Fig. 3.1 in a more efficient manner, as is discussed in this chapter. An additional goal of this chapter is to show how to use network theory concepts to manipulate *any* multirate digital system so as to achieve an efficient practical implementation. We consider only single-stage conversion in this chapter. However, in Chapter 5, we extend these concepts to include multistage implementations which can achieve greater efficiencies than single-stage designs when the conversion (either L or M or both) ratios are large.

Figure 3.1 Block diagram for a system to convert the sampling rate by a factor L/M.

In order to be able to manipulate network structures, the rules of signal-flow graphs must be understood and followed. As such, we begin this chapter by reviewing the basic concepts used to describe and modify a digital system. Since the vast majority of the computation in the implementation of a multirate system is in the digital filtering, we next review a number of the important structures that have been proposed for both finite impulse response (FIR) and infinite impulse response (IIR) digital filters. It is then shown how the filter structures can be combined with the time-varying components to define efficient structures for the implementation of various sampling rate conversion systems.

In Sections 3.1 to 3.4 we take a very utilitarian and informal point of view in terms of developing practical structures for multirate systems. Subtle concepts such as network transposition are informally described and used without proof in the development of these structures. Later, in Section 3.5 on advanced topics, we discuss in more detail some of the underlying network theoretic issues involved in formally proving these concepts.

3.1 SIGNAL-FLOW GRAPH REPRESENTATION OF DIGITAL SYSTEMS

Until now the networks we have been considering have been represented by block diagrams of the type shown in Fig. 3.1. The block diagram depicts the *generalized* operation of the system. The blocks that are used can be as simple (e.g., an adder, multiplier, etc.) or as complex (sampling rate changer) as desired. The block diagram generally has the property that the system cannot be implemented directly (in either software or hardware) without additional specifications on the blocks.

In order to define precisely the set of operations necessary to implement a digital system we use the signal-flow graph representation [3.1-3.3]. Signal-flow graphs provide a graphical representation of the explicit set of equations that are used to implement the system. As such signal-flow graphs can be manipulated similarly to the way in which mathematical equations can be manipulated and the results will be identical. Furthermore, finite-word-length properties of a network can be analyzed directly from signal-flow graphs.

In this section we briefly review the basic concepts of signal-flow graphs and show how they are used to describe structures. We also review a number of basic network concepts concerning signal-flow graphs, such as commutation and transposition, particularly as they relate to time-varying and multirate systems. These concepts are applied in an intuitive way in Sections 3.2 to 3.4 to assist in developing efficient structures for decimators and interpolators. In Section 3.5, after we have established a good intuitive understanding of these network concepts, we examine them again from a more formalistic network theoretic point of view and show how they can be used to derive other network properties.

3.1.1 Signal-Flow Graphs: Basic Principles

A signal-flow graph may be defined in terms of a set of branches and nodes. Branches define the signal operations in the structure such as delays, gains, sampling rate expanders and compressors, modulation, and so on. Nodes define the connection points of these branches in the structure. Figure 3.2 illustrates a number of typical branch operations that are commonly used in time-invariant and time-varying linear systems. The signal entering a branch is taken as the signal associated with the input node value of the branch. Input branches allow external signals to enter the network, and output branches are for terminal signals.

OPERATION	SYMBOL	TIME-DOMAIN DESCRIPTION	FREQUENCY DOMAIN DESCRIPTION
UNIT DELAY	$x(n) \xrightarrow{z^{-1}} y(n)$	$y(n)=x(n-1)$	$Y(e^{j\omega})=e^{-j\omega}X(e^{j\omega})$
M SAMPLE DELAY	$x(n) \xrightarrow{z^{-M}} y(n)$	$y(n)=x(n-M)$	$Y(e^{j\omega})=e^{-jM\omega}X(e^{j\omega})$
GAIN	$x(n) \xrightarrow{c} y(n)$	$y(n)=cx(n)$	$Y(e^{j\omega})=cX(e^{j\omega})$
GAIN AND DELAY	$x(n) \xrightarrow{cz^{-1}} y(n)$	$y(n)=cx(n-1)$	$Y(e^{j\omega})=ce^{-j\omega}X(e^{j\omega})$
SAMPLING RATE COMPRESSOR	$x(n) \to \boxed{\downarrow M} \to y(m)$	$y(m)=x(Mm)$	$Y(e^{j\omega})=\frac{1}{M}\sum_{\ell=0}^{M-1}X(e^{j(\omega-2\pi\ell)/M})$
SAMPLING RATE EXPANDER	$x(n) \to \boxed{\uparrow L} \to y(m)$	$y(m)=\begin{cases} x(m/L), & m=0,\pm L,\pm 2L,\cdots \\ 0 & \text{OTHERWISE} \end{cases}$	$Y(e^{j\omega})=X(e^{j\omega L})$
MODULATOR	$x(n) \to \otimes \to y(n),\ \phi(n)$	$y(n)=\phi(n)x(n)$	$Y(e^{j\omega})=X(e^{j\omega})*\phi(e^{j\omega})$
INPUT BRANCH	$x(n)$	—	—
OUTPUT BRANCH	$\xrightarrow{y(n)}$	—	—

$*$ DENOTES CONVOLUTION

Figure 3.2 Typical branch operations in signal-flow graphs.

Figure 3.3 illustrates the function of the nodes in the network. Associated with each node is a node value which is the sum of all the branch signals entering the node. Branches leaving the node have input branch signal values equal to this node value. It is assumed that all signals entering and leaving a *particular* node in a network have the *same* sampling rate. Otherwise, the signal-flow graph is inconsistent and unrealizable. It is permissible, however, for different nodes in a network to have different associated sampling rates and, in fact, this is the case for multirate systems such as decimators and interpolators.

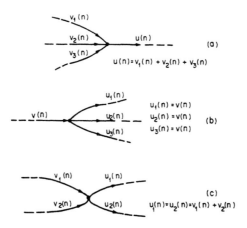

Figure 3.3 Illustration of node functions in a network; (a) summing, (b) branching, (c) summing and branching.

Figure 3.4 illustrates an example of a simple network described by a signal-flow graph. The input branch applies the external signal $x(n)$ to the network and the output of the network $y(n)$ is defined as one of the node values. From the signal-flow graph we can immediately write down the network equations as

$$y(n) = w(n) + c_2 w(n-1) \quad \text{(output node)} \qquad (3.1)$$

$$w(n) = x(n) + c_1 w(n-1) \quad \text{(input node)} \qquad (3.2)$$

Simple manipulation of Eqs. (3.1) and (3.2) leads to a difference equation of the network of the form

$$y(n) = x(n) + c_2 x(n-1) + c_1 y(n-1) \qquad (3.3)$$

explicitly showing that the signal-flow graph of Fig. 3.3 is a simple implementation of a general first-order difference equation.

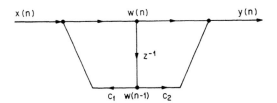

Figure 3.4 Signal-flow graph of a simple digital network.

3.1.2 Commutation of Branch Operations and Circuit Identities

Since a signal-flow graph corresponds to an explicit set of mathematical equations, the manipulation or reconfiguration of the signal-flow graph corresponds to the process of modifying the manner in which the computation for this set of equations is carried out. It is often easier, however, to modify or rearrange the signal-flow graph than it is to manipulate the corresponding set of equations, especially for multirate systems. The reasons we are interested in manipulating the structure are strictly computational: to find an equivalent form of the structure that is more efficient (in terms of multiplications, storage, etc.), or one that has better finite-word-length properties (i.e., lower round-off noise, less coefficient sensitivity, etc.).

An important concept in the manipulation of signal-flow graphs is the principle of commutation of branch operations. Two branch operations commute if the order of their cascade operation can be interchanged without affecting the input-to-output response of the cascaded system. Thus, interchanging commutable branches in a network is one way of modifying the network without affecting the desired input-to-output network response. In general, only some types of branches can be commuted and care must be observed in performing such operations (see Section 3.5 for a more formal definition of commutation). A few general rules can be stated, however, and we discuss some of them here.

One general class of branch operations which are commutable are those which are linear time-invariant operations. Such operations can be readily described in terms of their z-transforms. For example, the transform of a cascade of such operations corresponds to the product of the individual z-transforms. Since multiplication is a commutable operation, it follows that linear time-invariant operations are commutable. Figure 3.5 illustrates a number of examples of commutable operations for some common linear time-invariant branch operations. The commutability of linear time-invariant operations can also be extended to whole systems. For example, Fig. 3.6 illustrates the commutation of the block diagram of a cascade of two linear time-invariant networks denoted by their impulse responses $h_1(n)$ and $h_2(n)$.

Another relatively general class of commutable operations occurs for the case

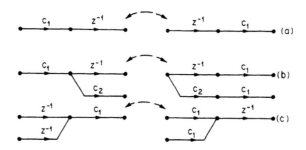

Figure 3.5 Examples of commutation of linear time-invariant branch operations.

Figure 3.6 Example of commutation of two linear time-invariant systems.

of a cascade of a (time-invariant) scalar and a linear but time-varying operation. This is a consequence of the linearity property. Figure 3.7 illustrates several examples of this class of commutable operations. Note that for the modulator one input signal path is considered as the input signal and the second input, $\phi(n)$, is a modulating function.

It is difficult to make general statements about the commutation of branches in which one or both of the branch operations are time-varying. However, for the specific case of sampling rate expanders and compressors, a number of highly useful commutable-type operations can be defined provided that we account for the change of sampling rate incurred by the sampling rate expander or compressor. Strictly speaking, however, the following operations are circuit identities rather than commutation rules.

The first class of identities are those associated with sampling rate compressors. Figure 3.8 illustrates a number of examples of this class. Figure 3.8(a) illustrates that an M-to-1 sampling rate compressor followed by a unit delay can be replaced by an M-sample delay followed by a sampling rate compressor. This identity follows directly from the definition of the sampling rate compressor; that is, letting

$$s(m) = x(Mm) \tag{3.4}$$

then

$$y(m) = s(m-1) = x(M(m-1)) = x(Mm - M) \tag{3.5}$$

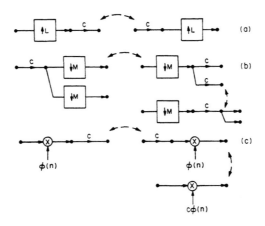

Figure 3.7 Examples of scalars commuting with linear time-varying operations.

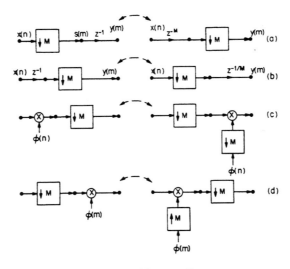

Figure 3.8 Identity systems involving sampling rate compressors.

Figure 3.8(b) illustrates that a delay followed by a sampling rate compressor can be replaced by a sampling rate compressor and a fractional sample delay $z^{-1/M}$. This identity follows from the relation

$$y(m) = x(mM - 1) = x(M(m - \frac{1}{M})) \tag{3.6}$$

Although this fractional delay of $1/M$ samples is not a realizable operation from a practical point of view, it is a conceptual device that is useful in intermediate stages of network manipulations. Examples of this are presented later in this chapter. Figure 3.8(c) and (d) illustrate additional circuit identities involving sampling rate compressors and modulators, namely that modulators commute with sampling rate compressors provided that the sampling rate of the modulating signal be compressed or expanded, depending on whether it precedes or follows the sampling rate compressor.

Figure 3.9 illustrates a class of identities involving sampling rate expanders which are similar to those of Fig. 3.8. Figure 3.9(c), in particular, illustrates a sequence of operations in which a unit delay is passed directly through a cascade of a sampling rate compressor followed by a sampling rate expander. This is accomplished through an intermediate stage which utilizes the concept of a fractional delay as discussed above.

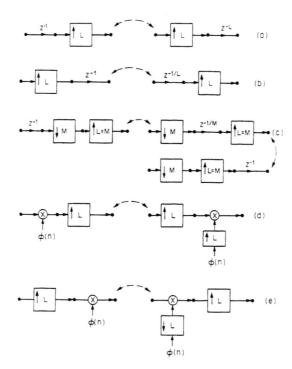

Figure 3.9   Identity systems involving sampling rate expanders.

Finally, Fig. 3.10 illustrates a number of circuit identities involving cascades of sampling rate expanders and compressors. Figure 3.10(a) shows that a compressor with a ratio of $M = L$ inverts the effects of an L-to-1 expander, leaving the signal unchanged. More formally, from Fig. 3.10(a) we have

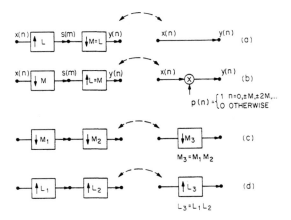

Figure 3.10   Identity operations involving cascades of sampling rate expanders and compressors.

$$s(m) = \begin{cases} x\left[\dfrac{m}{L}\right], & m = 0,\ \pm L,\ \pm 2L,\ \ldots \\ 0, & \text{otherwise} \end{cases} \tag{3.7}$$

and

$$y(n) = s(nL) = x(n) \tag{3.8}$$

Figure 3.10(b) shows that the reverse operation has a very different effect on the input signal (due to the effects of aliasing). If we first compress the sampling rate by a factor of M, we get

$$s(m) = x(nM) \tag{3.9}$$

If we next expand the sampling rate by a factor of $L = M$, we get

$$y(n) = \begin{cases} s\left[\dfrac{n}{M}\right], & n = 0,\ \pm M,\ \pm 2M,\ \ldots \\ 0, & \text{otherwise} \end{cases} \tag{3.10}$$

or

$$y(n) = \begin{cases} x(n), & n = 0,\ \pm M,\ \pm 2M,\ \cdots \\ 0, & \text{otherwise} \end{cases} \tag{3.11}$$

Equation (3.11) can be expressed as the modulation of the input signal by a unit pulse train with pulses spaced M samples apart, that is,

$$y(n) = x(n)p(n) \qquad (3.12a)$$

where

$$p(n) = \sum_{r=-\infty}^{\infty} u_0(n - rM) = \begin{cases} 1, & n = 0, \pm M, ... \\ 0, & \text{otherwise} \end{cases} \qquad (3.12b)$$

as shown at the right of Fig. 3.10(b). Thus the system of Fig. 3.10(a) is *not* commutable with the system of Fig. 3.10(b). Figure 3.10(c) and (d) show that a cascade of sampling rate compressors (or expanders) can be replaced by a single compressor (or expander) whose compression (expansion) factor is the product of the compression (expansion) factors of the cascaded stages.

In Chapter 5, 6, and 7 (Section 7.7.4) we consider a number of additional circuit equivalents involving cascades of decimators and interpolators and show how such operations can be applied in the design of efficient multirate systems. In Section 7.7.4, in particular, we show an exact circuit identity between that of a cascade of two or more interpolators (decimators) and a single-stage interpolator (decimator).

3.1.3 Transposition and Duality for Multirate Systems

The concepts of transposition and duality of networks are fundamental in network theory [3.1-3.4]. Basically, a dual system is one which performs an operation complementary to that of an original system. We have already seen several examples of systems and their duals in Chapter 2. For example, an interpolator performs the complementary operation to that of a decimator, and vice versa. Thus decimators and interpolators are dual systems. Similarly, as seen in Chapter 2, modulators and demodulators perform complementary operations and can be defined as dual systems.

In general, any linear time-invariant or time-varying network has a dual. Furthermore, given a network, its dual network can be obtained by the process of network transposition [3.1-3.4]. In this section we discuss network transposition in a nonrigorous manner. More formal definitions and discussion of duality and transposition are considered in Section 3.5. We consider first the transposition theorem as it relates to linear time-invariant networks with real coefficients and then consider the case of linear time-varying operations. These operations are used later in developing practical structures for decimation and interpolation in Section 3.3.

For the case of linear time-invariant digital networks with real coefficients, the transposition theorem states that if all the branches in a signal-flow graph are reversed in direction, and the roles of the input and output of the network are interchanged, the input-to-output system response of the (transposed) network is unchanged. Figure 3.11 shows an example of a simple linear time-invariant network and its transpose. Both networks are described by the same system function,

$$H(z) = \frac{1 + c_2 z^{-1}}{1 - c_1 z^{-1}} \tag{3.13}$$

It should be noted that in the transposition operation the roles of the nodes are reversed; that is, summing points become branching points and branching points become summing points. Linear time-invariant branch operations remain unchanged in the process of transposition (e.g., delay branches transpose to delay branches and coefficient branches transpose to coefficient branches); only the *direction* of the branches are reversed. This property extends to the entire linear time-invariant network according to the transposition theorem.

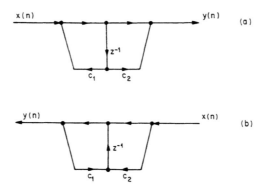

Figure 3.11 A simple linear time-invariant digital network and its transpose.

The transposition of time-varying networks (or branch operations) does not necessarily leave the system response unchanged. This can be seen from the fact that the input and output sampling rates of a multirate network and its transpose need not be the same. The transpose network implements the complementary operation of that of the original network. We consider first the transposition of sampling rate expanders and compressors and then extend these concepts to entire networks involving these types of branches. The discussion of transposition of modulators is deferred until later in this chapter because of the additional complications that arise in transposing these operations.

Figure 3.12 shows the results of transposition of a sampling rate compressor and a sampling rate expander. It is seen that the transpose of a sampling rate compressor is a sampling rate expander in which the expansion ratio is equal to the compression ratio M. Furthermore, since the transposition of a transposed branch returns the original branch, the transpose of a sampling rate expander is a sampling rate compressor.

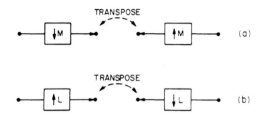

Figure 3.12 Transpositions of a sampling rate compressor and expander.

The transposition of a multirate network containing combinations of sampling rate compressors and expanders and linear time-invariant branches with real coefficients can be accomplished in a manner similar to that in the linear time-invariant case. The directions of all the branches in the network are reversed and the branch operations are replaced by their transpose operations. The roles of the inputs and outputs of the network are interchanged. Figure 3.13 shows that the transpose of a decimator is an interpolator (with $L = M$), and the transpose of a network that changes the sampling rate by a factor of L/M is a network that changes the sampling rate by a factor of M/L. Since the filter $h(n)$ is a linear time-invariant system, its transpose can also be represented by the same impulse response $h(n)$ (with the direction of inputs and outputs reversed).

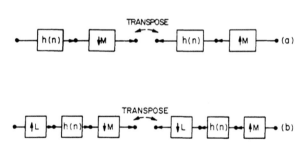

Figure 3.13 Transposes of a decimator and a generalized L/M sampling rate changer.

Other examples of networks and their transposes can be seen in Figs. 2.24 to 2.26 and 2.28 to 2.30. The left part of the networks (modulators) represent the initial networks and the right part of the networks (demodulators) represent their transposes.

3.2 REVIEW OF STRUCTURES FOR LINEAR TIME-INVARIANT FILTERS

Before considering network structures for multirate digital systems, we first review a number of important classes of structures for the linear time-invariant digital filters that are required in all sampling rate converters. These basic structural concepts coupled with the network identities and the operations of commutation and transposition discussed previously are then used in the next section to develop efficient structures for decimators and interpolators.

As seen from the previous discussion, a linear time-invariant digital system can be described as a system in which all the nodes in the network have the same sampling rate and all the branches in the network correspond to linear time-invariant operations.[1]

We consider two basic classes of structures in this section. The first class is associated with digital filters whose impulse response $h(n)$ is finite in duration [i.e., $h(n)$ is zero outside a finite interval of samples n] and they are denoted as finite impulse response (FIR) filters. The second class of filters is associated with impulse responses whose duration is infinite and they are referred to as infinite impulse response (IIR) digital filters.

3.2.1 FIR Direct-Form Structure

The most common structure for implementing time-invariant FIR systems is the direct-form structure [3.1, 3.5]. It corresponds to a direct implementation of the convolution equation

$$y(n) = \sum_{k=0}^{N-1} h(k)x(n-k) \tag{3.14}$$

where the N-point impulse response of the filter, $h(k)$, is assumed to be zero for $k < 0$ and $k > N - 1$. As shown in Eq. (3.14), the computation of an output sample $y(n)$ is obtained from the present and past $N - 1$ samples of the input signal $x(n)$, respectively, weighted by the filter coefficients $h(k)$. This sequence of operations is depicted in signal-flow graph form in Fig. 3.14.

In many applications the FIR filter is designed to have linear phase [3.1, 3.5]. Consequently, the impulse response is symmetric and satisfies the relation

$$h(k) = h(N-1-k) \tag{3.15}$$

This property can be utilized to reduce the number of multiplications in the direct-form structure by a factor of approximately 2. For example, when N is even, the application of the symmetry property of Eq. (3.15) to Eq. (3.14) leads to

$$y(n) = \sum_{k=0}^{N/2-1} h(k)[x(n-k) + x(n-(N-1-k))] \tag{3.16}$$

---

[1] It is possible, however, to realize a linear time-invariant system function with a time-varying digital system, as will be seen later.

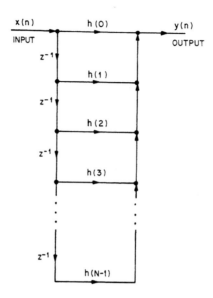

Figure 3.14 Direct form structure for an FIR digital filter.

Figure 3.15 illustrates the signal-flow graph for this computation. A similar form of this structure results when N is odd.

3.2.2 Transposed Direct-Form Structure for FIR Filters

By transposition of the direct-form FIR structures of Figs. 3.14 and 3.15, alternative direct-form structures are obtained. These structures, shown in Figs. 3.16 and 3.17, respectively, have the same system function as the original structures. Because of the transpose operation, the branch points in the shift registers become summing points in the transposed structures. Thus these shift registers become *accumulator shift registers* in the transposed structures. It will be seen later that for some cases, the transposed direct-form filters are more suitable for structural manipulations than the normal direct form structures. As such, it is important to be aware of both types of structures.

3.2.3 Structures for IIR Digital Filters

For a general, recursive implementation of a linear system with a denominator of order D and a numerator of order $N - 1$, the input $x(n)$ and output $y(n)$ satisfy the difference equation

$$y(n) = \sum_{k=1}^{D} a_k y(n - k) + \sum_{k=0}^{N-1} b_k x(n - k) \qquad (3.17)$$

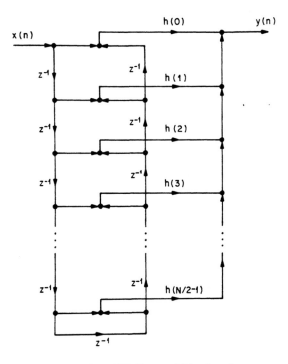

Figure 3.15 Modified direct form FIR filter exploiting impulse response symmetry.

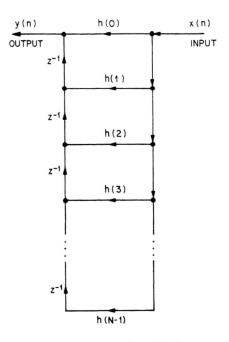

Figure 3.16 Transposed direct form FIR filter structure.

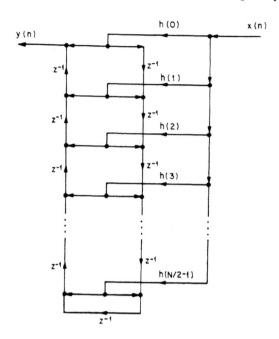

Figure 3.17 Transpose direct form structure that exploits symmetry in the impulse response.

The system function corresponding to the difference equation of Eq. (3.17) has the form

$$H(z) = \frac{Y(z)}{X(z)} = \frac{\displaystyle\sum_{k=0}^{N-1} b_k z^{-k}}{1 - \displaystyle\sum_{k=1}^{D} a_k z^{-k}} = \frac{N(z)}{D(z)} \tag{3.18}$$

where $N(z)$ and $D(z)$ are the numerator and denominator polynomials of $H(z)$, respectively. A signal-flow graph representation of Eq. (3.17) with $D = N - 1$ leads to the direct-form IIR structure shown in Fig. 3.18 [3.1]. The numerator term of Eq. (3.18) is realized by the "feed-forward" part of the structure and the denominator term is a consequence of the feedback branches in the structure.

The numerator and denominator polynomials of Eq. (3.18) can alternatively be factored into second-order polynomials giving an expression for $H(z)$ of the form

$$H(z) = A \cdot \prod_{i=1}^{P} \frac{1 + \beta_{1i} z^{-1} + \beta_{2i} z^{-2}}{1 - \alpha_{1i} z^{-1} - \alpha_{2i} z^{-2}} R(z) \tag{3.19}$$

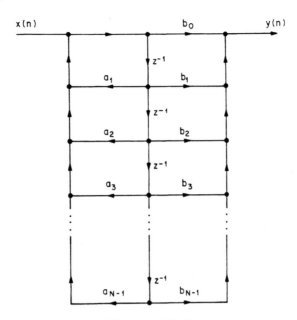

Figure 3.18 Direct form IIR filter structure.

where $P = D/2$ (assuming that D is an even integer), A is a gain constant, and $R(z)$ is a polynomial representing the higher-order numerator terms in Eq. (3.18) when $N > D + 1$. The structure that implements Eq. (3.19) [assuming that $R(z) = 1$] is the cascade form, as shown in Fig. 3.19. For IIR systems, the cascade structure is generally preferred over the direct-form structure because of its improved finite-word-length properties: reduced round-off noise, less coefficient sensitivity, and so on [3.1, 3.5].

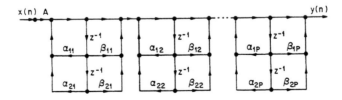

Figure 3.19 Cascade form IIR filter structure.

Although other structures exist for IIR filters (i.e., parallel form, lattice form, etc.) we are not concerned with them here because they have little relevance to the design problem we are dealing with in this book. In addition, transposed structures for each of the forms can readily be obtained. As in the FIR case, the use of a structural form or its transpose is dictated by the specific system being investigated.

3.3 STRUCTURES FOR FIR DECIMATORS AND INTERPOLATORS

In this section we combine the rules of Section 3.1 with the structures of Section 3.2 to show how efficient structures can be developed for realizing digital decimators and interpolators. We first consider decimation and interpolation structures with integral changes in the sampling rate. These structures are derived by combining the cascaded models of decimation and interpolation developed in Chapter 2 with the direct-form and transposed direct-form FIR filter structures in the preceding section. Next, a class of "polyphase" FIR structures for decimators and interpolators with integral rate changes is developed and their properties are discussed. Then we consider the design of FIR interpolators/decimators in which the sampling rate conversion ratio is a rational fraction of the form L/M.

3.3.1 Direct and Transposed Direct-Form FIR Structures for Decimators and Interpolators with Integer Changes in Sampling Rate

Consider the model of an M-to-1 decimator as developed in Chapter 2, and as shown in Fig. 3.20(a). According to this model, the filter $h(n)$ operates at the high

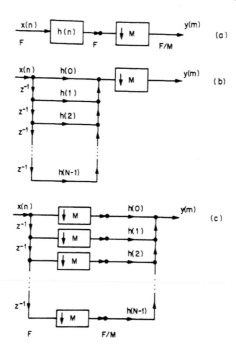

Figure 3.20 Generation of an efficient direct form structure of an M-to-1 decimator.

sampling rate F and $M - 1$ of every M output samples of the filter are discarded by the M-to-1 sampling rate compressor. In particular, if we assume that the filter $h(n)$ is an N-point FIR filter realized with a direct-form structure, the network of Fig. 3.20(b) results. The multiplications by $h(0)$, $h(1)$, ..., $h(N - 1)$ and the associated summations in this network must be performed at the rate F.

A more efficient realization of the foregoing structure can be achieved by noting that the branch operations of sampling rate compression and gain can be commuted as discussed in Section 3.1.2. By performing this series of commutative operations on the network, the modified network of Fig. 3.20(c) results. The multiplications and additions associated with the coefficients $h(0)$ to $h(N-1)$ now occur at the low sampling rate F/M, and therefore the total computation rate in the system has been reduced by a factor M. For every M samples of $x(n)$ that are shifted into the structure (the cascade of delays), one output sample $y(m)$ is computed. Thus the structure of Fig. 3.20(c) is seen to be a direct realization of the relation

$$y(m) = \sum_{n=0}^{N-1} h(n)x(Mm - n) \qquad (3.20)$$

which was derived in Section 2.3.2.

An alternative form of this structure which can exploit symmetry in $h(n)$ (for linear phase designs) can be derived by using the modified direct-form structure of Fig. 3.15 (for N even). This leads to the M-to-1 decimator structure shown in Fig. 3.21, which requires approximately a factor of 2 fewer multiplications than the structure of Fig. 3.20(c).

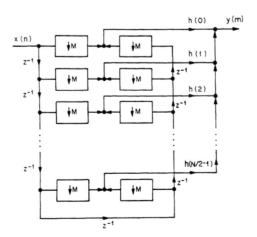

Figure 3.21 Direct form realization of an M-to-1 decimator that exploits symmetry in $h(n)$ for N even.

An efficient structure for the 1-to-L integer interpolator using an FIR filter can be derived in a similar manner. We begin with the cascade model for the interpolator shown in Fig. 3.22(a). In this case, however, if $h(m)$ is realized with the direct-form structure of Fig. 3.14, we are faced with the problem of commuting the 1-to-L sampling rate expander with a series of unit delays. As discussed earlier, this creates nonrealizable fractional delays. One way around this problem is to realize $h(m)$ with the transposed direct form FIR structure as shown in Fig. 3.22(b) [3.6]. The sampling rate expander can then be commuted into the network, resulting in the structure shown in Fig. 3.22(c). Since the coefficients $h(0)$, $h(1)$, ..., $h(N-1)$ are now commuted to the low-sampling-rate side of the network, this structure requires a factor of L times less computation than the structure in Fig. 3.22(b). By using the transposed direct-form structure of Fig. 3.17, an additional factor of approximately 2 in computational efficiency can be gained when $h(n)$ is a linear phase design. The resulting structure is shown in Fig. 3.23.

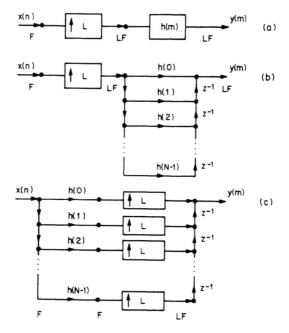

Figure 3.22 Steps in the generation of an efficient structure of a 1-to-L interpolator.

An alternative way of deriving the structures of Fig. 3.22(c) or 3.23 is by a direct transposition of the networks of Fig. 3.20 or 3.21, respectively (letting $L = M$). This is a direct consequence of the fact that decimators and interpolators are duals. A further property of transposition is that for the resulting network,

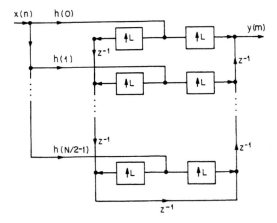

Figure 3.23 Efficient structure for a 1-to-L interpolator that exploits impulse response symmetry for N even.

neither the number of multipliers nor the rate at which these multipliers operate will change [3.4]. Thus if we are given a network that is minimized with respect to its multiplication rate, its transpose is also minimized with respect to its multiplication rate.

3.3.2 Polyphase FIR Structures for Decimators and Interpolators with Integer Changes in Sampling Rate

A second general class of structures that are of interest in multirate digital systems are the *polyphase* networks (sometimes referred to as N-path networks) [3.7]. We will find it convenient to first derive this structure for the L-to-1 interpolator and then obtain the structure for the decimator by transposing the interpolator structure.

In Section 2.3.3, it was shown that a general form for the input-to-output time domain relationship for the 1-to-L interpolator is

$$y(m) = \sum_{n=-\infty}^{\infty} g_m(n) x \left[\lfloor \frac{m}{L} \rfloor - n \right] \qquad (3.21)$$

where

$$g_m(n) = h(nL + m \oplus L), \text{ for all } m \text{ and } n \qquad (3.22)$$

is a periodically time-varying filter with period L and $\lfloor u \rfloor$ denotes the largest integer less than or equal to u. Thus to generate each output sample $y(m)$, $m = 0, 1, 2, ..., L - 1$, a different set of coefficients $g_m(n)$ are used. After L

outputs are generated, the coefficient pattern repeats; thus $y(L)$ is generated using the same set of coefficients $g_0(n)$ as $y(0)$, $y(L+1)$ uses the same set of coefficients $g_1(n)$ as $y(1)$, and so on.

Similarly, the term $\lfloor m/L \rfloor$ in Eq. (3.21) increases by one for every L samples of $y(m)$. Thus for output samples $y(L)$, $y(L + 1)$, ..., $y(2L - 1)$, the coefficients $g_m(n)$ are multiplied by samples $x(1 - n)$. In general, for output samples $y(rL)$, $y(rL + 1)$, ..., $y(rL + L - 1)$, the coefficients $g_m(n)$ are multiplied by samples $x(r - n)$. Thus it is seen that $x(n)$ in Eq. (3.21) is updated at the low sampling rate F, whereas $y(m)$ is evaluated at the high sampling rate LF.

An implementation of the 1-to-L interpolator based on the computation of Eq. (3.21) is shown in Fig. 3.24(a). The way in which this structures operates is as follows. The partitioned subsets, $g_0(n)$, $g_1(n)$, ..., $g_{L-1}(n)$, of $h(m)$ can be identified with L separate linear, time-invariant filters which operate at the low sampling rate F. To make this subtle notational distinction between the time-varying coefficients and the time-invariant filters, we will refer to the time-invariant filters respectively as $p_0(n)$, $p_1(n)$, ..., $p_{L-1}(n)$. Thus

$$p_\rho(n) = g_\rho(n) \quad \text{for } \rho = 0, 1, 2, ..., L - 1 \text{ and all } n \qquad (3.23)$$

These filters $p_\rho(n)$ will be referred to as the *polyphase filters*. Furthermore, by combining Eqs. (3.22) and (3.23) it is apparent that

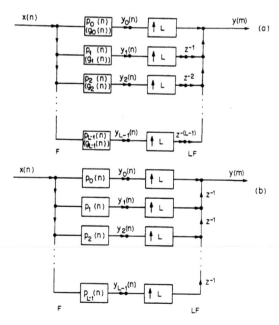

Figure 3.24 Polyphase structures for a 1-to-L interpolator.

$$p_\rho(n) = h(nL + \rho) \quad \text{for } \rho = 0, 1, 2, ..., L - 1 \text{ and all } n \qquad (3.24)$$

For each new input sample, $x(n)$, there are L output samples of $y(m)$, as seen in Fig. 3.24. The upper path (i.e., the $\rho = 0$ path of the polyphase filter), contributes sample values $y_0(n)$ for output samples $m = nL$, $n = 0, \pm1, \pm2, ...,$ of $y(m)$. That is,

$$y(nL) = y_0(n) \quad \text{for } n = 0, \pm1, \pm2, ... \qquad (3.25)$$

The output from the next path, $\rho = 1$, contributes sample values $y_1(n)$ for output samples $m = nL + 1$, for all n, that is,

$$y(nL + 1) = y_1(n) \quad \text{for } n = 0, \pm1, \pm2, ... \qquad (3.26)$$

In general, the output for the ρth path of the polyphase filter contributes sample values $y_\rho(n)$ for which

$$y(nL + \rho) = y_\rho(n) \qquad (3.27)$$

Thus for each input sample, $x(n)$, each of the L branches of the polyphase network contributes one nonzero output which corresponds to one of the L outputs of the network. The polyphase interpolation network of Fig. 3.24(a) has the property that the filtering is performed at the low sampling rate and thus it is an efficient structure. A simple manipulation of the structure of Fig. 3.24(a) leads to the equivalent network of Fig. 3.24(b), in which all the delays are single-sample delays.

From a practical point of view it is often convenient to implement the polyphase structures in terms of a *commutator model*. Since the outputs of the polyphase branches contribute samples of $y(m)$ for different time slots, the 1-to-L sampling rate expanders and delays in Fig. 3.24 can be replaced by a commutator, as shown in Fig. 3.25. The commutator rotates in a *counterclockwise* direction starting with the zeroth polyphase branch at time $m = 0$. For each input sample

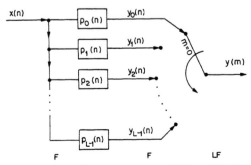

Figure 3.25 Counterclockwise commutator model for a 1-to-L interpolator.

$x(n)$ the commutator sweeps through the L polyphase branches to obtain L output samples of $y(m)$.

The individual polyphase filters $p_\rho(n)$, $\rho = 0, 1, 2, ..., L - 1$, have a number of interesting properties. This is a consequence of the fact that the impulse responses $p_\rho(n)$, $\rho = 0, 1, 2, ..., L - 1$, correspond to decimated versions of the impulse response of the prototype filter $h(m)$ [decimated by a factor of L according to Eq. (3.22) or Eq. (3.24)]. Figure 3.26 illustrates this for the case $L = 3$ and for an FIR filter $h(m)$ with $N = 9$ taps. The upper figure shows the samples of $h(m)$ where it is assumed that $h(m)$ is symmetric about $m = 4$. Thus $h(m)$ has a flat delay of four samples [3.1, 3.5]. The filter $p_0(n)$ has three samples, corresponding to $h(0)$, $h(3)$, $h(6) = h(2)$. Since the point of symmetry of the envelope of $p_0(n)$ is $n = 4/3$ it has a flat delay of 4/3 samples. Similarly, $p_1(n)$ has samples $h(1)$, $h(4)$, $h(7) = h(1)$ and because its zero reference $(n = 0)$ is offset by 1/3 sample (with respect to $m = 0$), it has a flat delay of one sample. Thus different fractional sample delays and consequently different phase shifts are associated with the different filters $p_\rho(n)$ as seen in Fig. 3.26(b). These delays are compensated for by the delays that occur at the high sampling rate LF in the network (see Fig. 3.24). The fact that different phases are associated with different paths of the network is, of course, the reason for the term *polyphase network*.

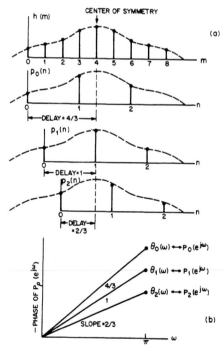

Figure 3.26 Illustration of the properties of polyphase network filters.

A second property of polyphase filters is shown in Fig. 3.27. The frequency response of the prototype filter $h(m)$ approximates the ideal lowpass characteristic $\widetilde{H}(e^{j\omega})$ shown in Fig. 3.27(a).[2] Since the polyphase filters $p_\rho(n)$ are decimated versions of $h(m)$ (decimated by L), the frequency response $0 \leqslant \omega \leqslant \pi/L$ of $\widetilde{H}(e^{j\omega})$ scales to the range $0 \leqslant \omega' \leqslant \pi$ for $\widetilde{P}_\rho(e^{j\omega'})$ as seen in Fig. 3.27, where $\widetilde{P}_\rho(e^{j\omega'})$ is the ideal characteristic that the polyphase filter $p_\rho(n)$ approximates. Thus the polyphase filters approximate *all-pass functions* and each value of ρ, $\rho = 0, 1, 2, ..., L - 1$, corresponds to a different phase shift. This property is discussed in greater detail in Chapter 4 in connection with filter design methods for polyphase filters.

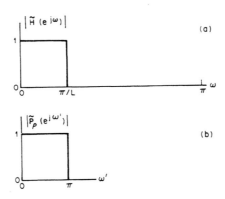

Figure 3.27   Ideal frequency response of the polyphase network filters.

The polyphase filters can be realized in a variety of ways. If the prototype filter $h(m)$ is an FIR filter of length N, the filters $p_\rho(n)$ will be FIR filters of length N/L. In this case it is often convenient to choose N to be a multiple of L so that all the polyphase filters are of equal length. These filters may be realized by any of the conventional methods for implementing FIR filters such as the direct form structure of Section 3.2 or the methods based on fast convolution [3.1, 3.5, 3.8]. If a direct form FIR structure is used for the polyphase filters, the polyphase structure of Fig. 3.24 will require the same multiplication rate as the direct form interpolator structure of Fig. 3.22. Exploiting symmetry in $h(m)$ is more difficult in this class of structures since, at most, only one of the $p_\rho(n)$ subfilters are symmetric. For large order filters of $p_\rho(n)$ (say $N/L > 30$), the methods of fast convolution can be used to reduce the computational requirements further [3.8].

By realizing each of the polyphase filters in the structures of Fig. 3.24 or 3.25 as a direct-form FIR filter it can also be observed that the same shift register for holding delayed values of $x(n)$ can be shared with each of the polyphase filters. Therefore the total storage requirements for data storage, as well as the

---

[2] Recall also that there is an additional gain of L required in the interpolator which we have ignored in this discussion (see Section 2.3.3).

computation rate, is reduced by a factor L since each of the polyphase filters is only $1/L$ as long as the original filter $h(m)$. This savings in required storage is an additional advantage of the polyphase structure.

By transposing the structure of the polyphase 1-to-L interpolator of Fig. 3.24(b), we get the polyphase M-to-1 decimator structure of Fig. 3.28, where L is replaced by M. Again the filtering operations of the polyphase filters occur at the low-sampling-rate side of the network and they can be implemented by any of the conventional structures discussed in Section 3.2.

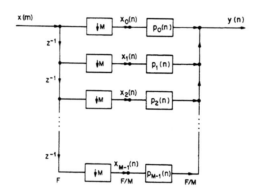

Figure 3.28   Polyphase structure for an M-to-1 decimator.

In the discussion above regarding the 1-to-L interpolator we have identified the coefficients of the polyphase filters, $p_\rho(n)$, with the coefficient sets $g_m(n)$ of the time-varying filter model. In the case of the M-to-1 decimator, however, this identification cannot be made directly. According to the time-varying filter model discussed in Chapter 2, the coefficients $g_m(n)$ for the M-to-1 decimator are

$$g_m(n) = g(n) = h(n), \quad \text{for all } n \text{ and } m \tag{3.28}$$

Alternatively, according to the transpose network of Fig. 3.28, the coefficients of the M-to-1 polyphase decimator are

$$p_\rho(n) = h(nM+\rho), \quad \text{for } \rho = 0, 1, 2, ..., M-1, \text{ and all } n \tag{3.29}$$

where ρ denotes the ρth polyphase filter. Thus the polyphase filters $p_\rho(n)$ for the M-to-1 decimator are *not* equal to the time-varying coefficients $g_m(n)$ of the decimator.

As in the case of the polyphase interpolator, the polyphase decimator structure can be more practically realized with the aid of a commutator model. This model is shown in Fig. 3.29. The commutator rotates *counterclockwise*, starting at the $\rho = 0$ branch at time $m = 0$. In effect, the commutator takes M input samples of the signal $x(m)$ and distributes them to the polyphase branches in the reverse sequence $\rho = M - 1, M - 2, ..., 1, 0$. When each of the polyphase filters has received a new input at time instants $n = mM$, the polyphase filters are computed and their outputs are summed to give one output sample of $y(n)$.

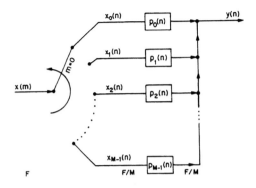

Figure 3.29 Counterclockwise commutator model for an M-to-1 decimator.

The counterclockwise commutator structure for the decimator described above can also be derived directly from the input-to-output relation developed in Chapter 2,

$$y(n) = \sum_{k=-\infty}^{\infty} h(k)x(nM - k) \qquad (3.30)$$

The summation in Eq. (3.30) can be separated into two summations by defining

$$k = rM + \rho \qquad (3.31)$$

and summing over the two variables $\rho = 0, 1, 2, ..., M - 1$ and $r = 0, \pm1, \pm2, ...$ This gives

$$y(n) = \sum_{\rho=0}^{M-1} \sum_{r=-\infty}^{\infty} h(rM + \rho)x((n - r)M - \rho) \qquad (3.32)$$

By recognizing that the input signals to the polyphase branches in Fig. 3.29 are

$$x_\rho(n) = x(nM-\rho), \quad \text{for } \rho = 0, 1, 2, ..., M - 1 \qquad (3.33)$$

and the coefficients of the polyphase filters are defined as in Eq. (3.29), Eq. (3.32) can be written in the form

$$y(n) = \sum_{\rho=0}^{M-1} \sum_{r=-\infty}^{\infty} p_\rho(r)x_\rho(n-r)$$

$$= \sum_{\rho=0}^{M-1} p_\rho(n) * x_\rho(n) \tag{3.34}$$

where $*$ denotes convolution. The structure of Fig. 3.29 is a direct implementation of Eq. (3.34).

As in the case of the polyphase interpolator structure, a savings of a factor of M in the storage requirements for internal filter variables can be achieved in the FIR polyphase decimator structures by sharing them among all of the polyphase filters. This can be achieved by realizing the polyphase filters $p_\rho(n)$ in Fig. 3.29 as transpose direct-form structures. Then the accumulator-shift register in each filter can be shared by summing the partial products of each respective tap of the polyphase filters with the respective taps of the other polyphase filters before storing them in the shift register.

3.3.3 Polyphase Structures Based on Clockwise Commutator Models

An alternative formulation of the polyphase structure can be developed by replacing ρ by $-\rho$ in the derivations above. It leads to a *different* but equivalent set of polyphase filters based on fractional phase advances rather than phase delays. It also results in commutator structures with *clockwise* rather than counterclockwise rotation. The reader should be cautioned that the subtle differences between these two formulations can lead to considerable confusion, especially since they have both been used in the literature.

To distinguish the clockwise formulation from the counterclockwise formulation, we will use the overbar notation. Thus the coefficients for the alternative (clockwise) polyphase formulation are defined as

$$\bar{p}_\rho(n) = h(nM - \rho) \text{ for } \rho = 0, 1, 2, ..., M - 1 \tag{3.35}$$

and they are related to the coefficients of the counterclockwise polyphase formulation according to the relation

$$\bar{p}_0(n) = p_0(n) \qquad \text{for all } n \tag{3.36a}$$

and

$$\bar{p}_\rho(n) = p_{m-\rho}(n - 1) \text{ for } \rho = 1, 2, ..., M - 1 \tag{3.36b}$$

The polyphase branch signals, $\bar{x}_\rho(n)$, for the alternative formulation are defined as

$$\bar{x}_\rho(n) = x(nM + \rho), \text{ for } \rho = 0, 1, 2, ..., M - 1 \tag{3.37}$$

Then we can derive, for the decimator,

$$y(n) = \sum_{k=-\infty}^{\infty} h(k)x(nM-k)$$

$$= \sum_{\rho=0}^{M-1} \sum_{r=-\infty}^{\infty} h(rM-\rho)x((n-r)M+\rho)$$

$$= \sum_{\rho=0}^{M-1} \sum_{r=-\infty}^{\infty} \bar{p}_\rho(r)\bar{x}_\rho(n-r)$$

$$= \sum_{\rho=0}^{M-1} \bar{p}_\rho(n) * \bar{x}_\rho(n) \qquad\qquad (3.38)$$

where we have applied the relation $k = rM - \rho$ in the derivation above. This derivation leads to the clockwise commutator model of the polyphase structure shown in Fig. 3.30(a) or the equivalent structure shown in Fig. 3.30(b). As seen by the sample advances rather than sample delays, these structures are, strictly speaking, not realizable. However, by allowing a one-sample delay at the output,

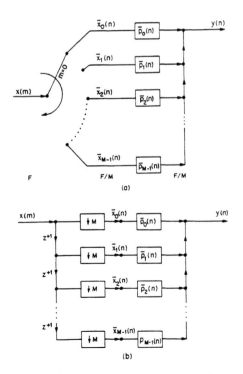

Figure 3.30 Clockwise commutator model and its equivalent structure for the alternative formulation of the polyphase network.

the commutator structure can be conveniently implemented. Inputs of $x(m)$ are applied by the commutator to the polyphase branches sequentially for $\rho = 0, 1, 2, ..., M - 1$. The output $y(n)$ can then be computed and be available for the next time slot $n + 1$ (i.e., a one-sample delay is incurred at the output).

By transposing the structures of Fig. 3.30, similar alternative structures can be derived for the interpolator. As in the case of the decimator these structures can be implemented only by allowing an additional one-sample delay at the low sampling rate.

3.3.4 FIR Structures with Time-Varying Coefficients for Interpolation/Decimation by a Factor of L/M

In the preceding three sections we have considered implementations of decimators and interpolators using the direct-form and polyphase structures for the case of integer changes in the sampling rate. Efficient realizations of these structures were obtained by commuting the filtering operations to occur at the low sampling rate. For the case of a network that realizes a change in sampling rate by a factor of L/M, it is difficult to achieve such efficiencies. The difficulty is illustrated in Fig. 3.31. If we realize the 1-to-L interpolation part of the structure using the techniques described earlier, we are faced with the problem of commuting the M-to-1 sampling rate compressor into the resulting network [Fig. 3.31(a)]. If we realize the decimator part of the structure first, the 1-to-L sampling rate expander must be commuted into the structure [Fig. 3.31(b)]. In both cases difficulties arise because of problems of generating unrealizable fractional delays. Thus we are faced with a network that cannot be implemented efficiently by simply using the techniques of commutation and transposition.

Efficient structures, however, exist for implementing a sampling rate converter with a ratio in sampling rates of L/M, and in this section we discuss one such class

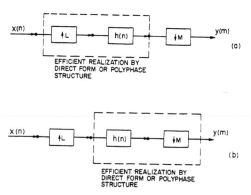

Figure 3.31 Possible realizations of an L/M sampling rate converter.

of FIR structures with time-varying coefficients [3.9]. This structure can be derived from the time domain input-to-output relation of the network, developed in Chapter 2,

$$y(m) = \sum_{n=-\infty}^{\infty} g_m(n)x\left[\lfloor \frac{mM}{L} \rfloor - n\right]$$

(3.39)

where

$$g_m(n) = h(nL + mM \oplus L), \text{ for all } m \text{ and all } n$$

(3.40)

and $h(k)$ corresponds to the lowpass (or bandpass) FIR prototype filer. It will be convenient for our discussion to assume that the length of the filter $h(k)$ is a multiple of L, that is,

$$N = QL$$

(3.41)

where Q is an integer. Then all the coefficient sets $g_m(n)$, $m = 0, 1, 2, ..., L - 1$, contain *exactly* Q coefficients. Furthermore, $g_m(n)$ is periodic in m with period L, as discussed previously; that is,

$$g_m(n) = g_{m+rL}(n), \quad r = 0, \pm 1, \pm 2, ...$$

(3.42)

Therefore, Eq. (3.39) can be expressed as

$$y(m) = \sum_{n=0}^{Q-1} g_{m \oplus L}(n)x\left[\lfloor \frac{mM}{L} \rfloor - n\right]$$

(3.43)

Equation (3.43) shows that the computation of an output sample $y(m)$ is obtained as a weighted sum of Q sequential samples of $x(n)$ starting at the $x(\lfloor mM/L \rfloor)$ sample and going backwards in n sequentially. The weighting coefficients are periodically time-varying so that the $m \oplus L$ coefficient set $g_{m \oplus L}(n)$, $n = 0, 1, 2, ..., Q - 1$, is used for the mth output sample. Figure 3.32 illustrates this timing relationship for the $n = 0$ term in Eq. (3.43) and for the case $M = 2$ and $L = 3$. The table in Fig. 3.32(a) shows the index values of $y(m)$, $x(\lfloor mM/L \rfloor)$ and $g_{m \oplus L}(0)$ for $m = 0$ to $m = 6$. Figure 3.32(b) illustrates the relative timing positions of the signals $y(m)$ and $x(n)$ drawn on an absolute time scale. By comparison of the table and the figure it can be seen that the value $x(\lfloor mM/L \rfloor)$ always represents the most recent available sample of $x(n)$ [i.e., $y(0)$ and $y(1)$ are computed on the basis of $x(0 - n)$]. For $y(2)$ the most recent available value of $x(n)$ is $x(1)$, for $y(3)$ it is $x(2)$, and so on.

Based on Eq. (3.43) and the foregoing description of how the input, output, and coefficients enter into the computation, the structure of Fig. 3.33 is suggested

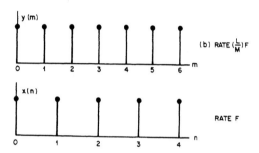

y (m)	$x\left(\left\lfloor\dfrac{mM}{L}\right\rfloor\right)$	$g_{m\oplus L}(0)$	(a)
m	$\left\lfloor\dfrac{2m}{3}\right\rfloor$	$m\oplus L$	
0	0	0	
1	0	1	
2	1	2	
3	2	0	
4	2	1	
5	3	2	
6	4	0	

Figure 3.32 Timing relationships between $y(m)$ and $x(n)$ for the case $M=2$, $L=3$.

for realizing an L/M sampling rate converter. The structure consists of:

1. A Q-sample "shift register" operating at the input sampling rate F, which stores sequential samples of the input signal.
2. A Q-tap direct-form FIR structure with time-varying coefficients $[(g_{m\oplus L}(n),\ n=0, 1, ..., Q-1)]$, which operates at the output sampling rate $(L/M)F$.
3. A series of digital "hold-and-sample" boxes that couple the two sampling rates. The input side of the box "holds" the most recent input value until the next input value comes along. The output side of the box "samples" the input values at times $n=mM/L$. For times when mM/L is an integer (i.e., input and output sampling times are the same), the input changes first and the output samples the changed input.

It should be clear that the structure of Fig. 3.33 is an efficient one for implementing an (L/M) sampling rate converter since the filtering operations are all performed at the output sampling rate with the minimum required number of coefficients used to generate each output.

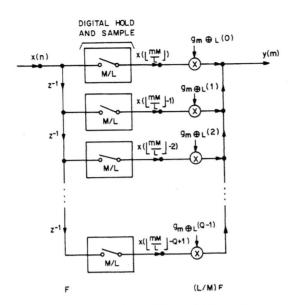

Figure 3.33 Efficient structure for realizing an L/M sampling rate converter.

Figure 3.34 shows a diagram of a program configuration to implement this structure in a block-by-block manner. The program takes in a block of M samples of the input signal, denoted as $x(n')$, $n' = 0, 1, 2, ..., M - 1$, and computes a block of L output samples $y(m')$ $m' = 0, 1, 2, ..., L - 1$. The primed variable notation in this context is used to denote sample times within the blocks rather than absolute sample times. For each output sample time m', $m' = 0, 1, ..., L - 1$, the Q samples from the state-variable buffer are multiplied respectively with Q coefficients from one of the coefficient sets $g_{m'}(n')$ and the products are accumulated according to Eq. (3.43) to give the output $y(m')$. Each time the quantity $\lfloor m'M/L \rfloor$ increases by one, one sample from the input buffer is shifted into the state-variable buffer. (This information can be stored in a control array.) Thus after L output values are computed, M input samples have been shifted into the state-variable buffer and the process can be repeated for the next block of data. In the course of processing one block of data (M input samples and L output samples) the state-variable buffer is sequentially addressed L times and the coefficient storage buffer is sequentially addressed once. A program that performs this computation can be found in Ref. 3.10.

3.4 IIR STRUCTURES FOR DECIMATORS AND INTERPOLATORS WITH INTEGER CHANGES IN SAMPLING RATE

In the preceding section we have assumed that the prototype filter in the cascade model for the decimator or interpolator is an FIR filter and we have used the

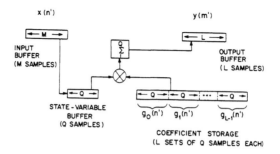

Figure 3.34 Block diagram of a program structure to implement the signal-flow graph of Fig. 3.33 in a block-by-block manner.

properties of FIR filter structures to achieve efficient designs. In this section we consider the case where this filter is an IIR filter and discuss the implication of this choice on the design of the decimator or interpolator structures. We begin by briefly considering an example where a general IIR filter design, with a system response of the form of Eq. (3.18), is used. We then show that a more efficient class of IIR decimator and interpolator structures can be developed if the form of the IIR filter, $H(z)$, is restricted to a more specific class of designs in which the denominator polynomial of $H(z)$ contains only powers of z^M (where M is the decimation or interpolation ratio). This form is developed in connection with the IIR polyphase structures in Section 3.4.1. It is also used in Section 3.4.2 to develop more efficient direct-form and time-varying coefficient form IIR structures for decimators and interpolators. Later in Chapter 4 we consider filter design techniques that allow this specific class of IIR filters to be designed.

Consider first the case where the IIR filter has the more general form given by Eq. (3.18). We assume, for convenience, that the orders of the numerator and denominator are equal, i.e. $D = N - 1$. Figure 3.35(a) illustrates an example where this filter is applied to the model of an M-to-1 decimator. If we further assume that this filter is realized as a direct form structure (as in Fig. 3.18), the sampling rate compressor can be commuted with the numerator coefficients as shown in Fig. 3.35(b). Thus the nonrecursive part of the filter structure can be realized efficiently at the low sampling rate (as in the case of pure FIR filters). However, the sampling rate compressor cannot be commuted with the denominator coefficients since the delays would become fractional delays, and hence would not be realizable. Thus while the nonrecursive part of the network can operate at the low sampling rate, the recursive part must still operate at the high sampling rate, giving a structure that is not totally efficient (i.e., it is a suboptimal design).

The above example illustrates a key difficulty in developing efficient IIR decimator or interpolator structures by directly applying conventional IIR filter designs to the prototype filter $H(z)$. Also, as mentioned above, more efficient structures are possible when the IIR filter design is restricted to a form such that

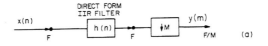

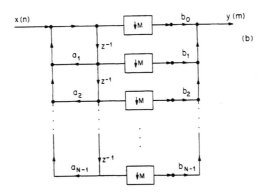

Figure 3.35 Direct-form IIR structure for an M-to-1 decimator.

the recursive part of the structure *can* be realized at the low sampling rate. Thus it is apparent that the problem of designing an efficient IIR decimator or interpolator structure and the filter design problem are more interdependent than in the FIR case. For this reason we consider in this section the *joint* issues of efficient structure design and the class of filter designs necessary to realize such structures. Later in Chapter 4, on filter design techniques, we discuss methods of designing filters which belong to this class (including ways in which conventional IIR designs can be transformed, in a nonoptimal way, to this form).

3.4.1 Polyphase Structures for IIR Decimators and Interpolators

It is convenient to first consider the above joint issues of structure design and IIR filter design in connection with the polyphase structures. Recall that the polyphase structures (for the counterclockwise model) are based on a decomposition of the impulse response of the prototype filter $h(n)$ into a set of M sub-filters or polyphase filters $p_\rho(n)$ of the form

$$p_\rho(n) = h(\rho+nM), \quad \rho = 0, 1, ..., M - 1 \tag{3.44}$$

where M is the decimation or interpolation ratio. Given the decomposition above, the polyphase filters can be readily applied in the structures already developed in Fig. 3.24 or 3.25, for the interpolator, (where $M = L$) or Fig. 3.28 or 3.29, for the decimator. Since $h(n)$ in this case is an IIR filter, it follows that the resulting polyphase filters, $p_\rho(n)$, are also IIR filters and we will denote their z-transforms as $P_\rho(z)$.

Assuming that $H(z)$, the z-transform of $h(n)$, has the form of a ratio of polynomials in z, it does not necessarily follow that the resulting polyphase filters $P_\rho(z)$ can be expressed as ratios of polynomials *unless* $H(z)$ has a more specific form. To show this form we assume that the filters $P_\rho(z)$ *do* have a rational form and then derive the resulting form that $H(z)$ must have in order for this to be true. Accordingly, we assume apriori that the filters $P_\rho(z)$ can be expressed as

$$P_\rho(z) = \frac{N_\rho(z)}{D_\rho(z)} \tag{3.45}$$

where $N_\rho(z)$ and $D_\rho(z)$ are polynomials in z.

To develop the relation for $H(z)$, we recognize that the filters $p_\rho(n)$, as a set, represent all the samples of $h(n)$. Thus given $p_\rho(n)$, $\rho = 0, 1, 2, ..., M$, it is possible to determine $h(n)$ by expanding the sampling rates of these sequences by a factor M and summing appropriately delayed versions of them. Let $\hat{p}_\rho(n)$ denote the sampling rate expanded sequences

$$\hat{p}_\rho(n) = \begin{cases} p_\rho\left[\dfrac{n}{M}\right], & n = 0, \pm M, \pm M, ... \\ 0, & \text{otherwise} \end{cases} \tag{3.46}$$

Then from Eq. (3.44) it can be shown that

$$h(n) = \hat{p}_0(n) + \hat{p}_1(n-1) + ... + \hat{p}_{M-1}(n - M + 1)$$

$$= \sum_{\rho=0}^{M-1} \hat{p}_\rho(n - \rho) \tag{3.47}$$

Furthermore letting $\hat{P}_m(z)$ denote the z-transform of $\hat{p}_\rho(n)$, it can be shown (see Section 2.3.3) that

$$\hat{P}_\rho(z) = P_\rho(z^M) \tag{3.48}$$

By transforming Eq. (3.47) and applying Eq. (3.48), $H(z)$ can be shown to have the form

$$H(z) = \sum_{\rho=0}^{M-1} z^{-\rho}\hat{P}_\rho(z)$$

$$= \sum_{\rho=0}^{M-1} z^{-\rho}P_\rho(z^M) \tag{3.49}$$

Applying Eq. (3.45) then gives

$$H(z) = \sum_{\rho=0}^{M-1} \frac{z^{-\rho} N_\rho(z^M)}{D_\rho(z^M)} \tag{3.50a}$$

or by expanding the terms in Eq. (3.50a), we get

$$H(z) = \frac{\sum_{\rho=0}^{M-1}\left[z^{-\rho} N_\rho(z^M) \prod_{\substack{i=0 \\ i\neq\rho}}^{M-1} D_i(z^\rho) \right]}{\prod_{\rho=0}^{M-1} D_\rho(z^M)} = \frac{N(z)}{D(z)} \tag{3.50b}$$

Equation (3.50b) illustrates the form that $H(z)$ takes when the polyphase filters $P_\rho(z)$ are rational functions. Observe that the numerator is a polynomial in z but the denominator is a polynomial in z^M.

Furthermore, it is often convenient to assume that the denominators, $D_\rho(z)$, of the polyphase filters are all identical [3.7, 3.11] so that their computation can be shared (as will be seen shortly). Then the form of $H(z)$ in Eq. (3.50a) simplifies to

$$H(z) = \frac{N(z)}{D(z)} = \frac{\sum_{\rho=0}^{M-1} z^{-\rho} N_\rho(z^M)}{\hat{D}(z^M)} \tag{3.51a}$$

where

$$D(z) = D_0(z^M) = D_1(z^M) = ... = D_{M-1}(z^M) = \hat{D}(z^M) \tag{3.51b}$$

From this expression it should be clear that the order of the denominator of $H(z)$ is M times the order of the denominators of the polyphase filters and the order of the numerator of $H(z)$ is M times the order of the numerators of the polyphase filters plus $M - 1$.

Because of the specific form that $H(z)$ must take in order for the polyphase filters $P_\rho(z)$ to be rational, special design procedures are required for the design of an efficient prototype filter for $H(z)$ rather than the classical IIR designs. This topic is taken up in greater detail in Chapter 4, which covers filter design issues. In the remainder of this section we assume that $H(z)$ has this form and, more specifically, that it has the form given by Eq. (3.51a).

Given a prototype filter design $H(z)$ that satisfies the form of Eq. (3.51a), it is a straightforward matter to obtain the coefficients for the polyphase filters $P_\rho(z)$. Let $H(z)$ have the form

$$H(z) = \frac{N(z)}{D(z)} = \frac{\sum_{k=0}^{N-1} b_k z^{-k}}{1 - \sum_{r=1}^{R} a_{rM} z^{-rM}} \tag{3.52}$$

and $P_\rho(z)$ have the form

$$P_\rho(z) = \frac{N_\rho(z)}{\hat{D}(z)} = \frac{\sum_{k=0}^{N_r-1} d_{k,\rho} z^{-k}}{1 - \sum_{r=1}^{R} c_r z^{-r}}, \quad \rho = 0, 1, 2, ..., M - 1 \tag{3.53}$$

where $N_\rho - 1$ is the order of the numerator and R is the order of the denominator. Then it can be clearly seen from Eq. (3.51b) that

$$c_r = a_{rM}, \quad r = 1, 2, ..., R \tag{3.54}$$

Also with the aid of Eq. (3.51a) it can be shown that

$$N(z) = \sum_{\rho=0}^{M-1} z^{-\rho} N_\rho(z^M) \tag{3.55}$$

The similarity of this form to the form of Eq. (3.49) and its derivation from Eqs. (3.46) and (3.47) reveals that the coefficients of the numerators of the polyphase filters $P_\rho(z)$ are simply decimated versions of the coefficients of the numerator of the prototype filter $H(z)$ in the same manner as they are for the case of FIR filters. Thus

$$d_{k,\rho} = b_{kM+\rho} \quad k = 0, 1, ..., N_\rho - 1; \rho = 0, 1, ..., M - 1 \tag{3.56}$$

An IIR polyphase structure for an M-to-1 decimator can now be defined by directly applying the IIR polyphase filters $P_\rho(z)$, $\rho = 0, 1, 2, ..., M - 1$, into the structure of Fig. 3.28. The resulting structure is shown in Fig. 3.36(a) where it is assumed that the numerator (nonrecursive) and denominator (recursive) parts of the structure are implemented separately in a cascaded manner. If the denominators of all the polyphase filters are identical, as in Eq. (3.51b), then the recursive parts of the polyphase filters can be commuted into a more efficient form, as shown in Fig. 3.36(b). This form of the structure should be applied with some caution, however, since separating the recursive and nonrecursive parts of the structure in this way may lead to severe problems of internal scaling and round-off noise.

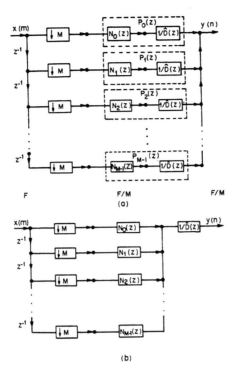

Figure 3.36 Polyphase structure for a M-to-1 IIR decimator.

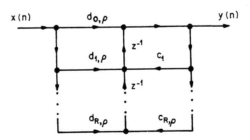

Figure 3.37 Transposed direct form structure for IIR filter.

One method of implementing the filters $P_\rho(z)$ is by the transposed direct-form structure shown in Fig. 3.37 (the transpose of the form of the structure in Fig. 3.18), where $P_\rho(z)$ is given by Eq. (3.53) (assuming that $N_\rho = R + 1$, $\rho = 0, 1, 2, ..., M-1$). This particular form of the structure for $P_\rho(z)$ has the advantage that the delays for all the polyphase filters can be shared, leading to the efficient IIR structure shown in Fig. 3.38 [3.7].

By taking the transpose of the structure of Fig. 3.38, an efficient polyphase

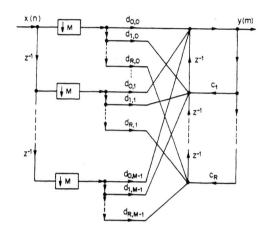

Figure 3.38   Efficient IIR polyphase structure for an M-to-1 decimator.

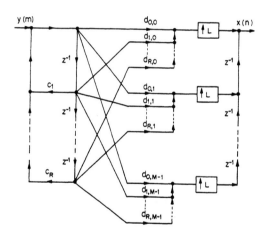

Figure 3.39   Efficient polyphase IIR structure for a 1-to-L interpolator.

IIR structure for a 1-to-L interpolator can be defined as given by Fig. 3.39, where M is replaced by L.

3.4.2 Direct-Form Structures and Structures with Time-Varying Coefficients for IIR Decimators and Interpolators

In the discussion above we have used the concept of polyphase structures for decimators and interpolators for a specific class of prototype filters $H(z)$. Next we will show that the polyphase structures are not the only way to implement efficient

IIR decimators or interpolators, but in fact they can just as easily be implemented with appropriate direct-form IIR structures or structures with time-varying coefficients.

This can be easily seen by looking again at the IIR decimator structure of Fig. 3.36. This structure can be interpreted as an integer FIR decimator cascaded with an IIR filter as shown by Fig. 3.40(a) [assuming that $H(z)$ has the form given by Eq. (3.51a)]. The IIR filter is the filter corresponding to the decimated denominator term of $H(z)$ [according to Eq. (3.51b)] and the prototype filter for the FIR interpolator is simply the numerator term of the IIR filter $H(z)$. In principle this FIR interpolator can be implemented by any of the methods discussed in Section 3.3.

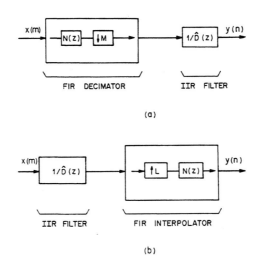

Figure 3.40 Cascade models for an IIR decimator and an IIR interpolator.

By transposing the structure of Fig. 3.40(a), the corresponding form for the IIR interpolator results, as shown by Fig. 3.40(b). Again the FIR interpolator can be implemented by any of the methods discussed in Section 3.3. For example, Fig. 3.41 illustrates the case where it is implemented by the structure with time-varying coefficients (see Fig. 3.33) and where the delays have been shared with the IIR filter $1/\hat{D}(z)$. With appropriate modifications this structure can also be used for conversions by factors of L/M.

3.4.3 Comparison of Structures for Decimation and Interpolation

In this chapter we have presented a variety of single-stage structures for decimation and interpolation. In addition, in Chapter 5 we show how multistage cascades of these structures can lead to additional gains in computational efficiency when

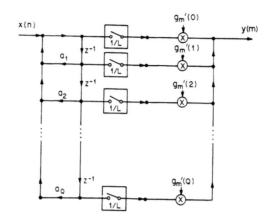

Figure 3.41 IIR structure for a 1-to-L interpolator with time-varying coefficients.

conversion ratios are large. A natural question to ask at this point is: Which of these methods is most efficient? The answer, unfortunately, is nontrivial and is highly dependent on the application being considered. Some insight and direction, however, can be provided by observing some general properties of the classes of structures noted above.

We may loosely categorize the structures above into three general classes: direct-form structures, polyphase structures, and structures with time-varying coefficients. In addition, each of these classes can be further subdivided based on the type of digital filter used into FIR and IIR classes.

Direct-form structures have the advantage that they can be easily modified to exploit symmetry in the numerator term of the system function to gain an additional reduction in computation by a factor of approximately 2. The polyphase structures have the advantage that the filters $p_\rho(n)$ can be easily realized with efficient techniques such as the fast convolution methods based on the FFT [3.8]. As such, they have been found to be useful for filter banks [3.7]. The structures with time-varying coefficients are particularly useful when considering conversions by factors of L/M.

There are many other factors that determine the overall efficiency of these structures. Most of these considerations, however, are filter design considerations and hence we must defer further comparisons of single-stage structures for decimators and interpolators until we have discussed the filter design issues in some detail. We will return to this topic at the end of Chapter 4.

3.5 ADVANCED NETWORK CONCEPTS OF LINEAR MULTIRATE AND TIME-VARYING STRUCTURES

In the preceding discussion we have taken a relatively informal approach in developing practical structures for multirate digital networks. We have relied

heavily on subtle network theoretic concepts, such as transposition, without establishing firm definitions of them, in order to present the practical issues of structures in a relatively straightforward way. In this section we reconsider many of these network concepts in a more rigorous framework and show how they are defined and verified.

We begin by considering a more general system representation of multirate and time-varying linear networks. We then use this framework to establish conditions under which networks are commutable. Next, we discuss concepts of duality and transposition. In the process of deriving the theorem for transposition we also consider a mathematical framework for linear time-varying signal-flow graphs and we develop an important network theorem called Tellegen's theorem. Finally, we discuss several issues concerning the transposition of complex networks. The material in this section is strongly influenced by Ref. 3.4.

3.5.1 System Representation of Linear Time-Varying and Multirate Networks

In Chapter 2 we showed that multirate digital systems belong to the class of periodically linear time-varying systems and that they can be described by a time-varying system response $g_m(n)$. This form of the system response is particularly useful from a practical point of view in developing the polyphase structures and the structures with time-varying coefficients. However, it is not the most general form for characterizing linear time-varying systems. Therefore, in this section we first need to describe a more general formulation of the system response of a time-varying system. This formulation is based on a discrete-time version of the continuous time-varying model used by Zadeh [3.4, 3.12, 3.13]. It is also referred to as the Green's function weighting pattern response.

Given a linear time-varying network with input $x(n)$, the output $y(m)$ can be expressed by the superposition sum

$$y(m) = \sum_{n=-\infty}^{\infty} k(m, n) x(n) \qquad (3.57)$$

The *system response* or *Green's function*, $k(m, n)$, corresponds to the response of the network at the (output) time m to a unit sample applied at (input) time n, where the sampling rates of $x(n)$ and $y(m)$ may, in general, be different. For time-invariant systems the system response $k(m, n)$ is dependent only on the difference $m - n$; that is,

$$k(m, n) = k(0, m - n) \quad \text{time-invariance} \qquad (3.58)$$

and Eq. (3.57) then takes the form of a convolution.

There is an important distinction to be made, however, between the system response $k(m, n)$ and the more conventionally used time-varying impulse response denoted as $h(m, n)$ [3.14, 3.15]. The impulse response $h(m, n)$ is defined as the response of the system at time m to a unit sample applied n samples earlier, assuming that $x(n)$ and $y(m)$ have the same sampling rates. Then

$$y(m) = \sum_{n=-\infty}^{\infty} h(m, n)x(m - n)$$
$$= \sum_{n=-\infty}^{\infty} h(m, m - n)x(n) \tag{3.59}$$

For time-invariant systems, $h(m, n)$ is independent of m, that is,

$$h(m, n) = h(0, n) = h(n) \quad \text{time-invariance} \tag{3.60}$$

and Eq. (3.59) then reduces to the conventional definition of the time-invariant impulse response $h(n)$. By comparing Eqs. (3.57) to (3.59), it can be seen that when $x(n)$ and $y(m)$ have the same sampling rates, $k(m, n)$ and $h(m, n)$ are related by

$$h(m, n) = k(m, m - n) \tag{3.61a}$$

or

$$k(m, n) = h(m, m - n) \tag{3.61b}$$

The major difficulty encountered in using $h(m, n)$ is that it cannot conveniently be used to characterize time-varying systems whose input and output sampling rates are *different*. The system response $k(m, n)$, however, is completely general in this respect and therefore we use it exclusively in this section.

The system response $k(m, n)$ in Eq. (3.57) describes a time domain relationship or mapping between $x(n)$ and $y(m)$. In a similar manner, a general frequency domain relationship can be defined. The Fourier transform pairs for the signals $x(n)$ and $y(m)$ are

$$X(e^{j\omega}) = \sum_{n=-\infty}^{\infty} x(n)e^{-j\omega n} \tag{3.62a}$$

$$x(n) = \frac{1}{2\pi} \int_{-\pi}^{\pi} X(e^{j\omega})e^{j\omega n} d\omega \tag{3.62b}$$

and

$$Y(e^{j\omega'}) = \sum_{m=-\infty}^{\infty} y(m)e^{-j\omega' m} \tag{3.63a}$$

$$y(m) = \frac{1}{2\pi} \int_{-\pi}^{\pi} Y(e^{j\omega'})e^{j\omega'm} d\omega' \qquad (3.63b)$$

where ω and ω' are the frequency domains associated with the signals $x(n)$ and $y(m)$, respectively. Applying Eq. (3.57) to (3.63a) and expressing $x(n)$ in terms of Eq. (3.62b) gives the frequency domain relationship between $X(e^{j\omega})$ and $Y(e^{j\omega})$ as

$$Y(e^{j\omega'}) = \frac{1}{2\pi} \int_{-\pi}^{\pi} \sum_{m=-\infty}^{\infty} \sum_{n=-\infty}^{\infty} k(m,n)e^{-j\omega'm}e^{j\omega n}X(e^{j\omega}) d\omega \qquad (3.64)$$

By defining

$$K(e^{j\omega'}, e^{j\omega}) \triangleq \frac{1}{2\pi} \sum_{m=-\infty}^{\infty} \sum_{n=-\infty}^{\infty} k(m,n)e^{-j\omega'm}e^{j\omega n} \qquad (3.65)$$

Eq. (3.65) can be expressed in the form

$$Y(e^{j\omega'}) = \int_{-\pi}^{\pi} K(e^{j\omega'}, e^{j\omega})X(e^{j\omega}) d\omega \qquad (3.66)$$

The function $K(e^{j\omega'}, e^{j\omega})$ is referred to as the *transmission function* or *bi-frequency system function* [3.4, 3.12-3.16] of the system described by $k(m,n)$. It defines the frequency domain mapping from the signal space $X(e^{j\omega})$ to the signal space $Y(e^{j\omega'})$ in a similar manner that $k(m,n)$ defines a time domain mapping between these two signal spaces. This relationship is graphically illustrated in Fig. 3.42.

For the case of linear time-invariant systems the mapping defined by Eq. (3.66) is clearly a one-to-one mapping such that $\omega' = \omega$. This can be shown by

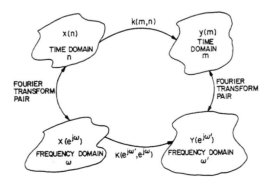

Figure 3.42 Graphic illustration of the time and frequency domain mappings defined by the system responses $k(n, m)$ and $K(e^{j\omega'}, e^{j\omega})$.

applying Eq. (3.61) to Eq. (3.65) to get

$$K(e^{j\omega'}, e^{j\omega}) = \frac{1}{2\pi} \sum_{m=-\infty}^{\infty} \sum_{n=-\infty}^{\infty} h(m, m-n)e^{-j\omega'm}e^{j\omega n} \tag{3.67}$$

By the substitution of variables $r = m - n$ and applying Eq. (3.60), this gives

$$K(e^{j\omega'}, e^{j\omega}) = \frac{1}{2\pi} \sum_{m=-\infty}^{\infty} \sum_{r=-\infty}^{\infty} h(m, r)e^{-j\omega'm}e^{j\omega(m-r)}$$

$$= \left[\sum_{r=-\infty}^{\infty} h(0, r)e^{-j\omega r} \right] \left[\frac{1}{2\pi} \sum_{m=-\infty}^{\infty} e^{-j(\omega'-\omega)m} \right]$$

$$= H(e^{j\omega}) \sum_{l=-\infty}^{\infty} \delta(\omega - \omega' + 2\pi l) \quad \text{time−invariance} \tag{3.68}$$

In this derivation we have applied the identity [3.17]

$$\frac{1}{2\pi} \sum_{m=-\infty}^{\infty} e^{j\theta m} = \sum_{l=-\infty}^{\infty} \delta(\theta + 2\pi l) \tag{3.69}$$

which defines a Dirac pulse train in θ with period 2π. Equation (3.68) defines the one-to-one mapping from $X(e^{j\omega})$ to $Y(e^{j\omega'})$ as illustrated in Fig. 3.43, where the principal value of the mapping is expressed by the solid line. According to this mapping, each signal component in ω is weighted by $H(e^{j\omega})$ and mapped to the location $\omega' = \omega$. Applying Eq. (3.68) to Eq. (3.66) then leads to the well-known relation for linear time-invariant systems

$$Y(e^{j\omega'}) = H(e^{j\omega})X(e^{j\omega'})\Big|_{\omega'=\omega}$$

$$= H(e^{j\omega'})X(e^{j\omega'}) \tag{3.70}$$

For the case of time-varying systems the mapping defined by $K(e^{j\omega'}, e^{j\omega})$ becomes more complex. When many frequency components in $X(e^{j\omega})$ map to the same frequency location in $Y(e^{j\omega'})$ the mapping is a many-to-one mapping which results in frequency domain *aliasing*. Alternatively, when one frequency component in $X(e^{j\omega})$ maps to many frequency locations in $Y(e^{j\omega'})$, the mapping is a one-to-many mapping which results in frequency domain *imaging*. In general, both aliasing and imaging may occur within the same mapping for complicated multirate systems (in time or frequency).

As seen in Chapter 2, an M-to-1 sampling rate compressor is an excellent example of a time-varying system which is characterized by a many-to-one

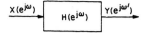

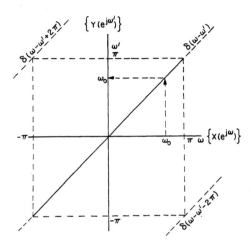

Figure 3.43 Illustration of the mapping defined by $K(e^{j\omega'}, e^{j\omega})$ for linear time-invariant systems.

mapping. Its time domain input-to-output relation is

$$y(m) = x(mM) \quad \text{for all } m \tag{3.71}$$

Comparing Eq. (3.71) with Eq. (3.57) gives

$$k(m, n) = u_0(mM - n) \quad M\text{-to-1 compressor} \tag{3.72}$$

where $u_0(m)$ is the unit sample function

$$u_0(m) = \begin{cases} 1, & m = 0 \\ 0, & \text{otherwise} \end{cases} \tag{3.73}$$

By applying the expression for $k(m, n)$ in Eq. (3.72) to Eq. (3.65) and using the identity in Eq. (3.69), the transmission function for the M-to-1 sampling rate compressor can be derived in the form

$$K(e^{j\omega'}, e^{j\omega}) = \frac{1}{2\pi} \sum_{m=-\infty}^{\infty} \sum_{n=-\infty}^{\infty} u_0(mM - n)e^{-j\omega'm}e^{j\omega n}$$

$$= \frac{1}{2\pi} \sum_{m=-\infty}^{\infty} e^{j(\omega M - \omega')m}$$

$$= \sum_{l=-\infty}^{\infty} \delta(\omega M - \omega' + 2\pi l) \quad M\text{-to-1 compressor} \tag{3.74}$$

This mapping is illustrated in Fig. 3.44 for $M = 3$. The principal values of the mapping are illustrated by the solid lines and it is clear that aliasing is occurring in this process.

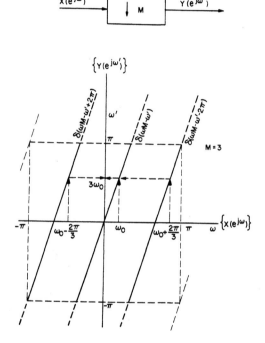

Figure 3.44 Illustration of a 3-to-1 mapping defined by $K(e^{j\omega'}, e^{j\omega})$ for a 3-to-1 sampling rate compressor.

In a similar manner, the 1-to-L sampling rate expander is a good example of a system whose transmission function $K(e^{j\omega'}, e^{j\omega})$ corresponds to a one-to-many (1-to-L) mapping. In Chapter 2 we saw that for the expander

$$y(m) = \begin{cases} x\left(\dfrac{m}{L}\right), & m = 0, \pm L, \pm 2L, \dots \\ 0, & \text{otherwise} \end{cases} \tag{3.75}$$

and therefore

$$k(m, n) = u_0(m - nL) \quad \text{1--to--}L \text{ expander} \tag{3.76}$$

Applying Eq. (3.75) to Eq. (3.65) then gives

$$K(e^{j\omega'}, e^{j\omega}) = \frac{1}{2\pi} \sum_{m=-\infty}^{\infty} \sum_{n=-\infty}^{\infty} u_0(m - nL)e^{-j\omega'm}e^{j\omega n}$$

$$= \frac{1}{2\pi} \sum_{n=-\infty}^{\infty} e^{j(\omega-\omega'L)n}$$

$$= \sum_{l=-\infty}^{\infty} \delta(\omega - \omega'L + 2\pi l) \quad \text{1--to--}L \text{ expander} \tag{3.77}$$

Figure 3.45 shows the mapping for this function and it is clear that for every

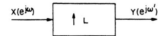

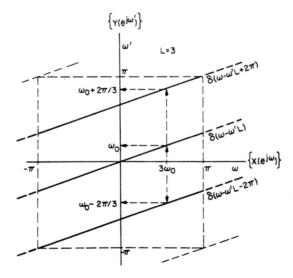

Figure 3.45   Illustration of a 1-to-3 mapping defined by $K(e^{j\omega'}, e^{j\omega})$ for a 1-to-3 sampling rate expander.

spectral component of $X(e^{j\omega})$ three images $(L = 3)$ of this signal occur in $Y(e^{j\omega'})$.

Another important class of time-varying systems are modulators. A modulator is characterized by the input-to-output relation

$$y(m) = \phi(n)x(n)\big|_{n=m} \qquad (3.78)$$

where the sampling rates of the input $x(n)$ and the output $y(m)$ are identical, and $\phi(n)$ is the modulation function. Comparison of Eqs. (3.57) and (3.78) yields

$$k(m, n) = u_0(m - n)\phi(n) \quad \text{modulator} \qquad (3.79)$$

where $u_0(m)$ is the unit sample function. The transmission function for the modulator can be derived as

$$K(e^{j\omega'}, e^{j\omega}) = \frac{1}{2\pi} \sum_{m=-\infty}^{\infty} \sum_{n=-\infty}^{\infty} u_0(m - n)\phi(n)e^{-j\omega'm}e^{j\omega n}$$

$$= \frac{1}{2\pi} \sum_{m=-\infty}^{\infty} \phi(n)e^{-j(\omega'-\omega)n}$$

$$= \frac{1}{2\pi} \Phi(e^{j(\omega'-\omega)}) \quad \text{modulator} \qquad (3.80)$$

where $\Phi(e^{j\omega})$ is the Fourier transform of $\phi(n)$.

An important modulator function seen in Chapter 2 is the complex modulator

$$\phi(n) = e^{j\omega_0 n} \qquad (3.81a)$$

whose transmission function is

$$K(e^{j\omega'}, e^{j\omega}) = \frac{1}{2\pi} \sum_{n=-\infty}^{\infty} e^{j\omega_0 n}e^{-j(\omega'-\omega)n}$$

$$= \sum_{l=-\infty}^{\infty} \delta(\omega_0 + \omega - \omega' + 2\pi l) \qquad (3.81b)$$

Figure 3.46 shows the mapping function that results from this operation where the double lines in the signal-flow graph indicate complex signals.

3.5.2 Cascading Networks and Commutation of Network Elements

Two important network operations that we have used extensively in Chapter 2 and in this chapter for building multirate systems are the operations of cascading and

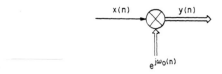

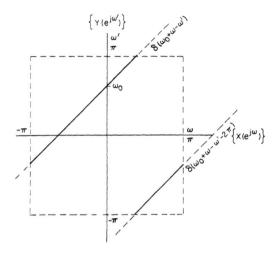

Figure 3.46 Mapping for a complex modulator $\phi(n) = e^{j\omega_0 n}$.

commuting of network elements. In this section we show how the general system response description of networks, $k(m, n)$ or $K(e^{j\omega'}, e^{j\omega})$, can conveniently be used to define these operations mathematically, and to define more rigorously the conditions under which time-varying network elements can be commuted.

Figure 3.47(a) shows an example of a cascade of two networks described by $k_1(r, n)$ and $k_2(m, r)$. If the overall response of this system is represented by the system response $k(m, n)$, then $k(m, n)$ can be derived in terms of $k_1(r, n)$ and $k_2(m, r)$ as follows:

$$w(r) = \sum_{n=-\infty}^{\infty} k_1(r, n)x(n) \tag{3.82}$$

$$y(m) = \sum_{r=-\infty}^{\infty} k_2(m, r)w(r)$$

$$= \sum_{n=-\infty}^{\infty} \sum_{r=-\infty}^{\infty} k_2(m, r)k_1(r, n)x(n) \tag{3.83a}$$

and also, by definition,

$$y(m) = \sum_{n=-\infty}^{\infty} k(m, n)x(n) \tag{3.83b}$$

By inspection of Eqs. (3.83) it can be seen that $k(m, n)$ can be expressed as

$$k(m, n) = \sum_{r=-\infty}^{\infty} k_2(m, r)k_1(r, n) \tag{3.84}$$

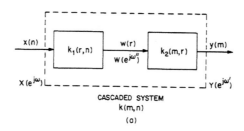

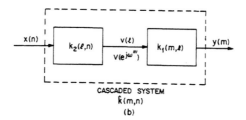

Figure 3.47 (a) Cascade connection of two network elements; (b) a cascade connection in which the two elements are commuted.

In a similar manner the overall transmission function $K(e^{j\omega'}, e^{j\omega})$ of this system can be conveniently expressed in terms of the individual transmission functions $K_1(e^{j\omega''}, e^{j\omega})$ and $K_2(e^{j\omega'}, e^{j\omega''})$ as

$$K(e^{j\omega'}, e^{j\omega}) = \int_{-\pi}^{\pi} K_2(e^{j\omega'}, e^{j\omega''})K_1(e^{j\omega''}, e^{j\omega}) \, d\omega'' \tag{3.85}$$

The derivation is completely analogous to that for Eq. (3.84). The form of Eqs. (3.84) and (3.85) show that cascading of networks implies a succession of mappings from the input signal space to an intermediate signal space and then to the output signal space.

If the two network responses $k_1(r, n)$ and $k_2(m, r)$ are interchanged as shown in Fig. 3.47(b), the overall system response is, in general, different. In this case the overall responses can be shown to be of the forms

$$\hat{k}(m, n) = \sum_{l=-\infty}^{\infty} k_1(m, l) k_2(l, n) \qquad (3.86a)$$

and

$$\hat{K}(e^{j\omega'}, e^{j\omega}) = \int_{-\pi}^{\pi} K_1(e^{j\omega'}, e^{j\omega'''}) K_2(e^{j\omega'''}, e^{j\omega}) \, d\omega''' \qquad (3.86b)$$

Thus the order in which the mappings occur is not necessarily commutable. As discussed in Section 3.1.2, two network elements $k_1(\cdot, \cdot)$ and $k_2(\cdot, \cdot)$ are *commutable* if and only if the two systems in Fig. 3.47 have the same system responses, that is, if and only if

$$k(m, n) = \hat{k}(m, n) \quad \text{commutable} \qquad (3.87a)$$

or equivalently if and only if

$$K(e^{j\omega'}, e^{j\omega}) = \hat{K}(e^{j\omega'}, e^{j\omega}) \quad \text{commutable} \qquad (3.87b)$$

Several examples of commutable network elements have been discussed in Section 3.1.2 and it can be verified that they satisfy the foregoing conditions.

In particular, an important class of systems that are commutable are those that are linear and time-invariant. If $K_1(\cdot, \cdot)$ and $K_2(\cdot, \cdot)$ are linear time-invariant elements, the principal value of their mappings can be defined as

$$K_1(e^{j\omega'}, e^{j\omega}) = H_1(e^{j\omega})\delta(\omega - \omega') \qquad (3.88a)$$

and

$$K_2(e^{j\omega'}, e^{j\omega}) = H_2(e^{j\omega})\delta(\omega - \omega') \qquad (3.88b)$$

where the variables ω and ω' are restricted to the range of principle values (i.e., $-\pi$ to π). It can then be verified that their commuted system responses are equal, that is,

$$K(e^{j\omega'}, e^{j\omega}) = \int_{-\pi}^{\pi} H_2(e^{j\omega''}) H_1(e^{j\omega}) \delta(\omega'' - \omega') \delta(\omega - \omega'') \, d\omega''$$

$$= H_2(e^{j\omega}) H_1(e^{j\omega}) \delta(\omega - \omega') \qquad (3.88c)$$

and

$$\hat{K}(e^{j\omega'}, e^{j\omega}) = \int_{-\pi}^{\pi} H_1(e^{j\omega'''}) H_2(e^{j\omega}) \delta(\omega''' - \omega') \delta(\omega - \omega''') \, d\omega'''$$

$$= H_1(e^{j\omega}) H_2(e^{j\omega}) \delta(\omega - \omega')$$

$$= K(e^{j\omega'}, e^{j\omega}) \qquad (3.88d)$$

3.5.3 Network Duality

In Section 3.1.3 we inferred in a nonrigorous manner that networks and their transposes are dual systems and that they perform complementary operations. These concepts of duality and transposition can be defined more rigorously with the aid of the general definitions of transmission functions and system responses for time-varying and multirate networks that were given above. In this section we consider the relationship between networks and their duals and present several definitions of duality. In the next section we outline a proof that transpose networks are duals.

The concepts of network duality and transposition have been formulated in several ways [3.1-3.4, 3.16, 3.18] and they are also closely related to concepts of adjoint networks [3.19, 3.20]. In this section we focus primarily on two definitions of duality and transposition: generalized duality (and transposition) and Hermitian duality (and transposition). Both definitions lead to the same results for the case of real-valued systems of the type discussed in Sections 3.1 to 3.4. Therefore, no distinction between these two formulations is necessary in developing the structures in those sections. Differences between these two formulations arise, however, when we consider networks with complex signals, or complex elements, such as the quadrature systems considered in Chapter 2. We will not consider systems constructed as adjoint systems since adjoint operators on causal systems lead to noncausal systems that are of less practical interest. The interested reader is referred to Ref. 3.19 for further details regarding adjoint systems and their properties.

Consider the network with a transmission function $K(e^{j\omega'}, e^{j\omega})$ and a system response $k(m, n)$. Then the *generalized dual* of this network is a network with the transmission function $K^T(e^{j\omega'}, e^{j\omega})$ such that

$$K^T(e^{j\omega'}, e^{j\omega}) \triangleq K(e^{j\omega}, e^{j\omega'}) \quad \text{generalized duality} \qquad (3.89a)$$

or equivalently with the system response $k^T(m, n)$ such that

$$k^T(m,n) \triangleq k(-n, -m) \quad \text{generalized duality} \qquad (3.89b)$$

The fact that Eqs. (3.89a) and (3.89b) are equivalent can be verified as follows:

$$K^T(e^{j\omega'}, e^{j\omega}) = \frac{1}{2\pi} \sum_{m=-\infty}^{\infty} \sum_{n=-\infty}^{\infty} k^T(m, n) e^{-j\omega'm} e^{j\omega n}$$

$$= \frac{1}{2\pi} \sum_{m=-\infty}^{\infty} \sum_{n=-\infty}^{\infty} k(-n, -m) e^{-j\omega'm} e^{j\omega n}$$

$$= \frac{1}{2\pi} \sum_{n=-\infty}^{\infty} \sum_{m=-\infty}^{\infty} k(n, m) e^{-j\omega n} e^{j\omega'm}$$

$$= K(e^{j\omega}, e^{j\omega'}) \tag{3.90}$$

We can further specify that two networks, represented by transmission functions $K(e^{j\omega'}, e^{j\omega})$ and $K^T(e^{j\omega'}, e^{j\omega})$, which are topologically related by the process of signal-flow graph reversal, and by replacing all branch operations by their generalized dual operations, are *generalized transpose networks*. The proof of this transposition property is outlined in the next section. We make the distinction, therefore, that transpose networks have dual transmission functions; however, networks that are duals are not necessarily transposes. For example, a 1-to-L interpolator realized with a polyphase structure may be the dual of an L-to-1 decimator realized with a direct-form structure; however, topologically they are not transpose networks.

The definition of generalized duality in Eq. (3.89a) suggests a reversal in the direction of the mapping implied by the transmission function. That is, the roles of ω and ω' are interchanged. A one-to-many mapping becomes a many-to-one mapping (and vice versa) and the roles of aliasing and imaging are interchanged. By comparing the transmission functions of sampling rate compressors and expanders [Eqs. (3.74) and (3.77), respectively] with the definition of generalized duality [Eq. (3.89a)], it is readily seen that they are duals. This is also seen by comparing their mappings in Figs. 3.44 and 3.45. The fact that linear-time invariant systems are their own duals is also readily seen by comparing Eq. (3.68) with Eqs. (3.89) and noting that their transmission functions are unchanged by interchanging the roles of ω and ω'.

The *Hermitian dual* of a network with a transmission function $K(e^{j\omega'}, e^{j\omega})$ is defined as a network with a transmission function

$$K^H(e^{j\omega'}, e^{j\omega}) \triangleq K^*(e^{-j\omega}, e^{-j\omega'}) \quad \text{Hermitian duality} \tag{3.91a}$$

or equivalently with the system response

$$k^H(m, n) \triangleq k^*(-n, -m) \quad \text{Hermitian duality} \tag{3.91b}$$

where * denotes the complex conjugate. In a similar manner to that of the generalized transpose, the *Hermitian transpose* of a network is a network that is topologically related by the process of signal-flow graph reversal and with all branch functions replaced by their Hermitian dual functions.

As seen from Eqs. (3.91), the Hermitian dual of a network is one in which the roles of ω and ω' are interchanged and replaced by their negatives and the resulting function is conjugated. For real systems or systems that are conjugate symmetric (e.g., real, time-invariant systems), operations of negation of the frequencies and conjugation cancel and the Hermitian dual (transpose) and the generalized dual (transpose) are identical. Thus all the transpose networks discussed in Sections 3.1 to 3.4 satisfy properties of both generalized and Hermitian duality.

Differences in the generalized and Hermitian duals arise when considering complex systems or complex signals. One important example is that of the modulator. For a modulator with modulation function $\phi(n)$, it was seen earlier that the system response and transmission function are, respectively,

$$k(m, n) = u_0(m - n)\phi(n) \quad \text{modulator} \tag{3.92a}$$

and

$$K(e^{j\omega'}, e^{j\omega}) = \frac{1}{2\pi}\Phi(e^{j(\omega'-\omega)}) \quad \text{modulator} \tag{3.92b}$$

where $\Phi(e^{j\omega})$ is the Fourier transform of $\phi(n)$. Therefore, the generalized dual of the modulator is characterized by

$$k^T(m, n) = u_0(m - n)\phi(-m) \tag{3.93a}$$

and

$$K^T(e^{j\omega'}, e^{j\omega}) = \frac{1}{2\pi}\Phi(e^{j(\omega-\omega')}) \tag{3.93b}$$

and the Hermitian dual is characterized by

$$k^H(m, n) = u_0(m - n)\phi^*(-m) \tag{3.94a}$$

and

$$K^H(e^{j\omega'}, e^{j\omega}) = \frac{1}{2\pi}\Phi^*(e^{j(\omega'-\omega)}) \tag{3.94b}$$

Figure 3.48 shows the resulting dual or transpose network elements described by Eqs. (3.92) to (3.94), respectively.

For the case of a complex modulator with the modulation function

$$\phi(n) = e^{j\omega_0 n} \tag{3.95}$$

the generalized transpose is a modulator with modulation function

$$\phi^T(n) = e^{-j\omega_0 n} \tag{3.96}$$

and the Hermitian transpose is a modulator with modulation function

$$\phi^H(n) = e^{j\omega_0 n} \tag{3.97}$$

Therefore, it is seen that the generalized transpose of a complex modulator is a demodulator, whereas the Hermitian transpose is a modulator whose modulation function is unchanged. More generally, we can extend this to the case of a signal-flow graph for the implementation of a DFT (discrete Fourier transform) and show that the generalized dual of a DFT structure is a structure that realizes an inverse DFT, whereas the Hermitian dual leads to a structure that realizes the original DFT [3.4]. Thus the concept of duality for complex systems takes on very different meanings depending on whether the generalized dual or the Hermitian dual is used.

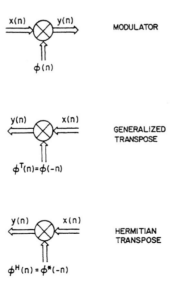

Figure 3.48 Generalized transpose and Hermitian transpose for a modulator.

3.5.4 Network Transposition and Tellegen's Theorem

The transposition theorem states that the process of signal-flow graph reversal and replacing branch operations in a network by their dual operations leads to a network whose overall system response is the dual to that of the original network. This theorem applies to both the generalized and Hermitian definitions of duality. As seen in Sections 3.1 to 3.4, this is an extremely useful theorem for manipulating and developing practical signal processing structures for multirate systems.

The proof of the transposition theorem is nontrivial and must be developed through a series of steps. These steps are outlined in this section together with a discussion of several related network theorems. We first consider a more rigorous mathematical formulation for uniquely describing signal-flow graphs. Then a powerful and general network theorem called Tellegen's theorem is derived. Tellegen's theorem is then used in proving the transposition theorem. Finally, an important and practical corollary to the transposition theorem is considered and some additional applications of Tellegen's theorem are mentioned.

The representation of a signal-flow graph in terms of a mathematical framework can be developed by considering the equations associated with each node and branch in the network [3.1-3.4]. Since different nodes in a multirate network may have different sampling rates, we define n_r and ω_r, respectively, as the discrete-time index and the frequency associated with node r. Also associated with node r is a node signal value $w_r(n_r)$ and its transform $W_r(e^{j\omega_r})$.

We will find it convenient to distinguish between source branches and network branches. *Source branches* represent the injection of external signals into the network [see Fig. 3.49(a)]. We will assume that there is at most one source branch associated with each node r and that its branch signal value is $x_r(n_r)$. For a single-input network only one node will have a nonzero source entering it. The output of a network is taken as one of the node signal values in the network.

Network branches originate from a network node and terminate at a network node [see Fig. 3.49(b)]. The input signal to a branch is the signal value of the node where the branch originates. Each network branch operates on its input signal to produce an output signal. Therefore, we can associate with a branch from node q to node r a branch system response $f_{rq}(n_r, n_q)$, as in Eq. (3.57), which describes this operation, and an output signal $v_{rq}(n_r)$ such that

$$v_{rq}(n_r) = \sum_{n_q=-\infty}^{\infty} f_{rq}(n_r, n_q) w_q(n_q) \tag{3.98}$$

For convenience (and without lack of generality) we will assume that only one branch (in each direction) connects each pair of nodes.

Nodes act as summing points in the network. Assuming that there are Q nodes in the network, we can then write Q equations (one for each node) of the

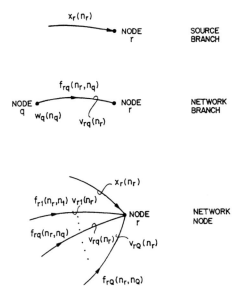

Figure 3.49 Signal-flow graph representation for; (a) a source branch; (b) a network branch; and (c) a network node.

form

$$w_r(n_r) = x_r(n_r) + \sum_{q=1}^{Q} v_{rq}(n_r) \tag{3.99a}$$

$$= x_r(n_r) + \sum_{q=1}^{Q} \sum_{n_q=-\infty}^{\infty} f_{rq}(n_r, n_q) w_q(n_q), \quad r = 1, ..., Q \tag{3.99b}$$

Alternatively, as in Eqs. (3.64) to (3.66), we can express the node signals in terms of their transforms as

$$W_r(e^{j\omega_r}) = X_r(e^{j\omega_r}) + \sum_{q=1}^{Q} V_{rq}(e^{j\omega_r}) \tag{3.100}$$

$$= X_r(e^{j\omega_r}) + \sum_{q=1}^{Q} \int_{-\pi}^{\pi} F_{rq}(e^{j\omega_r}, e^{j\omega_q}) W_q(e^{j\omega_q}) \, d\omega_q, \quad r = 1, ..., Q$$

The set of equations in Eqs. (3.99) or (3.100) then represents a complete mathematical formulation of the network. For any nonexistent branch in the network we can simply assign its system response and transmission function to be zero.

Tellegen's theorem relates variables of two different networks S and $\hat{S}$ having the same topology. The theorem has played an important role in understanding both analog and digital networks [3.1-3.4, 3.21] and is often used in deriving other network theorems. As in the case of duality and transposition there are many different forms and variations of Tellegen's theorem. We consider only two forms here, one associated with generalized transposition and the other associated with Hermitian transposition. The first form can be derived by considering two networks S and $\hat{S}$ with the same number of nodes Q. Let $W_r(e^{j\omega_r})$ be the transform of the rth node signal value in network S and $\hat{W}_r(e^{j\omega_r})$ be the corresponding value for the rth node in network $\hat{S}$. Then we can write the obvious identity

$$W_r(e^{j\omega_r})\hat{W}_r(e^{j\omega_r}) - \hat{W}_r(e^{j\omega_r})W_r(e^{j\omega_r}) = 0 \qquad (3.101)$$

By summing and integrating over r and ω_r, respectively, we get the expression

$$\sum_{r=1}^{Q}\int_{-\pi}^{\pi}\left[W_r(e^{j\omega_r})\hat{W}_r(e^{j\omega_r}) - \hat{W}_r(e^{j\omega_r})W_r(e^{j\omega_r})\right]d\omega_r = 0 \qquad (3.102)$$

Applying Eq. (3.100), once to $\hat{W}_r(e^{j\omega_r})$ and once to $W_r(e^{j\omega_r})$, and separating terms leads to the form

$$\sum_{r=1}^{Q}\sum_{q=1}^{Q}\int_{-\pi}^{\pi}\int_{-\pi}^{\pi}W_r(e^{j\omega_r})\hat{F}_{rq}(e^{j\omega_r},e^{j\omega_q})\hat{W}_q(e^{j\omega_q})\,d\omega_r\,d\omega_q$$

$$-\sum_{r=1}^{Q}\sum_{q=1}^{Q}\int_{-\pi}^{\pi}\int_{-\pi}^{\pi}\hat{W}_r(e^{j\omega_r})F_{rq}(e^{j\omega_r},e^{j\omega_q})W_q(e^{j\omega_q})\,d\omega_r\,d\omega_q$$

$$+\sum_{r=1}^{Q}\int_{-\pi}^{\pi}\left[W_r(e^{j\omega_r})\hat{X}_r(e^{j\omega_r}) - \hat{W}_r(e^{j\omega_r})X_r(e^{j\omega_r})\right]d\omega_r = 0 \quad (3.103)$$

Finally, we observe that the second term in this expression remains unchanged if we reverse r and q. By performing this interchange of variables and combining terms, we get the desired form for *Tellegen's theorem*:

$$\sum_{r=1}^{Q}\sum_{q=1}^{Q}\int_{-\pi}^{\pi}\int_{-\pi}^{\pi}W_r(e^{j\omega_r})\hat{W}_q(e^{j\omega_q})\left[\hat{F}_{rq}(e^{j\omega_r},e^{j\omega_q}) - F_{qr}(e^{j\omega_q},e^{j\omega_r})\right]d\omega_r\,d\omega_q$$

$$+\sum_{r=1}^{Q}\int_{-\pi}^{\pi}\left[W_r(e^{j\omega_r})\hat{X}_r(e^{j\omega_r}) - \hat{W}_r(e^{j\omega_r})X_r(e^{j\omega_r})\right]d\omega_r = 0 \quad (3.104)$$

We consider next the case where the network $\hat{S}$ is the generalized transpose of the network S. Then, according to the definition of generalized transposition, for every branch in network S from node r to node q with the transmission function $F_{qr}(e^{j\omega_q}, e^{j\omega_r})$, there exists a branch in network $\hat{S}$ from node q to node r with the generalized dual transmission function

$$\hat{F}_{rq}(e^{j\omega_r}, e^{j\omega_q}) = F_{qr}(e^{j\omega_q}, e^{j\omega_r}) \tag{3.105}$$

Applying this definition to Tellegen's theorem in Eq. (3.104) shows that the first term becomes zero when S and $\hat{S}$ are generalized transposes, and we get

$$\sum_{r=1}^{Q} \int_{-\pi}^{\pi} \left[W_r(e^{j\omega_r})\hat{X}_r(e^{j\omega_r}) - \hat{W}_r(e^{j\omega_r})X_r(e^{j\omega_r}) \right] d\omega_r = 0 \tag{3.106}$$

If we assume that network S has a single input, $X_a(e^{j\omega_a})$, at node a and a single output $Y(e^{j\omega_b}) = W_b(e^{j\omega_b})$ at node b, then network $\hat{S}$ has a single input, $\hat{X}_b(e^{j\omega_b})$, at node b and a single output $\hat{Y}(e^{j\omega_a}) = \hat{W}_a(e^{j\omega_a})$ as depicted in Fig. 3.50. Then Eq. (3.106) has only two terms and can be expressed as

$$\int_{-\pi}^{\pi} Y(e^{j\omega_b})\hat{X}_b(e^{j\omega_b}) \, d\omega_b - \int_{-\pi}^{\pi} \hat{Y}(e^{j\omega_a})X_a(e^{j\omega_a}) \, d\omega_a = 0 \tag{3.107}$$

The input-to-output relations for these two networks can be described in terms of their transmission functions as

$$Y(e^{j\omega_b}) = \int_{-\pi}^{\pi} K(e^{j\omega_b}, e^{j\omega_a})X_a(e^{j\omega_a}) \, d\omega_a \tag{3.108a}$$

and

$$\hat{Y}(e^{j\omega_a}) = \int_{-\pi}^{\pi} \hat{K}(e^{j\omega_a}, e^{j\omega_b})\hat{X}_b(e^{j\omega_b}) \, d\omega_b \tag{3.108b}$$

Applying Eqs. (3.108) to Eq. (3.107) then gives

$$\int_{-\pi}^{\pi} \int_{-\pi}^{\pi} X_a(e^{j\omega_a})\hat{X}_b(e^{j\omega_b}) \left[K(e^{j\omega_b}, e^{j\omega_a}) - \hat{K}(e^{j\omega_a}, e^{j\omega_b}) \right] d\omega_a \, d\omega_b = 0 \tag{3.109}$$

Equation (3.109) must be satisfied for all possible signals $X_a(e^{j\omega_a})$ and $\hat{X}_b(e^{j\omega_b})$. This can occur only if

$$\hat{K}(e^{j\omega_a}, e^{j\omega_b}) = K^T(e^{j\omega_a}, e^{j\omega_b}) = K(e^{j\omega_b}, e^{j\omega_a}) \tag{3.110}$$

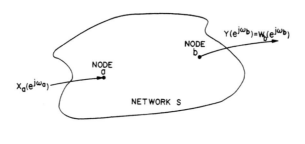

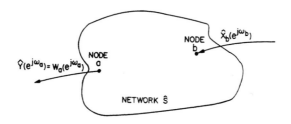

Figure 3.50 (a) A network S with input and output at nodes a and b respectively; and (b) its transpose with input and output at nodes b and a respectively.

which states that the two networks are generalized duals. This concludes the proof of the transposition theorem for generalized transpose networks.

A similar proof can be developed for the case of Hermitian transpose networks and it is based on a second form of Tellegen's theorem, which is obtained from the identity

$$W_r(e^{j\omega_r})\,\hat{W}_r^*(e^{-j\omega_r}) - \hat{W}_r^*(e^{-j\omega_r})\,W_r(e^{j\omega_r}) = 0 \qquad (3.111)$$

By following a similar approach to that for the case of the generalized form of Tellegen's theorem, we get the Hermitian form of Tellegen's theorem [3.4]:

$$\sum_{r=1}^{Q}\sum_{q=1}^{Q}\int_{-\pi}^{\pi}\int_{-\pi}^{\pi} \hat{W}_r^*(e^{-j\omega_r})\,W_q(e^{j\omega_q})\left[\hat{F}_{qr}^*(e^{-j\omega_q},\,e^{-j\omega_r}) - F_{rq}(e^{j\omega_r},\,e^{j\omega_q})\right]d\omega_q\,d\omega_r$$

$$+\sum_{r=1}^{Q}\int_{-\pi}^{\pi}\left[W_r(e^{j\omega_r})\hat{X}_r^*(e^{-j\omega_r}) - \hat{W}_r^*(e^{-j\omega_r})X_r(e^{j\omega_r})\right]d\omega_r = 0 \quad (3.112)$$

which can be used to prove the transposition theorem for Hermitian duality. In a like manner similar derivations can be developed for adjoint systems [3.19].

An important corollary to the transposition theorem is that if a particular structure that realizes a transmission function $K(e^{j\omega'},\,e^{j\omega})$ is minimized with respect to multiplication rate (e.g., is canonical), its transpose is a canonical structure for implementing the dual transmission function [3.4]. This fact was used implicitly earlier in this chapter as a tool in the development of the efficient decimator and interpolator structures.

Other important applications of the concepts of Tellegen's theorem and network transportation are in the areas of network sensitivity analysis and in defining properties of stationary linear time-varying systems. In the area of sensitivity analysis, these methods have been used to derive expressions for relating the sensitivity of the input-to-output relation of a multirate network to variations in its internal parameters (e.g., effects of round-off noise and coefficient quantization). In the area of defining properties of time-varying systems, they have been useful in defining consequences of causality and stationarity in time-varying systems. The reader is referred to Refs. [3.1-3.4, 3.18-3.22] for further details.

3.5.5 Transposition of Complex Networks

As we have seen from the discussion above, when dealing with real networks and real signals, the definitions of generalized and Hermitian transposes are identical and we can transpose branches in the networks in a straightforward manner as we have done in Sections 3.1 to 3.4 to obtain the dual networks. When dealing with structures that implicitly have complex signals or complex branches, some subtle differences arise depending on whether we use the generalized or the Hermitian definition of transposition. For example, in Section 3.5.3 we saw that the generalized transpose of a complex modulator is a complex demodulator, whereas its Hermitian transpose leaves it unchanged.

Problems often arise when structures that inherently involve complex signals are represented in a form such that their real and imaginary signal paths are expressed as separate "real" signal paths. For example, consider a simple network with a complex input and output and a complex system response denoted as $k(m, n)$. Then we can express the system response in terms of its real and imaginary parts as

$$k(m, n) = k_r(m, n) + jk_i(m, n) \qquad (3.113)$$

Figure 3.51(a) shows the complex representation for this network, where $x(n)$ and $y(m)$ are the complex inputs and outputs and double lines refer to complex signal paths. Alternatively, we can express this network in terms of its real and imaginary signal paths in terms of the "real" network shown in Fig. 3.51(b), where $x_r(n)$ and $x_i(n)$ are the real and imaginary components, respectively, of $x(n)$ and similarly, $y_r(m)$ and $y_i(m)$ are the real and imaginary components, respectively, of $y(m)$.

If we take the generalized transpose of this network, we get a system whose system response is

$$k^T(m, n) = k(-n, -m)$$

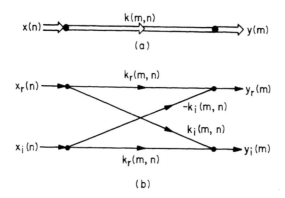

Figure 3.51 (a) Complex network; (b) its real representation.

$$= k_r(-n, -m) + jk_i(-n, -m) \qquad (3.114)$$

and the roles of the input and output nodes are interchanged. The complex network and its "real" representation for this system are shown in Fig. 3.52(a) and (b), respectively.

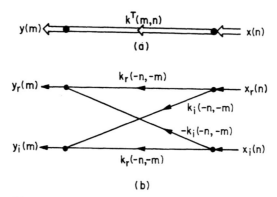

Figure 3.52 (a) Generalized transpose of the network in Fig. 3.51; (b) its real representation.

Alternatively, if we take the Hermitian transpose of the network, we get a system with the system response

$$k^H(m, n) = k^*(-n, -m)$$
$$= k_r(-n, -m) - jk_i(-n, -m) \qquad (3.115)$$

Figure 3.53 shows the corresponding complex and real representation of this network. In comparison to the generalized transpose network shown in Fig. 3.52, it

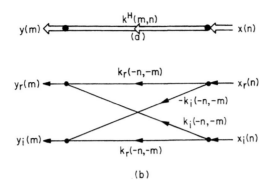

(a)

(b)

Figure 3.53 (a) Hermitian transpose of the network in Fig. 3.51; (b) its real representation.

is seen that the principal difference is in the signs of the cross branches in the real networks.

Given a "real" representation of a network, or structure, that is inherently a complex system, it is tempting to simply transpose this real network branch-by-branch as if it were in fact a real network (since the transpose of each branch is the same whether we use the generalized or Hermitian definitions of transposition). The question is: What does this process lead to in terms of a final complex network? The answer can be seen by branch-by-branch transposing the real network of Fig. 3.51(b) and comparing it to the real networks of Figs. 3.52(b) and 3.53(b). Figure 3.54 shows this transposition and it can be seen that it is identical to the "real" structure of Fig. 3.53(b). Thus we can conclude from this example that a branch-by-branch transposition of the real structure of an inherently complex network will lead to a complex network that is the Hermitian transpose of the original network. If we want the generalized transpose, we must consider the process of transposing the complex network directly.

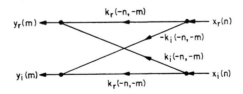

Figure 3.54 Branch-by-branch transposition of the real network of Fig. 3.51(b).

A good example is that of the quadrature modulator and demodulator discussed in Chapter 2 (see Fig. 2.24). If we branch-by-branch transposed the real representation of the quadrature modulator, each modulator function $\phi(n)$ would become $\phi(-n)$. Thus a $\cos(\omega_0 n)$ modulator would become a $\cos(\omega_0 n)$ modulator and a $\sin(\omega_0 n)$ modulator would become a $\sin(-\omega_0 n) = -\sin(\omega_0 n)$ modulator.

The sign on the sin $(\omega_0 n)$ modulator would be incorrect according to this process because the overall complex quadrature demodulator is the generalized transpose of the complex quadrature modulator, not the Hermitian transpose. In conclusion, some caution must be exercised when dealing with systems that inherently involve complex signals.

3.6 SUMMARY

This chapter has been directed at the topic of structures and network concepts for multirate digital systems. The first part of the chapter covered basic network concepts in an informal way. Three principal classes of practical structures were then developed for FIR and IIR decimators and interpolators: direct-form structures, polyphase structures, and structures with time-varying coefficients.

The second part of the chapter was directed at more theoretical aspects of network theory. Network concepts such as transposition that were informally developed and heavily used in the first part of the chapter were derived through the use of a much more rigorous mathematical framework for multirate and time-varying networks.

REFERENCES

3.1 A. V. Oppenheim and R. W. Schafer, *Digital Signal Processing.* Englewood Cliffs, NJ: Prentice-Hall Inc., 1975.

3.2 R. E. Crochiere and A. V. Oppenheim, "Analysis of Linear Digital Networks," *Proc. IEEE*, Vol. 63, No. 4, pp. 581-595, April 1975.

3.3 A. Fettweis, "A General Theorem for Signal-Flow Networks, with Applications," *Arch. Elek. Übertrag.*, Vol. 25, pp. 557-561, December 1971; also in *Digital Signal Processing*, (L. R. Rabiner and C. M. Rader, Eds.). New York: IEEE Press, 1972, pp. 126-130.

3.4 T. A. Claasen and W. F. Mecklenbräuker, "On the Transposition of Linear Time-Varying Discrete-Time Networks and Its Application to Multirate Digital Systems," *Philips J. Res.*, Vol. 23, pp. 78-102, 1978.

3.5 L. R. Rabiner and B. Gold, *Theory and Application of Digital Signal Processing.* Englewood Clifs, NJ: Prentice-Hall Inc., 1975.

3.6 M. G. Bellanger and G. Bonnerot, "Premultiplication Scheme for Digital FIR Filters with Application to Multirate Filtering," *IEEE Trans. Acoust. Speech Signal Process.*, Vol. ASSP-26, No. 1, pp. 50-55, February 1978.

3.7 M. G. Bellanger, G. Bonnerot, and M. Coudreuse, "Digital Filtering by Polyphase Network: Application to Sample Rate Alteration and Filter

Banks," *IEEE Trans. Acoust. Speech Signal Process.*, Vol. ASSP-24, No. 2, pp. 109-114, April 1976.

3.8   R. A. Meyer and C. S. Burrus, "Design and Implementation of Multirate Digital Filters," *IEEE Trans. Acoust. Speech Signal Process.*, Vol. ASSP-24, No. 1, pp. 53-58, February 1976.

3.9   R. E. Crochiere and L. R. Rabiner, "Optimum FIR Digital Filter Implementations for Decimation, Interpolation, and Narrow-Band Filtering," *IEEE Trans. Acoust. Speech Signal Process.*, Vol. ASSP-23, No. 5, pp. 444-456, October 1975.

3.10   R. E. Crochiere, "A General Program to Perform Sampling Rate Conversion of Data by Rational Ratios." in *Programs for Digital Signal Processing*, New York: IEEE Press, 1979, pp. 8.2-1 to 8.2-7.

3.11   H. G. Martinez and T. W. Parks, "A Class of Infinite-Duration Impulse Response Digital Filters for Sampling Rate Reduction," *IEEE Trans. Acoust. Speech Signal Process.*, Vol. ASSP-27, No. 2, pp. 154-162, April 1979.

3.12   L. A. Zadeh, "Time-Varying Networks," *Proc. IRE*, Vol. 49, pp. 1488-1503, October 1961.

3.13   L. A. Zadeh, "Frequency Analysis of Variable Networks," *Proc. IRE*, Vol. 32, pp. 291-299, March 1950.

3.14   T. Kailath, "Channel Characterization: Time-Variant Dispersive Channels," *Lectures on Communication System Theory* (E. J. Baghdaddy, Ed.). New York: McGraw-Hill, 1961, pp. 95-123.

3.15   M. R. Portnoff, "Time-Frequency Representation of Digital Signals and Systems Based on Short-Time Fourier Analysis," *IEEE Trans. Acoust. Speech Signal Process.*, Vol. ASSP-28, No. 1, pp. 55-69, February 1980.

3.16   A. Gersho and N. DeClaris, "Duality Concepts in Time-Varying Linear Systems," *IEEE Int. Conv. Rec.*, Part 1, pp. 344-356, 1964.

3.17   A. Papoulis, *The Fourier Integral and Its Application*. New York: McGraw-Hill, 1965.

3.18   L. B. Jackson, "On the Interaction of Roundoff Noise and Dynamic Range in Digital Filters," *Bell Syst. Tech. J.*, Vol. 49, No. 2, pp. 159-184, February 1970.

3.19   T. A. Claasen and W. F. Mecklenbräuker, "On the Description, Transposition, Adjointness and Other Functional and Structural Relations of Linear Discrete-Time Time-Varying Networks," *Proc. 4th Int. Symp. Math. Theory Networks Syst.*, Delft, The Netherlands, July 4-6, 1979.

3.20 L. A. Zadeh and C. A. Desoer, *Linear System Theory: The State Space Approach.* New York: McGraw-Hill, 1963, p. 346.

3.21 P. Penfield, Jr., R. Spence, and S. Duinker, *Tellegen's Theorem and Electric Networks*, Cambridge, Mass.: MIT Press, 1970.

3.22 T. A. Claasen and W. F. Mecklenbräuker, "On Stationary Linear Time-Varying Systems," *IEEE Trans. Circuits Systems*, Vol. CAS-29, No. 3, pp. 169-184, March 1982.

4

Design of Digital Filters for
Decimation and Interpolation

4.0 INTRODUCTION

In the preceding two chapters we have discussed the fundamental principles of sampling rate conversion and presented several examples of networks that implement this process. Chapter 2 described the basic conceptual model for sampling rate conversion by a factor of L/M (where L and M are integers), as shown in Fig. 4.1(a). As we have seen, this model is a cascade of three operations: a sampling rate expansion by a factor L, a linear time-invariant lowpass filtering operation with a filter $h(k)$, and a sampling rate compression by a factor M. It was also shown that this model could be used in conjunction with bandpass signals, in which case $h(k)$ becomes a bandpass rather than a lowpass filter.

Chapter 3 showed that a direct implementation of the model of Fig. 4.1(a) is not an efficient realization and that more efficient methods are possible: the direct-form structures, the polyphase structures, and structures with time-varying coefficients. All of these structures were developed either through signal-flow graph manipulations on the model of Fig. 4.1(a) or through interpretations of the mathematical input-to-output relations which describe this model. Thus, from an input-to-output point of view, all of these structures can be modeled by the conceptual model of Fig. 4.1. In fact, the coefficients in all these structures are obtained directly or indirectly from the prototype filter $h(k)$.

In the discussion in previous chapters we have assumed that the filter $h(k)$ approximates some ideal lowpass (or bandpass) characteristic. As such, the effectiveness of these systems is directly related to the type and quality of design of this digital filter. The purpose of this chapter is to review digital filter design techniques and discuss those methods that are especially applicable to the design of the digital filter in decimation and interpolation systems.

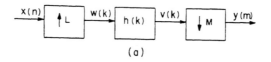

(a)

(b)

Figure 4.1 Basic prototype model for sampling rate conversion by a factor L/M.

4.1 DIGITAL FILTER DESIGN FUNDAMENTALS

4.1.1 Basic Considerations and Properties

Figure 4.1(b) shows a block diagram of a digital filter with input $w(k)$, output $v(k)$, and impulse response $h(k)$. For linear shift-invariant digital filters, the convolution equation

$$v(k) = \sum_{r=-\infty}^{\infty} h(r)w(k-r) \tag{4.1}$$

specifies the relation between $w(k)$, $v(k)$ and $h(k)$. Two major classes of digital filters have been defined: *finite impulse response* (FIR) and *infinite impulse response* (IIR) digital filters [4.1, 4.2]. FIR digital filters have the property that the impulse response, $h(k)$, satisfies the relation

$$h(k) = 0, \ k \geqslant N_2, \ k < N_1 \tag{4.2}$$

for finite values of both N_1 and N_2 [i.e., $h(k)$ is generally nonzero for the finite range $N_1 \leqslant k \leqslant N_2 - 1$]. The impulse response duration for such a filter is $N = N_2 - N_1$ samples. For realizability we require that $N_1 \geqslant 0$; however, for convenience we often consider the nonzero range of $h(k)$ to be the symmetric range $-(N-1)/2 \leqslant k \leqslant (N-1)/2$.

The z-transform of an FIR filter is the finite polynomial $H(z)$ given as

$$H(z) = \sum_{k=N_1}^{N_2-1} h(k)z^{-k} \tag{4.3}$$

which explicitly shows that FIR filters are "all-zero" filters, where the zeros are the roots of $H(z)$. Although several alternative implementations can be considered, Eq. (4.1) can be used directly to obtain $v(k)$ from $w(k)$ and $h(k)$ with the finite limits on k imposed by Eq. (4.2).

IIR digital filters have the property that the impulse response does not satisfy Eq. (4.2) for finite values of *both* N_1 and N_2. For IIR digital filters the convolution equation, although valid, is *not* an appropriate means for implementing the system. Instead, the structure of the network leading to the IIR filter is exploited to give a linear difference equation realization of the form

$$v(k) = \sum_{r=1}^{D} a_r v(k-r) + \sum_{r=0}^{N-1} b_r w(k-r) \tag{4.4}$$

Taking the z-transform of Eq. (4.4) and combining terms gives the transfer function, $H(z)$, of the filter as a finite ratio of polynomials in z, that is,

$$H(z) = \frac{V(z)}{W(z)} = \frac{\sum_{r=0}^{N-1} b_r z^{-r}}{1 - \sum_{r=1}^{D} a_r z^{-r}} \tag{4.5}$$

where D is the order of the denominator polynomial and $N-1$ is the order of the numerator polynomial. Equation (4.5) explicitly shows the pole-zero structure of an IIR filter: The poles are the roots of the denominator, and the zeros are the roots of the numerator.

The filter design problem is essentially one of determining suitable values of $h(k)$ (for FIR filters), or the a_r, b_r coefficients of Eq. (4.5) (for IIR filters), to meet given performance specifications on the filter. Such performance specifications can be made on the time response, $h(k)$, or the frequency response, $H(e^{j\omega})$, of the filter, defined as

$$H(e^{j\omega}) = \sum_{k=-\infty}^{\infty} h(k) e^{-j\omega k} \tag{4.6}$$

$$= H(z)\big|_{z=e^{j\omega}} \tag{4.7}$$

The frequency response is, in general, a complex function of ω. Thus it is convenient to represent it in terms of its magnitude, $|H(e^{j\omega})|$, and phase, $\theta(\omega)$, as

$$H(e^{j\omega}) = |H(e^{j\omega})| e^{j\theta(\omega)} \tag{4.8}$$

where

$$|H(e^{j\omega})| = \sqrt{\text{Re}^2 [H(e^{j\omega})] + \text{Im}^2 [H(e^{j\omega})]} \qquad (4.9)$$

and

$$\theta(\omega) = \tan^{-1} \left[\frac{\text{Im} [H(e^{j\omega})]}{\text{Re} [H(e^{j\omega})]} \right] \qquad (4.10)$$

An important filter parameter is the group delay, $\tau(\omega)$, defined as

$$\tau(\omega) = \frac{-d\theta(\omega)}{d\omega} \qquad (4.11)$$

The group delay is a measure of time delay of a signal as a function of frequency (i.e., the dispersion of the signal) as it passes through the filter. Nondispersive filters have the property that $\tau(\omega)$ is a constant (i.e., a fixed delay) over the frequency range of interest.

4.1.2 Advantages and Disadvantages of FIR and IIR Filters for Interpolation and Decimation

The filter required for sampling rate conversion systems can be obtained as the solution of a classical filter design problem: a lowpass (or bandpass) filter with, possibly, some time domain constraints (to be discussed later). In theory any of a large number of filter design techniques can be brought to bear on this problem. In the next few sections we discuss several such techniques. However, first we would like to distinguish general class differences between FIR and IIR filters and discuss how they relate to the filter design problem for interpolators and decimators.

FIR filters have the following general properties [4.1, 4.2]:

1. Linear phase designs are easily achieved [i.e., $\tau(\omega)$ is constant for all ω].
2. Arbitrary frequency responses can readily be approximated arbitrarily closely for sufficiently long impulse responses.
3. FIR filters are always stable.
4. FIR filters have good quantization properties (i.e., round-off noise can be made small, coefficients can be rounded to reasonable word-lengths for most practical designs, etc.).
5. No limit cycles occur with FIR filters.
6. The number of filter coefficients required for sharp-cutoff filters is generally quite large.
7. Excellent design techniques are readily available for a wide class of filters.

In contrast, IIR filters have the properties:

1. Arbitrary magnitude characteristics can readily be approximated.

2. No exact linear phase designs are possible; some good approximations do exist.
3. Classical design techniques are available for standard lowpass and bandpass designs (such designs, however, are not necessarily the most efficient ones for decimators or interpolators, as discussed in Chapter 3).
4. IIR designs are generally very efficient (small number of poles and zeros), especially for sharp-cutoff filters.
5. IIR filters can readily be checked for stability; if the filter is unstable, simple modifications can be made to ensure stability with no change in the magnitude response of the filter.
6. Finite-word-length effects (e.g., round-off noise, limit cycles, coefficient quantization) are well understood and can readily be controlled.

The properties listed above say basically that both FIR and IIR filters are suitable for the problem at hand. A major consideration in favor of the use of FIR filters, in sampling rate conversion, is that FIR filters can achieve exact linear phase while IIR filters can only approximate it. This consideration is particularly important when it is necessary to preserve the "continuous envelopes" of the sampled waveforms that are to be decimated or interpolated.

A second consideration in comparing FIR and IIR designs is that of required computation (number of coefficients, required accuracy of coefficients and internal data variables, structural complexity and modularity, etc.) For linear time-invariant filtering applications (where linear phase response is not required), IIR filters have generally been regarded as being computationally more efficient (i.e., lower filter orders and therefore fewer coefficients) than FIR filters [4.3]. However, IIR filters often require higher accuracy (more bits) than FIR filters for the specification of the filter coefficients and especially for internal state variables (to overcome effects of limit cycles and round-off noise). When we consider multirate structures (as in Chapter 3) and practical issues associated with implementing these systems (in hardware or software) the advantages of IIR designs are no longer as clear and we often see the balance tipping in favor of FIR designs. This is because a count of the number of coefficients in the structures, or the number of multiplications/second necessary to implement them, is only one measure of complexity. Equally important are more subtle issues such as the simplicity of the structure, control of the data flow in the structure, and the modularity of the structure. These factors are, of course, not as easy to quantify as computation, and are also dependent on the application and the particular form of the implementation.

Because of the above considerations, particularly that of linear phase, more emphasis is given in this chapter to the techniques of FIR design than to IIR design. At the end of this chapter, and later in Chapter 5, several comparisons are made, on the basis of required multiplication rate, between some specific FIR and IIR designs. These examples and the discussion in Chapter 3 will help to clarify the picture of when FIR or IIR designs may be most appropriate.

4.2 FILTER SPECIFICATIONS FOR SAMPLING RATE CONVERSION SYSTEMS

Before proceeding to a discussion of filter design techniques for decimators and interpolators, it is important to consider the *ideal* frequency domain and time domain criteria that specify such designs. It is also important to consider, in more detail, the representation of such filters in terms of a single prototype filter or as a set of polyphase filters. Although both representations are equivalent, it is sometimes easier to view filter design criteria in terms of one representation or the other. Also, some filter design techniques are directed at the design of a single prototype filter such as in the classical filter design methods, whereas other filter design techniques are directed at the design of the polyphase filters. Thus we will consider both representations in this section.

4.2.1 The Prototype Filter and Its Polyphase Representation

Figure 4.2(a) reviews the basic prototype model and its polyphase representation[1] [Fig. 4.2(b)] for a 1-to-L interpolator. A similar polyphase structure can also be defined for decimators (see Chapter 3). The signals $y_\rho(m)$, $\rho = 0, 1, 2, ..., L - 1$, represent the respective outputs of the polyphase filters $p_\rho(n)$ (appropriately expanded in sampling rate and delayed by ρ samples). Figure 4.3 gives a time domain interpretation of these signals for $L = 3$ and it can be seen that each branch ρ of the polyphase structure provides interpolated samples at time $m = nL + \rho$ in the output signal $y(m)$: for example, the $\rho = 0$ branch provides the outputs $y(0)$, $y(L)$, $y(2L)$...; the $\rho = 1$ branch provides outputs $y(1)$, $y(L + 1)$, $y(2L + 1)$, ...; and so on. [The reader should be cautioned here not to confuse the signal $u_\rho(n)$ with the unit sample $u_0(n)$.]

As discussed in Chapter 3, the coefficients, or impulse responses, of the polyphase filters correspond to sampled (and delayed) versions of the impulse response of the prototype filter. For a 1-to-L interpolator there are L polyphase filters and they are defined as (see Fig. 4.4)

$$p_\rho(n) = h(nL + \rho), \quad \rho = 0, 1, ..., L - 1, \text{ and all } n \qquad (4.12)$$

Similarly, for an M-to-1 decimator there are M polyphase filters in the polyphase structure and they are defined as

$$p_\rho(n) = h(nM + \rho), \quad \rho = 0, 1, ..., M - 1, \text{ and all } n \qquad (4.13)$$

[1] In this chapter we will only consider the counterclockwise commutator polyphase model. A similar clockwise commutator polyphase model can also be defined, as discussed in Chapter 3, by replacing ρ by $-\rho$ in the following derivations. The details are left to the reader.

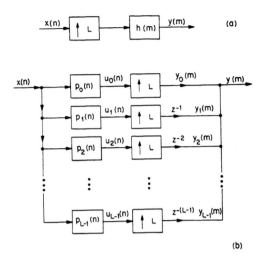

Figure 4.2 (a) A 1-to-L interpolator; (b) its polyphase structure.

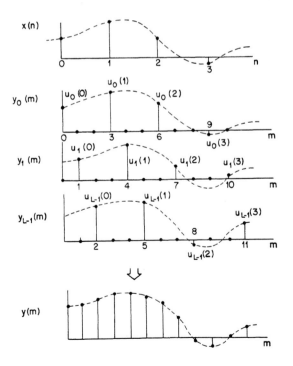

Figure 4.3 Illustration of signals in the polyphase interpolator for $L = 3$.

Also recall from the discussion in Section 3.3.2 that, in the case of interpolators, the polyphase filters $p_\rho(n)$ are the same as the periodically time-varying filters $g_m(n)$ (for $\rho = m \oplus L$) in the time-varying model, whereas in the case of decimators they are not [for decimators the values of $p_\rho(n)$ are equal to the values of $g_m(n)$ for the transposed system which is an interpolator].

Taken as a set, the samples $p_\rho(n)$ ($\rho = 0, 1, ..., L - 1$ for an interpolator, or $\rho = 0, 1, ..., M - 1$ for a decimator) represent all the samples of $h(k)$. Since the relationship between $p_\rho(n)$ and $h(k)$ is identical for both cases (1-to-L interpolators and M-to-1 decimators), we only consider the case of interpolators. The results for decimators can then simply be obtained by replacing L by M in the appropriate equations.

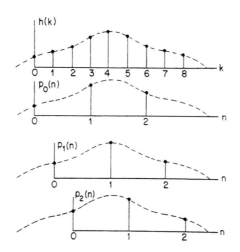

Figure 4.4 Relationship of the impulse responses of the prototype filter $h(n)$ and the polyphase filters $p_\rho(n)$ for $L = 3$.

The samples $h(k)$ can be recovered from $p_\rho(n)$ by sampling rate expanding the sequences $p_\rho(n)$ by a factor L. Each expanded set is then delayed by ρ samples and the L sets are then summed to give $h(k)$ (the reverse operation to that of Fig. 4.4). If we let $\hat{p}_\rho(k)$ represent the sampling rate expanded set

$$\hat{p}_\rho(k) = \begin{cases} p_\rho\left[\dfrac{k}{L}\right] & k = 0, \pm L, \pm 2L, ... \\[2ex] 0, & \text{otherwise} \end{cases} \qquad (4.14)$$

then $h(k)$ can be reconstructed from $\hat{p}_\rho(k)$ via the summation

$$h(k) = \sum_{\rho=0}^{L-1} \hat{p}_\rho(k - \rho) \tag{4.15}$$

The z-transform, $H(z)$, of the prototype filter can similarly be expressed in terms of the z-transforms of the polyphase filters, $P_\rho(z)$. It can be shown (see Section 3.4.1) that

$$H(z) = \sum_{\rho=0}^{L-1} z^{-\rho} P_\rho(z^L) \tag{4.16}$$

Finally, the z-transforms of the polyphase filters, $P_\rho(z)$, can be expressed in terms of $H(z)$ according to the following derivation. Let $g(n)$ be a shifted version of $h(n)$ such that

$$g(n) = h(n + \rho) \tag{4.17}$$

Then $p_\rho(n)$ can be expressed in terms of $g(n)$ as

$$p_\rho(n) = g(nL) \tag{4.18}$$

Equation (4.18) shows that $p_\rho(n)$ is simply a sampling rate compressed version of the sequence $g(n)$. Therefore, from Chapter 2, it can be shown that

$$P_\rho(z) = \frac{1}{L} \sum_{l=0}^{L-1} G(e^{-j2\pi l/L} z^{1/L}) \tag{4.19}$$

where $P_\rho(z)$ and $G(z)$ are the z-transforms of $p_\rho(n)$ and $g(n)$ respectively. Noting that

$$G(z) = z^\rho H(z) \tag{4.20}$$

and substituting this expression into Eq. (4.19) gives the desired z-transform relation

$$P_\rho(z) = \frac{1}{L} \sum_{l=0}^{L-1} e^{-j2\pi l\rho/L} z^{\rho/L} H(e^{-j2\pi l/L} z^{1/L}) \tag{4.21}$$

For the case of IIR filters we recall from Chapter 3 that it is desirable that $P_\rho(z)$ have the form of a ratio of polynomials in z. From Eq. (4.21) it is readily seen that this condition occurs only when the denominator polynomial of $H(z)$ can be expressed as (or transformed to) a polynomial in z^L.

By letting $z = e^{j\omega}$ and rearranging terms, Eq. (4.21) can be expressed as

$$P_\rho(e^{j\omega}) = \frac{1}{L}\sum_{l=0}^{L-1} e^{j(\omega-2\pi l)\rho/L} \sum_{k=-\infty}^{\infty} h(k)e^{-j(\omega-2\pi l)k/L}$$

$$= \frac{1}{L}\sum_{l=0}^{L-1} e^{j(\omega-2\pi l)\rho/L} H(e^{j(\omega-2\pi l)/L}), \quad \rho = 0, 1, ..., L-1 \qquad (4.22)$$

Equation (4.22) shows the relationships of the Fourier transforms of the polyphase filters to the Fourier transform of the prototype filter.

Equations (4.12) to (4.15), therefore, illustrate the time domain relationships between $h(k)$ and $p_\rho(n)$ and Eq. (4.16) and (4.22) show their frequency domain relationships.

4.2.2 Ideal Frequency Domain Characteristics for Interpolation Filters

In previous chapters we have assumed that the filter $h(k)$ approximates some ideal lowpass (or bandpass) characteristic. We elaborate on these "ideal" characteristics in somewhat more detail in the next two sections. In practice it is also necessary to specify a performance criterion to measure (in a consistent manner) how closely an actual filter design approximates this ideal characteristic. Since different design techniques are often based on different criteria, we will consider these criteria as they arise in later sections on filter design. Also, we consider, primarily, lowpass designs in this chapter; however, it should be noted that most of these design techniques can readily be extended to the case of bandpass filters.

Figure 4.5 briefly reviews the spectral interpretation of the 1-to-L interpolator (for $L = 5$). The input signal $x(n)$ is expanded in sampling rate by a factor L [by filling in $L-1$ zero-valued samples between each pair of samples of $x(n)$] to produce the signal $w(m)$. The spectrum of $w(m)$ contains not only the baseband spectrum of $x(n)$ but also periodic repetitions (images) of this spectrum, as illustrated in Fig. 4.5(c). The lowpass filter $h(m)$ attenuates these images to produce the desired output $y(m)$, and therefore it must approximate the ideal[2] lowpass characteristic

$$\tilde{H}(e^{j\omega'}) = \begin{cases} L, & |\omega'| < \dfrac{\pi}{L} \\ 0, & \text{otherwise} \end{cases} \qquad (4.23)$$

as illustrated in Fig. 4.5(d).

By applying Eq. (4.23) to Eq. (4.22) it is possible to derive the equivalent

---

[2] The "ideal" characteristic for a filter is denoted by a tilde over the variable throughout this chapter.

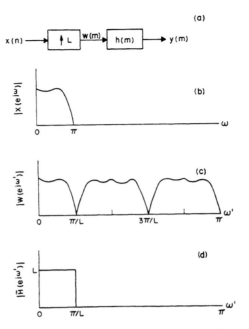

Figure 4.5 Review of the spectral interpretation for a 1-to-L interpolator $(L = 5)$.

ideal characteristics, $\widetilde{P}_\rho(e^{j\omega})$, that are implied in the polyphase filters.[3] Because of the constraint imposed by Eq. (4.23), only the $l = 0$ term in Eq. (4.22) is nonzero and thus it simplifies to the form

$$\widetilde{P}_\rho(e^{j\omega}) = \frac{1}{L}e^{j\omega\rho/L}\,\widetilde{H}(e^{j\omega/L})$$

$$= e^{j\omega\rho/L}, \quad \rho = 0, 1, 2, ..., L - 1 \qquad (4.24)$$

Equation (4.24) shows that the "ideal" polyphase filters, $\widetilde{p}_\rho(n)$, should approximate all-pass filters with linear phase shifts corresponding to fractional advances of ρ/L samples $(\rho = 0, 1, ..., L - 1)$ (ignoring any fixed delays that must be introduced in practical implementations of such filters). A further interpretation of the reason for this phase advance can be found in Section 3.3.2 in the discussion of polyphase structures.

In some cases it is known that the spectrum of $x(n)$ does not occupy its full bandwidth. This property can be used to advantage in the filter design and we will

---

[3] Since the conventional filter, $h(m)$, is implemented at the higher sampling rate, the frequency variable [in Eq. (4.23) for example] is ω', whereas since the polyphase filters are implemented at the lower sampling rate, the frequency variable is $\omega = \omega' L$.

see examples of this in Section 4.3.3 and in the next chapter with regard to cascaded (multistage) implementations of sampling rate conversion systems. If we define ω_c as the highest frequency of interest in $X(e^{j\omega})$, that is,

$$|X(e^{j\omega})| < \epsilon \quad \text{for} \quad \pi > |\omega| > \omega_c \tag{4.25}$$

where ϵ is a small quantity [relative to the peak of $|X(e^{j\omega})|$], then $W(e^{j\omega'})$ is an L-fold periodic repetition of $X(e^{j\omega})$ as shown in Fig. 4.6 (for $L = 5$). In this case, the ideal interpolator filter only has to remove the $(L - 1)$ images of the band of $X(e^{j\omega})$, where $|X(e^{j\omega})| > \epsilon$. Thus, in the frequency domain, the ideal interpolator filter satisfies the constraints

$$\widetilde{H}(e^{j\omega'}) = \begin{cases} L, & 0 \leqslant |\omega'| \leqslant \dfrac{\omega_c}{L} \\[3mm] 0, & \dfrac{2\pi r - \omega_c}{L} \leqslant |\omega'| \leqslant \dfrac{2\pi r + \omega_c}{L}, \quad r = 1, ..., L - 1 \end{cases} \tag{4.26}$$

as illustrated in Fig. 4.6(c). The bands from $(2\pi r + \omega_c)/L$ to $(2\pi(r + 1) - \omega_c)/L$ $(r = 0, 1, ...)$ are "don't care" (ϕ) bands in which the filter frequency response is essentially unconstrained. [In practice, however, $|H(e^{j\omega'})|$ should not be very large

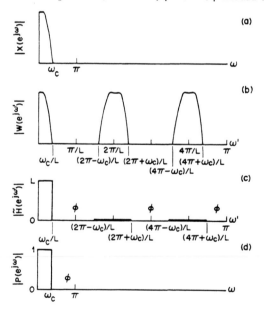

Figure 4.6 Illustration of ϕ bands in the specification of an interpolation filter $(L = 5)$.

in these ϕ bands, (e.g., not larger than L), to avoid amplification of any noise (or tails of $X(e^{j\omega})$) that may exist in these bands]. We will see later how these ϕ bands can have a significant effect on the filter design problem. Figure 4.6(d) shows the response of the ideal polyphase filter, which is converted from an all-pass to a lowpass filter with cutoff frequency ω_c. Of course, the phase response of each polyphase filter is unaltered by the don't-care bands for frequencies less than ω_c.

4.2.3 Ideal Frequency Domain Characteristics for Decimation Filters

Figure 4.7 briefly reviews the spectral interpretation for the M-to-1 decimator (for $M = 5$). The input signal $x(n)$ is lowpass filtered by $h(n)$ to eliminate aliasing which would occur in the sampling rate compression. Thus $H(e^{j\omega})$ should approximate the ideal lowpass characteristic

$$\widetilde{H}(e^{j\omega}) = \begin{cases} 1, & 0 \leqslant \omega \leqslant \dfrac{\pi}{M} \\ 0, & \text{otherwise} \end{cases} \tag{4.27}$$

where ω denotes the frequency at the input sampling rate. Alternatively, the

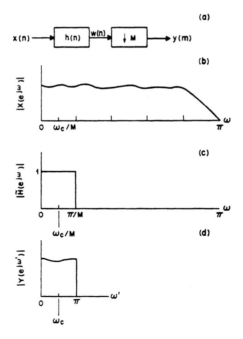

Figure 4.7 Review of the spectral interpretation for an M-to-1 decimator ($M = 5$).

polyphase filters should approximate the ideal all-pass characteristics

$$\widetilde{P}_\rho(e^{j\omega'}) = \frac{1}{M} e^{j\omega'\rho/M}, \quad \rho = 0, 1, ..., M - 1 \tag{4.28}$$

where ω' denotes the frequency at the output sampling rate.

If we are only interested in preventing aliasing, at the output of the decimator, in a band from 0 to ω_c, where $\omega_c < \pi$ [see Fig. 4.7(d)], and we are willing to tolerate aliased components (in the output) for frequencies above ω_c, then we again have a situation where don't-care bands are permitted in the filter design. The don't-care regions are the same as those in Eq. (4.26), as illustrated in Fig. 4.6(c) (with L replaced by M). In fact, all the frequency domain constraints that apply to the design of interpolation filters also apply to the design of decimation filters, a consequence of the property that they are transpose systems.

4.2.4 Time Domain Properties of Ideal Interpolation Filters

If we view the interpolation filter design problem in the time domain, an alternative picture of the "ideal" interpolation filter is obtained. By taking the inverse transform of the ideal filter characteristic defined by Eq. (4.23), we get the well-known $\sin(x)/x$ characteristic

$$\widetilde{h}(k) = \frac{\sin(\pi k/L)}{\pi k/L}, \quad k = 0, \pm 1, \pm 2, ... \tag{4.29}$$

In a similar manner we can determine the ideal time responses of the polyphase filters, either by taking the inverse transform of Eq. (4.28) or by sampling the time response $\widetilde{h}(k)$ according to Eq. (4.12). The net result is that the ideal time responses of the polyphase filters are

$$\widetilde{p}_\rho(n) = \frac{\sin[\pi(n + \rho/L)]}{\pi(n + \rho/L)}, \quad \rho = 0, 1, ..., L - 1, \text{ and all } n \tag{4.30}$$

A number of interesting observations can be made about the above ideal time responses noted above. First we see that they constrain every Lth value of $\widetilde{h}(k)$ such that

$$\widetilde{h}(k) = \begin{cases} 1, & k = 0 \\ 0, & k = rL, r = 0, \pm 1, \pm 2, ... \end{cases} \tag{4.31}$$

Alternatively, this implies the constraint that the zeroth polyphase filter have an impulse response that is a unit pulse, that is,

$$\widetilde{p}_0(n) = u_0(n), \quad \text{for all } n \tag{4.32}$$

In terms of the polyphase structure of Fig. 4.2 and its signal processing interpretation of Fig. 4.3, the foregoing constraint is easy to visualize. It simply implies that the output, $y_0(m)$, of the zeroth polyphase branch is identical to the input, $x(n)$, filled in with $L - 1$ zeros (i.e., these sample values are already known). The remaining $L - 1$ samples in between these values must be interpolated by the polyphase filters $p_\rho(n)$, $\rho = 1, 2, ..., L - 1$. Since these filters are theoretically infinite in duration, they must be approximated, in practice, with finite-duration filters. Thus the interpolation "error" between the outputs $y(m)$ of a practical system and an ideal system can be zero for $m = 0, \pm L, \pm 2L, ...$. However "in between" these samples, the error will always be nonzero.

By choosing a design that does not specifically satisfy the constraint of Eq. (4.31) or Eq. (4.32), a trade-off can be made between errors that occur at sample times $m = 0, \pm L, \pm 2L, ...$ and errors that occur between these samples.

Another "time domain" property that can be observed is that the ideal filter $\widetilde{h}(k)$ is symmetric about zero, that is,

$$\widetilde{h}(k) = \widetilde{h}(-k) \tag{4.33}$$

(Alternatively, for practical systems it may be symmetrical about some fixed nonzero delay.) This symmetry does not necessarily extend directly to polyphase filters since they correspond to sampled values of $\widetilde{h}(k)$ offset by some fraction of a sample. Their envelopes, however, are symmetrical (see Fig. 4.4).

The symmetry property described above does, however, imply a form of mirror-image symmetry between pairs of polyphase filters $\widetilde{p}_\rho(n)$ and $\widetilde{p}_{L-\rho}(n)$. Applying Eq. (4.33) to Eq. (4.12) gives

$$\widetilde{p}_\rho(n) = \widetilde{h}(-\rho - nL)$$

$$= \widetilde{h}(L - \rho - (n + 1)L) \tag{4.34}$$

Also noting that

$$\widetilde{p}_{L-\rho}(n) = \widetilde{h}(L - \rho + nL) \tag{4.35}$$

it can be seen that this symmetry is of the form

$$\widetilde{p}_\rho(n) = \widetilde{p}_{L-\rho}(-n - 1) \tag{4.36}$$

Figure 4.8 illustrates this for the case $L = 5$. Figure 4.8(a) shows $\widetilde{h}(k)$ and Fig. 4.8b shows $\widetilde{p}_0(n)$, which is seen to be a unit sample. Figure 4.8(c) and (d) show the polyphase filters $\widetilde{p}_2(n)$ and $\widetilde{p}_3(n)$ and they are seen to be mirror-image symmetric according to the definition of Eq. (4.36) [i.e., $\widetilde{p}_2(0) = \widetilde{p}_3(-1)$, $\widetilde{p}_2(1) = \widetilde{p}_3(-2)$, $\widetilde{p}_2(2) = \widetilde{p}_3(-3)$, ...]. When $\widetilde{h}(k)$ is symmetric about some fixed nonzero delay, the mirror-image symmetry relation of Eq. (4.36) changes accordingly; however, it still involves pairwise filters $\widetilde{p}_\rho(n)$ and $\widetilde{p}_{L-\rho}(n)$.

4.2.5 Time Domain Properties of Ideal Decimation Filters

In the case of decimators it is not possible to identify the outputs of specific polyphase branches with specific output samples of the network. All branches contribute to each output. Also, as seen by comparing Figs. 4.5 and 4.7, the filter in the decimator is used to remove arbitrary, unnecessary signal components, whereas in the interpolator it is used to remove periodic repetitions of the signal spectrum. Thus it is not as convenient to give meaningful time domain interpretations to the operation of the filter in a decimator.

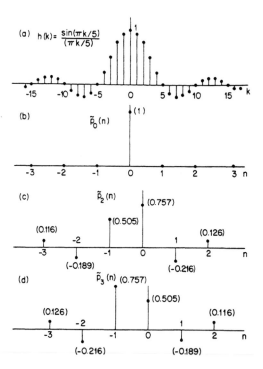

Figure 4.8 Ideal prototype filter and polyphase filters for $L = 5$.

The ideal time responses for $\widetilde{h}(k)$ and $\widetilde{p}_\rho(n)$ for decimators, however, are the same as those of Eqs. (4.29) and (4.30), respectively, with L replaced by M.

4.3 FILTER DESIGN PROCEDURES FOR FIR DECIMATORS AND INTERPOLATORS

In the remainder of this chapter we will discuss a number of filter design procedures that apply to the design of multirate systems. Since the filter design problem for such systems generally is a lowpass (or bandpass) design problem, nearly all of the work in digital signal processing filter theory can be brought to bear on this problem. We will not attempt to discuss all these methods in detail because they are well documented elsewhere [4.1, 4.2]; rather, we will try to point to the relevant issues involved in these designs that apply particularly to multirate systems. We focus on FIR design procedures in this section and IIR design procedures in Section 4.4.

For FIR filters we will discuss five main categories of filter design procedures:

1. Window designs [4.1, 4.2, 4.4]
2. Optimal, equiripple linear phase designs [4.5-4.9, 4.21]
3. Half-band designs [4.10-4.12]
4. Special FIR interpolator designs based on time domain filter specifications [4.13-4.15, 4.19, 4.20] or stochastic properties of the signal [4.22]
5. Classical interpolation designs: namely, linear and Lagrangian [4.23, 4.24].

4.3.1 FIR Filters Based on Window Designs

One straightforward approach to designing FIR filters for decimators and interpolators is by the well-known method of windowing or truncating the ideal prototype response $\tilde{h}(k)$. A direct truncation [i.e., a rectangular windowing of $\tilde{h}(k)$], however, leads to the Gibbs phenomenon, which manifests itself as a large (9%) ripple in the frequency behavior of the filter in the vicinity of filter magnitude discontinues (i.e., near the edges of the passband and stopband). Furthermore, the amplitude of this ripple does not decrease with increasing duration of the filter (it only becomes narrower in width). Thus a direct truncation, or a rectangular windowing, of $\tilde{h}(k)$ is rarely used in practice.

A more successful way of windowing the ideal characteristic, $\tilde{h}(k)$, is by more gradually tapering its amplitude to zero near the ends of the filter with a weighting sequence, $w(k)$, known as a window. The resulting filter design $h(k)$ is thus the product of the window $w(k)$ with the ideal response $\tilde{h}(k)$:

$$h(k) = \tilde{h}(k)w(k), \quad -\frac{N-1}{2} \leqslant k \leqslant \frac{N-1}{2} \qquad (4.37)$$

where we assume that $w(k)$ is a symmetric N-point (N odd) window. This process is illustrated in Fig. 4.9 for an $N = 19$ tap window. Figure 4.9(a) shows the ideal

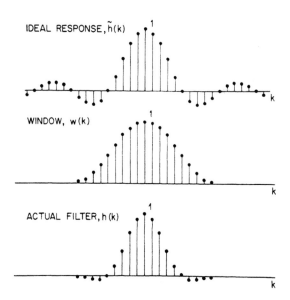

Figure 4.9 Illustration of FIR filter design by windowing.

impulse response, Fig. 4.9(b) shows the window sequence, and Fig. 4.9(c) shows the resulting filter impulse response.

A number of windows have been proposed in the literature for controlling the effects of the Gibbs phenomenon, and their properties are well understood [4.1, 4.2, 4.4]. Two commonly used types of windows are the "generalized" Hamming windows and the Kaiser windows. In the following discussion we assume that the desired duration of $h(k)$ is N samples, where N is assumed to be odd. With simple modifications, similar results can be derived for the case of N even.

The "generalized" Hamming window has the form

$$w_H(k) = \begin{cases} \gamma + (1 - \gamma) \cos \left[\dfrac{2\pi k}{N} \right], & -\dfrac{N-1}{2} \leqslant k \leqslant \dfrac{N-1}{2} \\ 0, & \text{otherwise} \end{cases} \tag{4.38}$$

where γ is in the range $0 \leqslant \gamma \leqslant 1$. If $\gamma = 0.54$, the window is called a *Hamming window*, and if $\gamma = 0.5$, it is called a *Hanning window* [the window in Fig. 4.9(b) is a Hanning window].

The *Kaiser window* has the form

$$w_K(k) = \begin{cases} \dfrac{I_0[\beta \sqrt{1 - (2k/(N-1))^2}]}{I_0[\beta]}, & -\dfrac{N-1}{2} \leqslant k \leqslant \dfrac{N-1}{2} \\ 0, & \text{otherwise} \end{cases} \tag{4.39}$$

where β is a constant that specifies a frequency-response trade-off between the peak height of the sidelobe ripples and the width or energy of the main lobe, and $I_0[x]$ is the modified zeroth-order Bessel function. The details of this trade-off are given in Refs. 4.1 and 4.4.

The window designs have the property that they preserve the zero-crossing pattern of $\tilde{h}(k)$ in the actual filter design $h(k)$. Thus if $\tilde{h}(k)$ is obtained from Eq. (4.29), then the time domain properties discussed in Section 4.2.4 apply to this class of filters.

The window method for designing digital filters for use in sampling rate conversion systems has a number of advantages and disadvantages. Among the advantages are that the method is simple, easy to use, and can readily be implemented in a direct manner (i.e., closed-form expressions are available for the window coefficients; hence the filter responses can be obtained simply from the ideal filter response). Another strong advantage is that the properties (in the frequency domain) of digital filters designed by the window method are relatively well understood. For example, in the case of lowpass filters, if we define the set of filter parameters [as illustrated in the classical tolerance scheme of Figure 4.10(a)]

$$\delta_p = \text{ripple (deviation) in the passband from the ideal response}$$

$$\delta_s = \text{ripple (deviation) in the stopband from the ideal response}$$

$$F_p = \frac{\omega_p}{2\pi} = \text{normalized passband edge frequency}$$

$$F_s = \frac{\omega_s}{2\pi} = \text{normalized stopband edge frequency}$$

then approximate relationships between N, δ_p, δ_s, F_p, and F_s can be obtained for several types of windows. For example, for the Kaiser window lowpass filters [4.4], we have

$$\delta = \delta_p = \delta_s \tag{4.40}$$

since the window function produces equal-amplitude ripples in both the passband and the stopband of the ideal lowpass filter. If we define a normalized transition width ΔF as

$$\Delta F = F_s - F_p = \frac{\omega_s - \omega_p}{2\pi} \tag{4.41}$$

then the design relationship between N, ΔF, and δ is of the form

$$N = \frac{-20 \log_{10} \delta - 7.95}{14.36 \Delta F} + 1 \tag{4.42}$$

For these designs, Eq. (4.42) shows that for the required filter length N is inversely proportional to the normalized filter transition width ΔF, and directly proportional to the logarithm of the required ripple. Using Eq. (4.42), the required filter length for matching any desired set of lowpass filter characteristics (i.e., δ, ΔF) can readily be computed. (Extensive use of such relationships will be made in Chapter 5 when we discuss multistage implementations of sampling rate conversion systems.)

Among the disadvantages of the window technique are that there is only limited control in choosing values of F_p, F_s, δ_p, and/or δ_s for most cases. The resulting filter designs are also suboptimal in that a smaller value of N can be found (using other design methods) such that all specifications on the filter characteristics are met or exceeded. As such, the window design method has been used primarily when ease of design has been the major consideration. In the next section we discuss a more general class of FIR filter designs which are optimal in a Chebyshev approximation sense.

4.3.2 Equiripple (Optimal) FIR Designs

The window technique of the preceding section represents a simple, straightforward approach to the design of the digital filter required in sampling rate conversion systems. However, considerably more sophisticated design techniques have been developed for FIR digital filters [4.1, 4.2, 4.6-4.9]. One such technique is the method of equiripple design based on Chebyshev approximation methods. The filters designed by this technique are optimal in the sense that the peak (weighted) approximation error in the frequency domain over the frequency range of interest is minimized. In this section we consider this approach for the simple lowpass design criterion. In the next section we show how it can be extended to designs that include don't care regions in the design criterion.

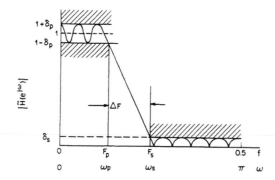

Figure 4.10   Illustration of the classical tolerance scheme for practical FIR lowpass filters.

To apply the method of equiripple design to the lowpass digital filter required for decimators or interpolators, consider again the classical tolerance scheme[4] in Fig. 4.10. Recall that in this scheme δ_p and δ_s denote the approximation error (ripple) of the actual filter in the passband and stopband respectively. Similarly, $F_p = \omega_p/2\pi$ and $F_s = \omega_s/2\pi$ denote the (normalized) passband and stopband edge frequencies respectively. If we further specify that the highest frequency of interest in a decimated signal (as in Section 4.2.3) or the total bandwidth of a signal to be interpolated (see Section 4.2.2) is $\omega_c \leqslant \pi$, then the passband edge frequency of the filter is

$$
\omega_p \leqslant \begin{cases} \dfrac{\omega_c}{L} & 1\text{-to-}L \text{ interpolator} \\[2ex] \dfrac{\omega_c}{M} & M\text{-to-}1 \text{ decimator} \\[2ex] \min\left[\dfrac{\omega_c}{L}, \dfrac{\omega_c}{M}\right] & \text{conversion by } L/M \end{cases} \tag{4.43}
$$

In particular if $\omega_c = \pi$ then the inequality in Eq. (4.43) applies to allow some margin for the transition band.

If no aliasing in the decimator or imaging in the interpolator is allowed then the stopband edge frequency is defined as

$$
\omega_s = \begin{cases} \dfrac{\pi}{L} & 1\text{-to-}L \text{ interpolator} \\[2ex] \dfrac{\pi}{M} & M\text{-to-}1 \text{ decimator} \\[2ex] \min\left[\dfrac{\pi}{L}, \dfrac{\pi}{M}\right] & \text{conversion by } L/M \end{cases} \tag{4.44a}
$$

Alternatively, if aliasing *is* allowed in the transition region from ω_c to π for the decimator or in the region from π to $2\pi - \omega_c$ for the interpolator, then the stopband edge can be chosen (as discussed in Sections 4.2.2 and 4.2.3) to exclude the don't care region nearest the passband. In this case ω_s becomes

---

[4] Note that we have omitted the scale factor of L for the interpolator filter. This factor can be ignored in the design phase and handled in the implementation phase.

$$\omega_s = \begin{cases} \dfrac{2\pi - \omega_c}{L} & 1\text{-to-}L \text{ interpolator} \\[4mm] \dfrac{2\pi - \omega_c}{M} & M\text{-to-}1 \text{ decimator} \qquad (4.44b) \\[4mm] \min\left[\dfrac{2\pi - \omega_c}{L}, \dfrac{2\pi - \omega_c}{M}\right] & \text{conversion by } L/M \end{cases}$$

Given the tolerance scheme of Fig. 4.10, it is a simple matter to set up a filter approximation problem based on Chebyshev approximation methods. Assume that the FIR filter to be designed is an N-point filter (N odd) that is symmetric around the origin, that is,

$$h(k) = h(-k), \quad k = 1, 2, ..., \frac{N-1}{2} \qquad (4.45)$$

Then the resulting frequency response, $H(e^{j\omega})$, is purely real, and can be expressed in the form

$$H(e^{j\omega}) = h(0) + 2 \sum_{k=1}^{(N-1)/2} h(k) \cos(\omega k) \qquad (4.46)$$

In practice, a delay of $(N-1)/2$ samples is used to make $h(k)$ realizable. Also, even values of N can be readily handled with simple modifications to the analysis.

Based on the analysis above, the filter approximation problem for the FIR interpolator or decimator can be expressed as

$$1 - \delta_p \leqslant H(e^{j\omega}) \leqslant 1 + \delta_p, \quad 0 \leqslant \omega \leqslant \omega_p$$

$$-\delta_s \leqslant H(e^{j\omega}) \leqslant \delta_s, \qquad \omega_s \leqslant \omega \leqslant \pi \qquad (4.47)$$

where we use Eqs. (4.43) and (4.44) for ω_p and ω_s in Eq. (4.47). To convert Eq. (4.47) into a more manageable form for solution via Chebyshev approximation methods, we set

$$\delta_p = K\delta_s \qquad (4.48)$$

where K is a fixed weighting factor that determines the relative importance of the errors δ_p and δ_s. This gives the modified approximation problem

$$1 - K\delta_s \leqslant H(e^{j\omega}) \leqslant 1 + K\delta_s, \quad 0 \leqslant \omega \leqslant \omega_p$$

$$-\delta_s \leqslant H(e^{j\omega}) \leqslant \delta_s, \qquad \omega_s \leqslant \omega \leqslant \pi \qquad (4.49)$$

The Chebyshev (minimax) approximation solution to Eq. (4.49) is given as

$$\delta_s = \begin{array}{c} \min \\ \{h(k)\} \end{array} \begin{array}{c} \max[|E(e^{j\omega})|] \\ \omega \in [0,\omega_p] \cup [\omega_s,\pi] \end{array} \tag{4.50}$$

where the weighted error function $E(e^{j\omega})$ is defined as

$$E(e^{j\omega}) = \begin{cases} \dfrac{H(e^{j\omega}) - 1}{K}, & 0 \leqslant \omega \leqslant \omega_p \\ |H(e^{j\omega})|, & \omega_s \leqslant \omega \leqslant \pi \end{cases} \tag{4.51}$$

Hence the solution to Eq. (4.50) provides the set of filter coefficients, $h(k)$, and δ_s directly, and δ_p can be determined from Eq. (4.48).

Several highly developed techniques have been presented in the literature for solving the Chebyshev approximation problem of Eq. (4.50) [4.6-4.8], including a well-documented, widely used computer program [4.9]. The solutions are based on either a multiple-exchange Remez algorithm, or a single-exchange linear programming solution. We will not concern ourselves with details of the various solution methods. Instead, we will discuss the empirical relationships that have been found among the five lowpass filter parameters for this class of filters. These design relationships are widely used in Chapter 5 in our discussion of filter optimization for a cascade of lowpass filters.

To establish the complex relationships among N, ω_p, ω_s, δ_p, and δ_s, we first define the measures D, $D_\infty(\delta_p, \delta_s)$, and $f(\delta_p, \delta_s)$ as

$$D = (N - 1)\,\Delta F = \frac{(N - 1)(\omega_s - \omega_p)}{2\pi} \tag{4.52}$$

$$D_\infty(\delta_p, \delta_s) = \log_{10} \delta_s [a_1(\log_{10}\delta_p)^2 + a_2 \log_{10}\delta_p + a_3]$$

$$+ [a_4(\log_{10}\delta_p)^2 + a_5 \log_{10}\delta_p + a_6] \tag{4.53}$$

$$f(\delta_p, \delta_s) = 11.012 + 0.512(\log_{10}\delta_p - \log_{10}\delta_s), \text{ for } |\delta_s| \leqslant |\delta_p| \tag{4.54}$$

where $a_1 = 0.005309$, $a_2 = 0.07114$, $a_3 = -0.4761$, $a_4 = -0.00266$, $a_5 = -0.5941$, and $a_6 = -0.4278$. The functions D, D_∞, and f are related by the empirically derived formula [4.3, 4.21]

$$D_\infty(\delta_p, \delta_s) = D + f(\delta_p, \delta_s)(\Delta F)^2 \tag{4.55}$$

By appropriate manipulations of the parameter relationships above, it is possible to solve for any one of the five lowpass filter parameters in terms of the other four.

For example, we can solve for N in terms of ω_p, ω_s, δ_p, and δ_s, giving

$$N = \frac{D_\infty(\delta_p, \delta_s)}{(\omega_s - \omega_p)/2\pi} - f(\delta_p, \delta_s) \frac{\omega_s - \omega_p}{2\pi} + 1 \qquad (4.56)$$

Although the design relationships of Eqs. (4.52) to (4.55) appear complex, they are fairly simple to apply. By way of example, Fig. 4.11 shows a series of plots of the quantity $D_\infty(\delta_p, \delta_s)$ as a function of δ_s for several values of δ_p. Through the use of either the analytic form [e.g., Eq. (4.56)] or from widely available design charts and tables, it is a simple matter to determine the value of N needed for an FIR lowpass filter to meet any set of design specifications [4.1, 4.3, 4.21].

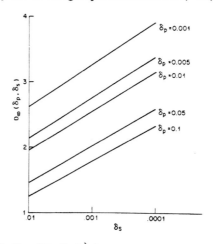

Figure 4.11   Plot of $D_\infty(\delta_p, \delta_s)$ for practical values of δ_p and δ_s.

4.3.3 Effects of the ϕ Bands for Equiripple Designs

The design relationships of Eqs. (4.52) to (4.55) are for lowpass filters. However, as discussed in Sections 4.2.2 and 4.2.3, digital filters for interpolation and decimation need not be strictly lowpass filters, but instead can also include ϕ, or don't-care bands, which can influence the filter design problem [4.23]. This is especially true when the total width of the ϕ bands is a significant portion of the total frequency band. For such cases no simple design formula such as Eq. (4.56) exists that relates the relevant filter parameters. Thus the simplest way of illustrating the effects of the ϕ bands on the filter design problem is by way of example.

Figure 4.12 shows an example of a practical tolerance scheme for a 1-to-L interpolator (where $L = 5$) that takes account of ϕ bands in the filter design. It is assumed, as discussed above, that the signal to be interpolated is bandlimited to the range 0 to ω_c. As seen from this figure, the range of the passband is unchanged from that in Eq. (4.43). However, the stopband is now divided into a series of stopbands separated by ϕ bands. These bands are defined as in Fig. 4.6 or Eq. (4.26). In the ϕ band regions the design is essentially unconstrained and in each stopband the design is constrained to have a peak error of less than δ_s. Note also that when $\omega_p < \omega_c/L$ the first ϕ band is wider than the others by an amount $\omega_c/L - \omega_p$. This additional width near the transition edge of the passband helps to relax the constraints (and the filter order) of the design. The method of equiripple design can be applied directly to this tolerance scheme in a similar manner to that for the simple lowpass case.

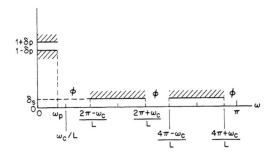

Figure 4.12 Illustration of a practical tolerance scheme for a 1-to-L interpolator with ϕ bands ($L = 5$).

Given the tolerance scheme above, consider the design of a set of interpolators with specifications

$$L = 5$$
$$\delta_p \leqslant 0.001$$
$$\delta_s \leqslant 0.0001$$
$$0 \leqslant \omega_c \leqslant \pi$$
$$\omega_p = \omega_c/L$$

where the parameter that we allow to vary is ω_c [see Fig. 4.12]. First we designed, for comparison, a series of lowpass filters with passband cutoff frequencies $\omega_p = \omega_c/5$, and stopband cutoff frequencies $\omega_s = (2\pi - \omega_c)/5$. Next we designed a series of multiband filters with a single passband (identical to that of the lowpass design), and a series of stopbands, separated by don't-care bands (see Fig. 4.12). If we compare the required impulse response duration of the lowpass filter to the required impulse duration of the multiband filter, we get a result of the type shown in Fig. 4.13(a), which gives the percentage decrease in N as a function of ω_c/π. The heavy dots shown in this figure are measured values (i.e., they are not theoretical computations), and the smooth curve shows the trend in the data. The numbers next to each heavy dot are the minimum required impulse response durations (in samples) for the lowpass and multiband designs, respectively.

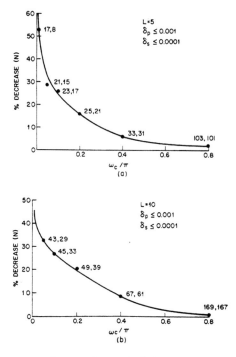

Figure 4.13 Percentage decrease in required filter order obtained by using a multiband design instead of a lowpass design as a function of ω_c/π for (a) $L = 5$; (b) $L = 10$.

The trends in the curve are quite clear. For ω_c/π close to 1.0, there is essentially no gain in designing and using a multiband filter instead of a standard lowpass filter, since the total width of the ϕ bands is small. However, as ω_c/π tends to zero, significant reductions in N are possible; for example, for $\omega_c/\pi \approx 0.05$, a 50% reduction in impulse response duration is possible. Figure 4.13(b) shows similar trends for a series of interpolators with a value of $L = 10$.

Figures 4.14 and 4.15 show typical frequency responses for lowpass and multiband interpolation filters based on the equiripple design method. For the example of Figure 4.14 the specifications were $L = 5$, $\delta_p = 0.001$, $\delta_s = 0.0001$, $\omega_p = 0.95\omega_c/L$, and $\omega_c = 0.5\pi$. The required values of N were 41 for the lowpass filter, and 39 for the multiband design. Thus for this case the reduction in filter order was insignificant. However, the change in filter frequency response [as seen in Fig. 4.14(b)] was highly significant. Figure 4.15 shows the same comparisons for designs with $L = 10$, $\delta_p = 0.001$, $\delta_s = 0.0001$, $\omega_p = 0.95\omega_c/L$, and $\omega_c = 0.1\pi$. In this case a 27% reduction in filter order (from 45 for the lowpass filter to 33 for the multiband filter) was obtained since the frequency span of the stopbands was small compared to the frequency span of the don't-care bands. In both of these examples we see that, in the ϕ bands, the frequency response of the resulting filter is truly unconstrained, and can, in fact, become extremely large (as compared to

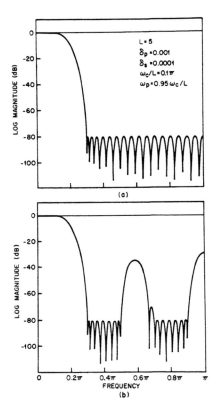

Figure 4.14 Comparison between (a) a lowpass filter; and (b) its equivalent multiband design for $L = 5$.

the passband response). The practical implication is that some care must be taken to ensure that the amplitude response in the ϕ bands stays below some well-specified level to guarantee that the noise in the input signal, in the ϕ bands, is not amplified so much that it becomes excessive in the output signal.

The foregoing examples have shown that when ω_c/L, the cutoff frequency of the passband, is relatively small (compared to π/L), significant reductions in computation can be obtained by exploiting the don't-care bands in the design of the interpolation (or decimation) digital filter. In the limit for very small values of ω_c/L, relative to the width of the ϕ bands, it can be observed that the above design procedure leads to filter designs having a comb filter characteristic. In fact, a considerable simplification occurs in the design when ω_c/L is small enough so that comb filter designs can be directly applied to this problem. In practice, such conditions rarely occur; that is, we are usually dealing with signals where ω_c/L is relatively large. Thus the question arises as to whether the multiband filter approach is of practical utility in the implementation of single-stage digital sampling rate conversion systems. However, as we will see in Chapter 5, the

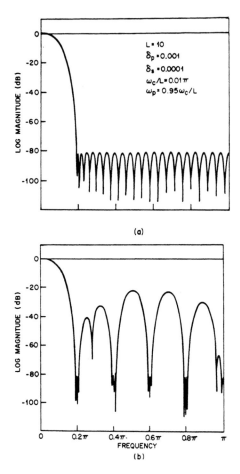

Figure 4.15 Comparison between (a) a lowpass filter; and (b) its equivalent multiband design for $L = 10$.

techniques discussed here are of considerable value in multistage implementations of decimators and interpolators involving large changes in sampling rates. As such, we will defer the discussion of comb filter designs until Chapter 5.

4.3.4 Equiripple FIR Filters for Interpolation with Time Domain Constraints

For the design of ideal interpolators, the frequency domain constraints that we have been discussing can readily be combined with the time domain constraints of Eq. (4.31) to yield optimal (in the Chebyshev sense) interpolation filters. This is because the constraints of Eq. (4.31) are *linear* in the filter variables $h(k)$. Thus Eqs. (4.50) and (4.31) can be solved simultaneously to minimize δ_s for a given set of specifications. However, the standard Remez multiple-exchange procedure for equiripple design *cannot* be used directly since the resulting filter coefficients do not satisfy a necessary condition for optimality (the Haar condition). Instead, considerably slower, but theoretically equivalent linear programming optimization techniques must be used to obtain the optimal filter [4.8]. The resulting filters are optimal in a minimax sense, but not equiripple in the frequency domain.

Since alternative and considerably simpler methods exist for designing optimal interpolators satisfying Eq. (4.31) we will defer a discussion of this important topic to Sections 4.3.6 to 4.3.8.

4.3.5 Half-Band FIR Filters — A Special Case of FIR Designs for Conversion by Factors of 2

Let us again consider the tolerance specifications of the ideal lowpass filter of Figure 4.10. If we consider the special case

$$\delta_s = \delta_p = \delta \tag{4.57}$$

$$\omega_s = \pi - \omega_p \tag{4.58}$$

then the resulting equiripple optimal solution to the approximation problem has the property that

$$H(e^{j\omega}) = 1 - H(e^{j(\pi-\omega)}) \tag{4.59}$$

That is, the frequency response of the optimal filter is symmetric around $\omega = \pi/2$, and at $\omega = \pi/2$ we get

$$H(e^{j\pi/2}) = 0.5 \tag{4.60}$$

(This, of course, implies that some aliasing or imaging occurs at this frequency.) It can also be readily shown that any symmetric FIR filter satisfying Eq. (4.59) also satisfies the ideal time domain constraints discussed in Section 4.2.4, namely

$$h(k) = \begin{cases} 1, & k = 0 \\ 0, & k = \pm 2, \pm 4, \dots \end{cases} \tag{4.61}$$

That is, every other impulse response coefficient (except for $k=0$) is *exactly* 0. Thus in implementing sampling rate conversion systems using such filters, an additional factor of two reduction in computation [over that obtained because of the symmetry of $h(k)$ or the L- or M-to-1 reduction discussed earlier] is obtained.

Filters designed using the constraints of Eqs. (4.57) and (4.58) have been called *half-band filters* [4.10-4.12], and their properties have been studied intensively. They can be designed in a variety of ways, including the window designs and equiripple designs discussed previously.

In the case of window designs, half-band filters can be obtained as windowed versions of the ideal response

$$\widetilde{h}(k) = \frac{\sin{(\pi k/2)}}{\pi k/2} \tag{4.62}$$

It can easily be shown that these filters preserve the half-band conditions of Eqs. (4.57) to (4.61).

In the case of equiripple designs, half-band filters can be obtained by carefully imposing the specifications of Eqs. (4.57) and (4.58) (and assuming N odd). The properties of Eqs. (4.59) to (4.61) will then follow. Because of these constraints there are only three design parameters to consider for half-band filters: ω_p, δ, and N where N is the number of taps in the filter (including zero-valued taps within the filter). From Eqs. (4.41) and (4.58) it is seen that

$$\Delta F = \frac{\pi - 2\omega_p}{2\pi} \tag{4.63}$$

Alternatively, ω_p is often expressed in the form

$$\omega_p = \frac{\alpha\pi}{2} \tag{4.64}$$

so that α defines the ratio of the passband width to the Nyquist band width ($\pi/2$) at the lower sampling rate. Then

$$\Delta F = \frac{1 - \alpha}{2} \tag{4.65}$$

Since the design equations (4.52) to (4.56) are approximations and are less accurate for small N, Oetken [4.13, 4.14] has measured curves that show exact relationships of the foregoing parameters for small values of $N (3 \leqslant N \leqslant 39)$. This set of design curves is presented in Fig. 4.16 and they show the trade-offs between ΔF (or α) and $20 \log_{10} \delta$ for values of $N = 3, 7, 11, ..., 39$ for half-band filters.

The half-band symmetric filter has a natural application in decimators or interpolators with sampling rate changes of 2. For such systems the 2-to-1 reduction in computation is quite significant. However, the half-band filter is even more important in a class of multistage implementations of decimators and interpolators in which cascaded stages of 2-to-1 changes in sampling rate are used to realize a large overall change in sampling rate. We discuss such systems in Chapter 5.

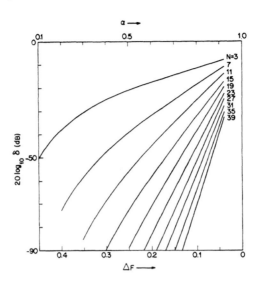

Figure 4.16 Design relations for half-band FIR filters.

Perhaps the only real drawback of the half-band filter is the requirement that $\delta_p = \delta_s$. For most practical systems we have $\delta_s \ll \delta_p$. However, since the design curves are relatively insensitive to δ, the computational price paid for designing a filter with δ_p much smaller than required is generally small, and almost always much less than the 2-to-1 speedup achieved by these filters.

4.3.6 Minimum Mean-Square-Error Design of FIR Interpolators — Deterministic Signals

Thus far we have considered the design of filters for interpolators and decimators from the point of view of designing the prototype filter $h(m)$ such that it satisfies a prescribed set of frequency domain specifications. In the remainder of Section 4.3 we consider an alternative point of view in designing filters (particularly for integer interpolators). In this approach the error criterion to be minimized is a function of the difference between the actual interpolated *signal* and its ideal value rather than a direct specification on the filter itself. We will see in this section and in following sections that such an approach leads to a number of filter design techniques [4.13-4.15, 4.19, 4.20, 4.22] which are capable of accounting directly for the spectrum of the signal being interpolated.

Figure 4.17(a) depicts the basic theoretical framework used for defining the interpolator error criterion. We wish to design the FIR filter $h(m)$ such that it can be used to interpolate the signal $x(n)$ by a factor of L with minimum interpolation error. To define this error we need to compare the output of this actual interpolator with that of an ideal (infinite duration) interpolator $\tilde{h}(m)$ whose characteristics were derived in Section 4.2. This signal error is defined as

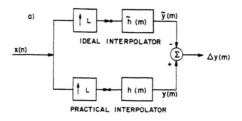

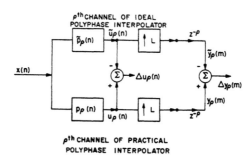

Figure 4.17   Framework for defining error criteria for interpolation filters.

$$\Delta y(m) = y(m) - \tilde{y}(m) \tag{4.66}$$

where $y(m)$ is the output of the actual interpolator and $\tilde{y}(m)$ is the ideal output.

In this section we consider interpolator designs that minimize the mean-square value of $\Delta y(m)$, defined as

$$E^2 = \|\Delta y(m)\|^2 = \lim_{K \to \infty} \frac{1}{2K+1} \sum_{m=-K}^{K} \Delta y(m)^2 \tag{4.67a}$$

which, by Parseval's theorem, can be expressed as

$$E^2 = \frac{1}{2\pi} \int_{-\pi}^{\pi} |Y(e^{j\omega'}) - \tilde{Y}(e^{j\omega'})|^2 \, d\omega' \tag{4.67b}$$

$$= \frac{1}{2\pi} \int_{-\pi}^{\pi} |\Delta Y(e^{j\omega'})|^2 \, d\omega' \tag{4.67c}$$

where ω' denotes the frequency at the high sampling rate. In Section 4.3.9 we consider designs that minimize the maximum value of $|\Delta Y(e^{j\omega'})|$ over a prescribed frequency range, and in Section 4.3.10 we consider designs that minimize the maximum value of $|\Delta y(m)|$ in the time domain.

The design problems described above are greatly simplified by considering them in the framework of the polyphase structures as illustrated in Fig. 4.2. Here it is seen that the signal $y(m)$ is actually composed of interleaved samples of the signals $u_\rho(n)$, $\rho = 0, 1, 2, ..., L - 1$, as shown by Fig. 4.3, where $u_\rho(n)$ is the output of the ρth polyphase filter. Thus the errors introduced by each polyphase branch are orthogonal to each other (since they do not coincide in time) and we can define the error in the ρth branch as the error between the actual output and the output of the ρth branch of an ideal polyphase interpolator as shown in Fig. 4.17(b), that is,

$$\Delta y_\rho(m) = y_\rho(m) - \tilde{y}_\rho(m) \qquad (4.68)$$

or, equivalently,

$$\Delta u_\rho(n) = u_\rho(n) - \tilde{u}_\rho(n) \qquad (4.69)$$

Because of this orthogonality property we can *separately* and *independently* design each of the polyphase filters for minimum error and arrive at an overall interpolator design which minimizes the error $\|\Delta y(m)\|$. Thus a large (multirate) filter design problem can be broken down into L smaller (time-invariant) filter design problems.

In the case of the mean-square-error criterion it can be seen that

$$\mathbf{E}^2 = \|\Delta y(m)\|^2 = \frac{1}{L} \sum_{\rho=0}^{L-1} \mathbf{E}_\rho^2 \qquad (4.70)$$

where

$$\mathbf{E}_\rho^2 = L\|\Delta y_\rho(m)\|^2 = \lim_{K \to \infty} \frac{L}{2K+1} \sum_{m=-K}^{K} \Delta y_\rho(m)^2$$

$$= \lim_{K \to \infty} \frac{1}{2K+1} \sum_{n=-K}^{K} \Delta u_\rho(n)^2$$

$$= \|\Delta u_\rho(n)\|^2 \qquad (4.71)$$

To minimize $\mathbf{E}^2$ we then need to design L independent polyphase filters $p_\rho(n)$, $\rho = 0, 1, 2, ..., L - 1$, which independently minimize the respective mean-square errors $\mathbf{E}_\rho^2$.

In order to set up the filter design problem analytically, we can express $\mathbf{E}_\rho^2$ in the form (using Parseval's theorem)

$$\mathbf{E}_\rho^2 = \|\Delta u_\rho(n)\|^2 = \frac{1}{2\pi} \int\limits_{-\pi}^{\pi} |\Delta U_\rho(e^{j\omega})|^2 \, d\omega \tag{4.72}$$

We observe from Fig. 4.17(b) that we can express $\Delta U_\rho(e^{j\omega})$ as

$$\Delta U_\rho(e^{j\omega}) = P_\rho(e^{j\omega})X(e^{j\omega}) - \widetilde{P}_\rho(e^{j\omega})X(e^{j\omega}) \tag{4.73}$$

and also recall from Section 4.2.2 that the ideal polyphase filter response is

$$\widetilde{P}_\rho(e^{j\omega}) = e^{j\omega\rho/L} \tag{4.74}$$

Combining Eqs. (4.72) to (4.74) then leads to the form

$$\mathbf{E}_\rho^2 = \|\Delta u_\rho(n)\|^2 = \frac{1}{2\pi} \int\limits_{-\pi}^{\pi} |P_\rho(e^{j\omega}) - e^{j\omega\rho/L}|^2 |X(e^{j\omega})|^2 \, d\omega \tag{4.75}$$

Equation (4.75) reveals that in the minimum mean-square-error design, we are in fact attempting to design a polyphase filter such that the integral of the squared difference between its frequency response $P_\rho(e^{j\omega})$ and a linear (fractional sample) phase delay $e^{j\omega\rho/L}$, weighted by the spectrum of the input signal $|X(e^{j\omega})|^2$, is minimized. Note also that the integral from $-\pi$ to π in Eq. (4.75) is taken over the frequency range of the input signal of the interpolator, not the output signal.

In practice, this error criterion is often modified slightly [4.13-4.16, 4.18] by specifying that $X(e^{j\omega})$ is bandlimited to the range $0 \leqslant \omega \leqslant \alpha\pi$ where $0 < \alpha < 1$, that is,

$$|X(e^{j\omega})| = 0, \text{ for } |\omega| \geqslant \alpha\pi \tag{4.76}$$

Then Eq. (4.75) can be expressed as

$$\mathbf{E}_{\rho,\alpha}^2 = \frac{1}{2\pi} \int\limits_{-\alpha\pi}^{\alpha\pi} |P_\rho(e^{j\omega}) - e^{j\omega\rho/L}|^2 |X(e^{j\omega})|^2 \, d\omega \tag{4.77}$$

where the subscript α will be used to distinguish this norm from the one in Eq. (4.75). Alternatively, we can consider the modification above as a means of specifying that we want the design of $P_\rho(e^{j\omega})$ to be minimized only over the frequency range $0 \leqslant \omega \leqslant \alpha\pi$, and that the range $\alpha\pi \leqslant \omega \leqslant \pi$ is allowed to be a transition region. Then α can be used as a parameter in the filter design procedure.

The solution to the minimization problem of Eq. (4.77) involves expressing the norm $\mathbf{E}_{\rho,\alpha}^2$ directly in terms of the filter coefficients $p_\rho(n)$. Then, since the problem is formulated in a classical mean-square sense, it will be seen that $\mathbf{E}_{\rho,\alpha}^2$ is a quadratic function of the coefficients $p_\rho(n)$, and thus it has a single, unique minimum for some optimum choice of coefficients. At this minimum point, the derivative of $E_{\rho,\alpha}^2$ with respect to all the coefficients $p_\rho(n)$ is zero. Thus the second step in the solution is to take the derivative of $E_{\rho,\alpha}^2$ with respect to the coefficients $p_\rho(n)$ and set it equal to zero. This leads to a set of linear equations in terms of the coefficients $p_\rho(n)$ and the solution to this set of equations gives the optimum choice of coefficients that minimize $E_{\rho,\alpha}^2$. This minimization problem is solved for each value of ρ, $\rho = 0, 1, 2, ..., L - 1$, and each solution provides the optimum solution for one of the polyphase filters. Finally, these optimum polyphase filters can be combined, as in Eqs. (4.14) to (4.16), to obtain the optimum prototype filter $h(m)$ which minimizes the overall norm

$$E_\alpha^2 = \frac{1}{L} \sum_{\rho=0}^{L-1} E_{\rho,\alpha}^2 \qquad (4.78)$$

We will now outline the foregoing steps in somewhat more detail. We will assume that the overall filter $h(m) = h(-m)$ has a total of

$$N = 2RL - 1 \qquad (4.79)$$

taps, where L is the interpolation factor and $2R$ is the number of taps in each polyphase filter except for $\rho = 0$ (we will consider the case for $\rho = 0$ separately). Thus $h(m)$ is zero for $m \leqslant -RL$ and for $m \geqslant RL$ and it follows that $p_\rho(n)$ is zero for $n < -R$ and $n \geqslant R$. Then $P_\rho(e^{j\omega})$ can be expressed as the FIR polynomial

$$P_\rho(e^{j\omega}) = \sum_{n=-R}^{R-1} p_\rho(n) e^{-j\omega n} \qquad (4.80)$$

Substituting Eq. (4.80) into Eq. (4.77) leads to the desired expression of $E_{\rho,\alpha}^2$ in terms of the coefficients $p_\rho(n)$,

$$E_{\rho,\alpha}^2 = \frac{1}{2\pi} \int_{-\alpha\pi}^{\alpha\pi} \left| \left[\sum_{n=-R}^{R-1} p_\rho(n) e^{-j\omega n} \right] - e^{j\omega\rho/L} \right|^2 |X(e^{j\omega})|^2 \, d\omega \qquad (4.81)$$

We can note immediately that in the case $\rho = 0$, the choice of the polyphase filter

$$P_0(e^{j\omega}) = 1 \qquad (4.82a)$$

or equivalently

$$p_0(n) = u_0(n) = \begin{cases} 1, & n = 0 \\ 0, & \text{otherwise} \end{cases} \qquad (4.82b)$$

results in zero error (i.e., $E_{0,\alpha}^2 = 0$), as anticipated from the discussion in Section 4.2.4 [see Eq. (4.32)]. All other polyphase filters will have minimized values of $E_{\rho,\alpha}^2$ which are greater than zero.

As discussed above, the second step in our solution (for $\rho = 1, 2, ..., L - 1$) is to take the derivatives of $E_{\rho,\alpha}^2$ with respect to the coefficients $p_\rho(n)$ and set them to zero; that is, we compute

$$\frac{\partial E_{\rho,\alpha}^2}{\partial p_\rho(n_0)} = 0 \quad \text{for } n_0 = - R, ..., 0, 1, ..., R - 1 \tag{4.83}$$

After careful manipulation, this leads to

$$\int_{-\alpha\pi}^{\alpha\pi} \left[\left(\sum_{n=-R}^{R-1} p_\rho(n) \cos\left(\omega(n_0 - n)\right) \right) - \cos\left(\omega(n_0 + \tfrac{\rho}{L})\right) \right] |X(e^{j\omega})|^2 \, d\omega = 0$$

$$\text{for } n_0 = - R, ..., 0, 1, ..., R - 1 \tag{4.84}$$

Defining $\phi(r)$ as

$$\phi(r) = \int_{-\alpha\pi}^{\alpha\pi} |X(e^{j\omega})|^2 \cos(\omega r) \, d\omega \tag{4.85}$$

and substituting Eq. (4.85) into Eq. (4.84) then gives

$$\sum_{n=-R}^{R-1} p_\rho(n)\phi(n - n_0) = \phi\left(n_0 + \frac{\rho}{L}\right), \quad n_0 = - R, ..., 0, 1, .., R - 1 \tag{4.86}$$

Equation (4.86) defines a set of $2R$ linear equations in terms of the coefficients $p_\rho(n)$, as expected. They can be expressed more compactly in the matrix form

$$\mathbf{\Phi}_T \mathbf{P}_\rho = \mathbf{\Phi}_\rho \tag{4.87}$$

where $\mathbf{P}_\rho$ is a column vector of the form

$$\mathbf{P}_\rho = [p_\rho(-R), ..., p_\rho(0), ..., p_\rho(R - 1)]^t \tag{4.88}$$

$\mathbf{\Phi}_T$ is a Toeplitz matrix of the form

$$\mathbf{\Phi}_T = \begin{bmatrix} \phi(0) & \phi(1) & \cdots & \phi(2R - 1) \\ \phi(1) & \phi(0) & & \\ \vdots & & & \vdots \\ \phi(2R - 1) & & \cdots & \phi(0) \end{bmatrix} \tag{4.89}$$

and Φ_ρ is a column vector of the form

$$\Phi_\rho = \left[\phi\left(-R + \frac{\rho}{L}\right), \, ..., \, \phi\left(\frac{\rho}{L}\right), \, \phi\left(1 + \frac{\rho}{L}\right), \, ..., \, \phi\left(R-1 + \frac{\rho}{L}\right) \right]^t \quad (4.90)$$

The solution of Eq. (4.87) is simply

$$\mathbf{P}_\rho = \Phi_T^{-1} \Phi_\rho \quad (4.91)$$

requiring the inversion of a $2R \times 2R$ matrix. Note also that the matrix Φ_T is independent of ρ and therefore only one matrix inversion must be performed to obtain the solution for all the polyphase filters. By carefully taking advantage of symmetries, this problem can be further reduced to a problem requiring the inversion of two $R \times R$ matrices. The details for this approach can be found in Ref. 4.14. This reference also contains a computer program that designs minimum mean-square interpolators according to the techniques described above, and it greatly simplifies the task of designing these filters.

4.3.7 Solution of the Matrix Equation

The process of solving Eq. (4.87) for $\mathbf{P}_\rho$ involves matrix inversion. Although in theory the matrix Φ_T is nonnegative definite, in practice it is usually positive definite. Therefore, in inverting Φ_T it is essential to keep the computation minimal to minimize the possibility of numerical instability caused by the accumulation of round-off error. General matrix inversion methods require on the order of $(2R)^3$ operations (multiplications and additions) to solve for the inverse. Using the symmetries in Φ_T, on the order of R^3 operations are required in the procedure discussed in the previous section. However, as Lu and Gupta [4.15] have pointed out, Eq. (4.87) can be solved in an even more efficient manner since Φ_T is a Toeplitz matrix (i.e., the elements along any diagonal of Φ_T are identical). Thus a wide class of Toeplitz matrix inversion methods can be used to solve Eq. (4.87). In particular, Durbin's method [4.16, 4.17] or the Burg method [4.18] can be used to solve for Φ_T^{-1} in $(2R)^2$ operations in the following way [4.15]. Consider the matrix equation

$$\Phi_T \mathbf{a} = \boldsymbol{\epsilon} \quad (4.92)$$

where

$$\mathbf{a} = \begin{bmatrix} 1 & 0 & \cdots & 0 \\ a_1^{(2R-1)} & 1 & & 0 \\ \vdots & \vdots & & \vdots \\ a_{2R-1}^{(2R-1)} & a_{2R-2}^{(2R-2)} & \cdots & 1 \end{bmatrix} \quad (4.93)$$

is a matrix of the "solution" vectors $a^{(2R-1)}$, $a^{(2R-2)}$, ..., $a^{(1)}$, and

$$
\epsilon = \begin{bmatrix} \epsilon^{(2R-1)} & 0 & 0 & \cdots & 0 \\ 0 & \epsilon^{(2R-2)} & 0 & \cdots & 0 \\ \vdots & & & & \\ 0 & 0 & 0 & \cdots & \epsilon^{(0)} \end{bmatrix} \tag{4.94}
$$

is a matrix of "residual analysis" errors. The well-known Durbin or Burg methods can be used directly to solve Eq. (4.92) for a and ϵ as follows. Since matrix a is lower triangular and matrix ϵ is a diagonal matrix, we first solve for the element $\epsilon^{(0)}$ from the product of the last row of $\mathbf{\Phi}_T$ and the last column of **a**, giving

$$
\epsilon^{(0)} = \phi(0) \tag{4.95}
$$

This solution is essentially the zeroth-order solution to the equivalent linear prediction problem.

We next solve for the coefficient and error residual of the first-order solution, namely $a_1^{(1)}$ and $\epsilon^{(1)}$, from the last two rows of $\mathbf{\Phi}_T$ and the last two columns of **a**, giving the equations

$$
\phi(1) + a_1^{(1)}\phi(0) = 0 \tag{4.96a}
$$

$$
\phi(0) + a_1^{(1)}\phi(1) = \epsilon^{(1)} \tag{4.96b}
$$

which can be solved for $a_1^{(1)}$ and $\epsilon^{(1)}$ giving

$$
a_1^{(1)} = -\frac{\phi(1)}{\phi(0)} \tag{4.97a}
$$

$$
\epsilon^{(1)} = \phi(0) - \frac{[\phi(1)]^2}{\phi(0)} \tag{4.97b}
$$

At each successive iteration, the order of the solution increases by 1 and the number of equations and unknowns increase by 1.

An efficient, recursive algorithm for solving for the **a** coefficients and ϵ terms is due to Durbin [4.16, 4.17], and can be stated as follows:

Initialization: $\epsilon^{(0)} = \phi(0)$

Recursion: $k_i = \dfrac{\phi(i) - \sum\limits_{j=1}^{i-1} a_j^{(i-1)}\phi(i-j)}{\epsilon^{(i-1)}}$, $1 \leqslant i \leqslant 2R - 1$

$$a_i^{(i)} = k_i$$

$$a_j^{(i)} = a_j^{(i-1)} - k_i a_{i-j}^{(i-1)}, \quad 1 \leqslant j \leqslant i - 1$$

$$\epsilon^{(i)} = (1 - k_i^2) \epsilon^{(i-1)}$$

This recursive algorithm provides an efficient method of solving for $\mathbf{a}$ and ϵ of Eq. (4.92), once Φ_T is specified. Now we must examine how we can use this solution to provide us with the solution to Eq. (4.87). To see how this is done, we first solve for Φ_T^{-1} in terms of $\mathbf{a}$ and ϵ^{-1}. If we premultiply Eq. (4.92) by $\mathbf{a}'$, we get

$$\mathbf{a}' \Phi_T \mathbf{a} = \mathbf{a}' \epsilon = \epsilon \qquad (4.98)$$

since $\mathbf{a}' = 1$ along the diagonal. We now take the inverse of Eq. (4.98), giving

$$\mathbf{a}^{-1} \Phi_T^{-1} (\mathbf{a}')^{-1} = \epsilon^{-1} \qquad (4.99)$$

and then we pre- and postmultiply both sides of Eq. (4.99) by $\mathbf{a}$ and $\mathbf{a}'$, giving

$$\Phi_T^{-1} = \mathbf{a} \epsilon^{-1} \mathbf{a}' \qquad (4.100)$$

Since ϵ is a diagonal matrix, its inverse is trivially obtained. We now substitute Eq. (4.100) into Eq. (4.91), giving

$$\mathbf{P}_\rho = \mathbf{a} \epsilon^{-1} \mathbf{a}' \Phi_\rho \qquad (4.101)$$

which is the desired solution in terms of known quantities.

4.3.8 Properties of the Minimum Mean-Square-Error Interpolators

The minimum mean-square-error interpolators designed using either of the procedures described above have a number of interesting properties [4.13 to 4.15]:

1. The resulting filters have the same symmetry properties as the ideal filters [Eqs. (4.33) and (4.36)], namely

$$h(m) = h(-m), \quad \text{for all } m \qquad (4.102)$$

and

$$p_\rho(n) = p_{L-\rho}(-n - 1) \qquad (4.103)$$

2. The minimum error $\min \mathbf{E}^2_{\rho,\alpha}$ for the polyphase filters also satisfies the symmetry condition

$$\min \mathbf{E}^2_{\rho,\alpha} = \min \mathbf{E}^2_{L-\rho,\alpha} \qquad (4.104)$$

This error increases monotonically as ρ increases (starting with $\mathbf{E}^2_{0,\alpha} = 0$) until $\rho = L/2$, at which point it decreases monotonically according to Eq. (4.104). Thus the greatest error occurs in interpolating sample values that are halfway between two given samples. Figure 4.18 shows a plot of the quantity $\min \mathbf{E}^2_{\rho,\alpha}$ normalized to its maximum value $\min \mathbf{E}^2_{L/2,\alpha}$ as a function of ρ for $\alpha = 0.5$, $R = 4$, and assuming that $|X(e^{j\omega})| = 1$. This normalized error is closely approximated by the sine-squared function [4.19], that is,

$$\frac{\min \mathbf{E}^2_{\rho,\alpha}}{\min \mathbf{E}^2_{L/2,\alpha}} \approx \sin^2\left[\frac{\rho\pi}{L}\right] \qquad (4.105)$$

3. If an interpolator is designed for a given signal with a large value of L, all interpolators whose lengths are fractions of L are obtained by simply sampling the original filter; that is, if we design an interpolator for $L = 100$, then for the same parameters α and R, we can derive from this filter the optimum mean-square-error interpolators for $L = 50, 25, 20, 10, 5$, and 2 by taking appropriate samples (or appropriate polyphase filters).

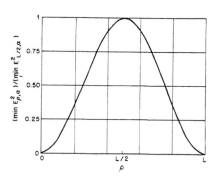

Figure 4.18 Plot of normalized minimum mean-squared interpolation error as a function of the shift ρ.

Figure 4.19 shows an example of the impulse response and frequency response for a minimum mean-square error interpolation filter with parameter values $\alpha = 0.5$, $R = 3$, $L = 8$, and assuming that $|X(e^{j\omega})| = 1$ in Eq. (4.85). Notice that this design procedure automatically takes into account the don't care regions discussed

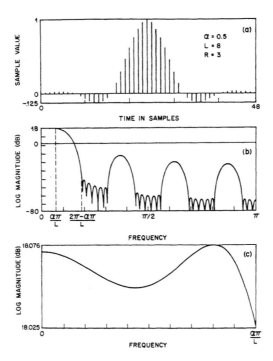

Figure 4.19 The impulse and frequency responses of a minimum mean-squared error interpolation filter with $\alpha = 0.5$ $R = 3$ and $L = 8$.

in Sections 4.2.2 and 4.3.3. Also a gain factor of L is included in this interpolation filter for reasons discussed in Chapter 2.

4.3.9 Design of FIR Interpolators with Minimax Error in the Frequency Domain

In the preceding section we considered the design of FIR interpolators based on minimizing the mean-square-error norm

$$\|\Delta Y(e^{j\omega'})\|^2 = \frac{1}{2\pi} \int_{-\alpha\pi/L}^{\alpha\pi/L} |\Delta Y(e^{j\omega'})|^2 \, d\omega' \qquad (4.106)$$

where $\Delta Y(e^{j\omega'})$ is the transform of $\Delta y(m)$ as defined in Fig. 4.17. It is seen that this norm is the integral of the error $|\Delta Y(e^{j\omega'})|^2$ over the frequency range $|\omega'| \leqslant \alpha\pi/L$, and thus the design based on this norm provides no direct control over errors at individual frequencies. For example, large errors may occur near the passband edge (i.e., at $\omega' = \alpha\pi/L$) in favor of smaller in-band errors.

In this section we consider another class of designs of interpolation filters in which the maximum of the error $|\Delta Y(e^{j\omega'})|$ over the frequency range of interest is minimized. This type of design gives a greater degree of control over the errors at specific frequencies [4.19]. In minimax designs of this type, the error $\Delta Y(e^{j\omega'})$ oscillates (ripples) in the frequency domain between positive and negative values of this maximum error. If the number of taps in the overall filter is

$$N = 2RL - 1 \qquad (4.107)$$

then each polyphase filter has $2R$ taps and therefore $2R$ degrees of freedom. These $2R$ degrees of freedom allow the error $|\Delta Y(e^{j\omega'})|$ to be exactly zero at $2R$ frequencies or R conjugate pairs of frequencies. Thus $\Delta Y(e^{j\omega'})$ will have $R + 1$ extremal values in the range of positive frequencies $0 \leqslant \omega' \leqslant \alpha\pi/L$.

To formulate the design problem described above, we can start by expressing $|\Delta Y(e^{j\omega'})|^2$ in the form

$$|\Delta Y(e^{j\omega'})|^2 = |H(e^{j\omega'}) - \widetilde{H}(e^{j\omega'})|^2 |X(e^{j\omega'L})|^2 \qquad (4.108)$$

which follows from Fig. 4.17(a). Next we express $H(e^{j\omega'})$ and $\widetilde{H}(e^{j\omega'})$ in terms of their polyphase filters. These expressions were derived in Section 4.2.1 and it follows from Eqs. (4.16) and (4.108) that

$$|\Delta Y(e^{j\omega'})|^2 = |\sum_{\rho=0}^{L-1} e^{-j\omega'\rho} \left[P_\rho(e^{j\omega'L}) - \widetilde{P}_\rho(e^{j\omega'L}) \right]|^2 |X(e^{j\omega'L})|^2 \qquad (4.109)$$

Also, in Section 4.2.2 we derived the expressions for the ideal filters $\widetilde{P}_\rho(e^{j\omega})$. Applying these expressions [Eq. (4.24)] to Eq. (4.109) and simplifying terms leads to

$$|\Delta Y(e^{j\omega'})|^2 = |\sum_{\rho=0}^{L-1} \left[P_\rho(e^{j\omega'L}) e^{-j\omega'\rho} - 1 \right]|^2 |X(e^{j\omega'L}|^2 \qquad (4.110)$$

In the discussion above we noted that $|\Delta Y(e^{j\omega'})|$ can be made exactly zero at R frequencies which we will specify as

$$0 \leqslant \omega_\lambda = \omega'_\lambda L \leqslant \alpha\pi, \quad \lambda = 1, 2, ..., R \qquad (4.111)$$

where the reader may recall that ω refers to frequencies specified with reference to the low sampling rate and ω' refers to frequencies specified with reference to the high sampling rate. From Eq. (4.110) it then follows that

$$P_\rho(e^{j\omega'_\lambda L}) = e^{j\omega'_\lambda \rho}, \quad \lambda = 1, ...R, \quad \rho = 0, 1, ..., L - 1 \qquad (4.112a)$$

or, equivalently,

$$P_\rho(e^{j\omega_\lambda}) = e^{j\omega_\lambda\rho/L}, \quad \lambda = 1, ..., R, \quad \rho = 0, 1, ..., L-1 \quad (4.112b)$$

Thus at these frequencies the polyphase filter responses $P_\rho(e^{j\omega_\lambda})$ are all equal to the ideal filter responses $\widetilde{P}_\rho(e^{j\omega_\lambda})$, $\lambda = 1, 2, ..., R$.

We can now express the polyphase filters $P_\rho(e^{j\omega})$ in terms of their coefficients according to Eq. (4.80). This leads to the set of conditions

$$\sum_{n=-R}^{R-1} p_\rho(n)e^{j\omega_\lambda n} = e^{j\omega_\lambda\rho/L}, \quad \lambda = 1, ..., R, \quad \rho = 0, 1, ..., L-1 \quad (4.113)$$

Equation (4.113) is now in the form of $2R$ linear equations in $2R$ variables for each value of ρ. The $2R$ equations occur by equating the real and imaginary components of Eq. (4.113) for the R frequencies ω_λ. They can be expressed more compactly in matrix form as

$$\boldsymbol{\Theta}_T\mathbf{P}_\rho = \boldsymbol{\Theta}_\rho \quad (4.114)$$

where $\mathbf{P}_\rho$ is the column vector

$$\mathbf{P}_\rho = [p_\rho(-R), ..., p_\rho(0), p_\rho(1), ..., p_\rho(R-1)]^t \quad (4.115)$$

$\boldsymbol{\Theta}_T$ is a $2R \times 2R$ matrix of the form

$$\boldsymbol{\Theta}_T = \begin{bmatrix} \cos(-R\omega_1) & \cos((-R+1)\omega_1) & \cdots & \cos((R-1)\omega_1) \\ -\sin(-R\omega_1) & -\sin((-R+1)\omega_1) & \cdots & -\sin((R-1)\omega_1) \\ \cos(-R\omega_2) & & \cdots & \\ \vdots & & & \\ -\sin(-R\omega_R) & & \cdots & -\sin((R-1)\omega_R) \end{bmatrix} \quad (4.116)$$

and $\boldsymbol{\Theta}_\rho$ is a column vector of the form

$$\boldsymbol{\Theta}_\rho = \left[\cos\left[\frac{\omega_1\rho}{L}\right], \sin\left[\frac{\omega_1\rho}{L}\right], \cos\left[\frac{\omega_2\rho}{L}\right], ..., \sin\left[\frac{\omega_R\rho}{L}\right] \right]^t \quad (4.117)$$

Assuming that we know the frequencies ω_λ, the solution to Eq. (4.114) is

$$\mathbf{P}_\rho = \boldsymbol{\Theta}_T^{-1}\boldsymbol{\Theta}_\rho, \quad \text{for } \rho = 0, 1, 2, ..., L-1 \quad (4.118)$$

and since $\boldsymbol{\Theta}_T$ is not dependent on ρ, it has to be inverted only once for the calculation of all the polyphase filters. Once the polyphase filters are determined, it is a simple matter to get the prototype filter $h(m)$ as discussed in Section 4.2.1.

The problem above has now been converted to a problem of finding the set of R frequencies ω_λ, $\lambda = 1, 2, ..., R$. The assumption that these frequencies are distinct is not essential to the validity of the result. Indeed, if we assume that $\omega_\lambda = 0$ for all λ and $X(e^{j\omega}) = $ constant for $|\omega| \leqslant \alpha\pi$, then all the zeros of $||\Delta Y(e^{j\omega'})||^2$ occur at $\omega' = 0$. In this case we get a maximally flat behavior of $||\Delta Y(e^{j\omega'})||^2$ such that its first $4R$ derivatives with respect to ω' are zero. The solution of Eq. (4.114) then leads to a well-known class of Lagrange interpolators [4.19]. The design of linear and Lagrange interpolators will be taken up in further detail in Sections 4.3.11 and 4.3.12.

For equiripple designs a procedure for finding the frequencies ω_λ has been proposed by Oetken [4.19]. First he showed that the error $||\Delta Y(e^{j\omega})||^2$ can be expressed as the sum of the errors due to each polyphase filter. Assuming a sinusoidal input with frequency ω_0, the error can be expressed as

$$||\Delta Y(e^{j\omega_0'})||^2 = \lim_{K \to \infty} \frac{1}{2K+1} \sum_{m=-K}^{K} \Delta y(m, \omega_0')^2 \qquad (4.119)$$

$$= \sum_{\rho=0}^{L-1} \lim_{K \to \infty} \frac{1}{2K+1} \sum_{m=-K}^{K} \Delta y_\rho(m, \omega_0)^2$$

$$= \sum_{\rho=0}^{L-1} ||\Delta Y_\rho(e^{j\omega_0})||^2 \qquad (4.120)$$

where $||\Delta Y_\rho(e^{j\omega_0})||^2$ denotes the interpolation error at the output of the ρth polyphase branch as seen in Fig. 4.17. Oetken then showed that the individual errors $||\Delta Y_\rho(e^{j\omega_0})||^2$ are almost exactly proportional to each other over the whole frequency range. This proportionality has the form [similar to that of Eq. (4.105)]

$$||\Delta Y_\rho(e^{j\omega_0})|| \approx ||\Delta Y_{L/2}(e^{j\omega_0})|| \sin\left(\frac{\rho\pi}{L}\right) \qquad (4.121)$$

where $||\Delta Y_{L/2}(e^{j\omega_0})||^2$ denotes the error for the polyphase filter $\rho = L/2$; that is, it is the filter which interpolates samples exactly halfway between the input samples. This form is a stronger condition than that of Eq. (4.105) in that it applies to individual frequencies as opposed to the mean square error integrated over the entire frequency range. The deviations to the approximation have been found to be smaller than 1% [4.19].

Because of the condition described above, the design problem may be converted to a simpler problem of designing one of the polyphase filters such that it has the desired minimax error. The zeros of this filter can be calculated to obtain the frequencies ω_λ, $\lambda = 1, 2, ..., R$, and they can be applied to Eqs. (4.113) to (4.118) to obtain the minimax solutions to the other polyphase filters. It is convenient to choose the polyphase filter $\rho = L/2$, which interpolates sample values halfway between input samples. Also, as in the case of mean-square-error designs, the design of the polyphase filters is independent of L (assuming that there are always $2R$ taps for each polyphase filter). Therefore, it is convenient to choose $L = 2$ so that the foregoing design requirements are those of a minimax half-band filter design. This design can be obtained using the techniques described in Section 4.3.5, with the appropriate weighting factor $|X(e^{j\omega})|$.

If we assume that $|X(e^{j\omega})| =$ constant, it can be shown that [4.19]

$$\max_{0 \leqslant \omega \leqslant \alpha\pi/L} |\Delta Y_{L/2}(e^{j\omega})| = \frac{2\hat{\delta}}{\sqrt{L}} \tag{4.122}$$

where $\hat{\delta}$ is the ripple of the half-band filter. The maximum error of $\|\Delta Y(e^{j\omega'})\|$ can then be shown to be

$$
\begin{aligned}
\delta &= \max_{0 \leqslant \omega' \leqslant \alpha\pi/L} |\Delta Y(e^{j\omega'})| \\
&\approx \max_{0 \leqslant \omega \leqslant \alpha\pi/L} \left[\|\Delta Y_{L/2}(e^{j\omega})\|^2 \sum_{\rho=0}^{L-1} \sin^2\left(\frac{\rho\pi}{L}\right) \right]^{1/2} \\
&\approx \left[\frac{4\hat{\delta}^2}{L} \frac{L}{2} \right]^{1/2} = \sqrt{2}\,\hat{\delta}
\end{aligned}
\tag{4.123}
$$

The design procedure outlined above can be summarized as follows:

1. Choose the cutoff frequency parameters α, and the maximum tolerable ripple δ, of the rms interpolation error $\|Y(e^{j\omega})\|$.
2. Design an equiripple half-band filter $h(m)$ with $4R - 1$ taps, where R is chosen such that its ripple $\hat{\delta}$ satisfies $\hat{\delta} \leqslant \delta/\sqrt{2}$ for the passband range $0 \leqslant \omega' \leqslant \alpha\pi/2$ (see Fig. 4.16).
3. Calculate the R roots of the polyphase filter $P_1(e^{j\omega})$ of this half-band filter to get the frequencies ω_λ, $\lambda = 1, 2, ..., R$
4. Calculate matrix Θ_T (Eq. 4.116) and the values for the polyphase filters P_ρ according to Eq. (4.118), for any interpolation rate L.

Figure 4.20 shows an example of a minimax interpolator design for $\alpha = 0.5$, $R = 3$, $L = 8$, and assuming that $|X(e^{j\omega})| = 1$. Figure 4.20(a) shows the individual errors $\|\Delta Y_\rho(e^{j\omega})\|$ [note that $\|\Delta Y_\rho(e^{j\omega})\| = \|\Delta Y_{L-\rho}(e^{j\omega})\|$] and Fig. 4.20(b) shows the total error $\|\Delta Y(e^{j\omega})\|$. Figure 4.20(c)-(e) shows the impulse response and frequency response of the final prototype filter $H(e^{j\omega})$. Note that although the error signal $\|\Delta Y(e^{j\omega})\|$ exhibits an equal-ripple behavior (as specified by the design criterion), $H(e^{j\omega})$ does not exhibit equal-ripple behavior.

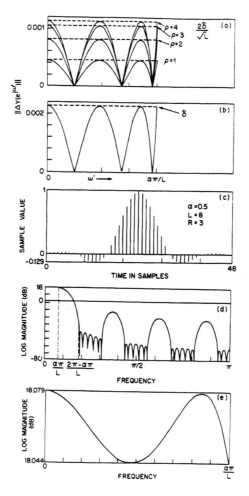

Figure 4.20 Impulse and frequency responses of a minimax design interpolation filter with $\alpha = 0.5$, $R = 3$, and $L = 8$.

4.3.10 Design of FIR Interpolators with Minimax Error in the Time Domain

Thus far we have considered the design of FIR interpolators based on a mean-square-error criterion (in time or frequency), and on a minimax error criterion in the frequency domain. In this section we consider another class of designs based on a minimax criterion in the time domain [4.20].

The problem can be formulated by again referring to Fig. (4.17). The minimax error in time can be expressed in the form

$$\hat{E} = \min_{h(m)} \max_{m} |\Delta y(m)| \tag{4.124a}$$

$$= \min_{p_\rho(n)} \cdot \max_n |\Delta u_\rho(n)|, \text{ for all } \rho \qquad (4.124\text{b})$$

Thus the design can again be reduced to the design of L polyphase filters $p_\rho(n)$, $\rho = 0, 1, ..., L - 1$, such that each polyphase filter minimizes its respective error $|\Delta u(n)|$.

From Fig. (4.17) we can note that $|\Delta u_\rho(n)|$ can be expressed in the form

$$|\Delta u_\rho(n)| = \left| \frac{1}{2\pi} \int_{-\pi}^{\pi} [P_\rho(e^{j\omega}) - \widetilde{P}_\rho(e^{j\omega})] X(e^{j\omega}) e^{j\omega n} \, d\omega \right| \qquad (4.125)$$

We will again assume that $|X(e^{j\omega})|$ is bandlimited, that is,

$$|X(e^{j\omega})| = 0 \text{ for } |\omega| \geq \alpha\pi \qquad (4.126)$$

Equivalently, we can define an ideally bandlimited function $F(e^{j\omega})$ such that

$$F(e^{j\omega}) = \begin{cases} 1, & |\omega| \leq \alpha\pi \\ 0, & |\omega| > \alpha\pi \end{cases} \qquad (4.127)$$

Thus $F(e^{j\omega})$ has an inverse transform, $f(n)$, of the form

$$f(n) = \frac{\sin(\alpha\pi n)}{\pi n} \qquad (4.128)$$

Equation (4.125) can then be expressed as

$$|\Delta u_\rho(n)| = \left| \frac{1}{2\pi} \int_{-\pi}^{\pi} [P_\rho(e^{j\omega}) F(e^{j\omega}) - \widetilde{P}_{\rho,f}(e^{j\omega})] X(e^{j\omega}) e^{j\omega n} \, d\omega \right| \quad (4.129)$$

where $\widetilde{P}_{\rho,f}(e^{j\omega})$ is defined as

$$\widetilde{P}_{\rho,f}(e^{j\omega}) \triangleq \widetilde{P}_\rho(e^{j\omega}) F(e^{j\omega}) \qquad (4.130\text{a})$$

$$= \begin{cases} e^{j\omega\rho/L}, & |\omega| \leq \alpha\pi \\ 0, & |\omega| > \alpha\pi \end{cases} \qquad (4.130\text{b})$$

From the discussion in Section 4.2.2 it can be shown that the inverse transform of $\widetilde{P}_{\rho,f}(e^{j\omega})$ is

$$\widetilde{P}_{\rho,f}(n) = \frac{\sin(\alpha\pi(n + \rho/L))}{\pi(n + \rho/L)} \qquad (4.131)$$

Inverse transforming Eq. (4.129) and applying Eq. (4.131) then gives the error $|\Delta u_\rho(n)|$ in the form

$$|\Delta u_\rho(n)| = |x(n) * [p_\rho(n) * f(n) - \tilde{p}_{\rho,f}(n)]| \qquad (4.132a)$$

$$= |x(n) * \left[\sum_{r=-R}^{R-1} p_\rho(r) f(n-r) - \tilde{p}_{\rho,f}(n) \right]| \qquad (4.132b)$$

where * denotes the direct convolution operator.

From Eq. (4.132b) it is seen that in order to minimize $\hat{E}$ we would ideally like to choose the design of the polyphase filter such that it forces the expression in brackets,

$$\sum_{r=-R}^{R-1} p_\rho(r) f(n-r) - \tilde{p}_{\rho,f}(n)$$

to zero for all values of n. However, since this expression is infinite in duration and there are only $2R$ degrees of freedom in $p_\rho(n)$ [assuming that $p_\rho(n)$ has $2R$ taps], only $2R$ terms of this expression can be made to be exactly zero. Parks and Kolba [4.20] showed that these terms are the terms $n = -R,...,0,1,...,R-1$, and that errors due to all other terms (which may be nonzero) are orthogonal to the error terms that have been forced to zero. Thus we are lead to the set of equations

$$\sum_{r=-R}^{R-1} p_\rho(r) \frac{\sin(\alpha\pi(n-r))}{\pi(n-r)} = \frac{\sin(\alpha\pi(n+\rho/L))}{\pi(n+\rho/L)}, \quad n = -R, ..., R-1 \quad (4.133)$$

This set of equations forms a set of $2R$ linear equations which can be solved in a manner similar to that discussed in the preceding two sections.

The form of Eq. (4.133) is particularly revealing in that it shows that the problem of minimizing $\hat{E}$ is reduced to one of approximating a $\sin(x)/x$ function centered at $x = n + \rho/L$ with a linear combination of $2R$ $\sin(x)/x$ functions located at the integer values of x nearest to $n + \rho/L$ (for each polyphase filter ρ).

Parks and Kolba [4.20] also showed that the solution obtained from Eq. (4.133) is identical to that of the mean-square-error design in Section 4.3.6 when it was assumed that the input spectrum was flat. Thus the same design program can be used here.

Note that in the development of Eq. (4.133) the properties of the signal $x(n)$ (other than the assumption that it is bandlimited to 2π) have been decoupled from the design problem [i.e., Eq. (4.133) is independent of $x(n)$]. If $x(n)$ is known, a priori, the problem can be modified to include this knowledge of $x(n)$ so that the design is specifically tailored to interpolating this one signal. This can be accomplished by expressing Eq. (4.132) in the form

$$|\Delta u_\rho(n)| = |\sum_{r=-R}^{R-1} p_\rho(r) f_x(n-r) - \widetilde{p}_{\rho,f_x}(n)|$$

(4.134)

where $f_x(n)$ and $\widetilde{p}_{\rho,f_x}(n)$ are defined as

$$f_x(n) = f(n) * x(n)$$

(4.135)

and

$$\widetilde{p}_{\rho,f_x}(n) = \widetilde{p}_{\rho,f}(n) * x(n)$$

(4.136)

The problem can now be reduced to the solution of the set linear equations

$$\sum_{r=-R}^{R-1} p_\rho(r) f_x(n-r) = \widetilde{p}_{\rho,f_x}(n)$$

(4.137)

The practicality of this approach depends, of course, on our ability to compute Eqs. (4.135) and (4.136). For example, if $x(n)$ is infinite in duration, these expressions cannot be determined directly (although it may be possible to get closed-form analytical expressions for them).

4.3.11 Linear Interpolation

Perhaps the most familiar of all data interpolators is the linear interpolator, in which the values interpolated between two consecutive data samples lie on a straight line connecting these two samples. It is a fairly simple procedure to derive the impulse response of the linear interpolation filter as follows. Since the linear interpolation filter encompasses exactly two values of the input signal, from Eq. (4.79) we get

$$N = 2L - 1$$

(4.138)

as the overall length of the filter (i.e., $R = 1$), where L is the interpolation factor. Thus $h(m)$ is defined for $-L < m < L$. Rather than solving for $h(m)$ directly, it is much simpler to solve for the polyphase filters $p_\rho(n)$ of Fig. 4.2(b), and get $h(m)$ via the reconstruction Eq. (4.15). Since $h(m)$ extends for $2L - 1$ samples, each polyphase filter has two taps, $p_\rho(-1)$ and $p_\rho(0)$. From Fig. 4.2(b) we can solve for $y(m)$ at sample $m = nL + \rho$ as

$$y(nL + \rho) = y_\rho(nL + \rho) = p_\rho(0)x(n) + p_\rho(-1)x(n+1)$$

(4.139)

Based on the definition of the linear interpolator, we get

$$y(nL + \rho) = x(n) + \Delta x(n)\left[\frac{\rho}{L}\right] \tag{4.140a}$$

$$= x(n) + [x(n+1) - x(n)]\left[\frac{\rho}{L}\right] \tag{4.140b}$$

$$= x(n)\left[1 - \frac{\rho}{L}\right] + x(n+1)\left[\frac{\rho}{L}\right] \tag{4.140c}$$

From Eqs. (4.139) and (4.140), we get

$$p_\rho(-1) = \frac{\rho}{L}, \qquad \rho = 0, 1, ..., L - 1 \tag{4.141a}$$

$$p_\rho(0) = 1 - \frac{\rho}{L}, \quad \rho = 0, 1, ..., L - 1 \tag{4.141b}$$

and from Eq. (4.15) we get

$$h(m) = \begin{cases} 1 - \dfrac{|m|}{L}, & -L < m < L \\ 0, & \text{otherwise} \end{cases} \tag{4.142}$$

as the impulse response of the linear interpolation filter.

Equation (4.142) shows that the impulse response of a linear interpolator has a triangular shape, as illustrated in Fig. 4.21(a) for a value $L = 5$. The frequency response of the linear interpolator is obtained as the Fourier transform of Eq. (4.142), giving

$$H(e^{j\omega}) = \frac{1}{L}\left[\frac{\sin(\omega L/2)}{\sin(\omega/2)}\right]^2 \tag{4.143}$$

The magnitude and log magnitude responses of a linear interpolator with $L = 5$ are shown in Fig. 4.21(b) and (c). It is seen from Eq. (4.143) and Fig. 4.21 that the peak sidelobes of the linear interpolator are attenuated only about 20 dB relative to the passband attenuation. Thus unless the original signal being interpolated is highly bandlimited (i.e., to a band much smaller than π/L), the linear interpolator will not be capable of sufficiently attenuating the images of the signal spectrum.

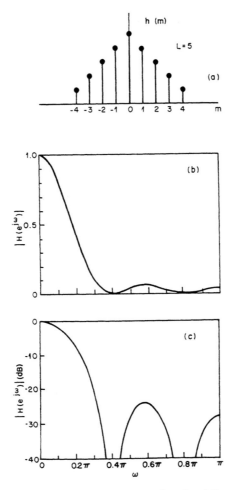

Figure 4.21 Impulse and frequency responses of an $L = 5$ linear interpolator.

Thus the applicability of linear interpolation is somewhat limited.[6]

4.3.12 Lagrange Interpolators

The inadequacies of linear interpolators were recognized by classical numerical analysts, who sought highly accurate methods of interpolating mathematical functions. Their solution was to interpolate using Q values of the original signal with a polynomial of degree $Q - 1$ that passed through the original signal values.

[6] Because of its simplicity and the fact that only two signal samples are involved in the computation, linear interpolation is widely used in practice. This section is intended to guide a potential user regarding problems that could develop when using linear interpolation.

The interpolated values were then samples of the polynomial. Figure 4.22 illustrates this technique for $Q = 2$, 3, and 4 with $L = 3$ and points out some of the issues involved with polynomial interpolation. For $Q = 2$ we get the linear interpolator discussed in the preceding section. For $Q = 3$ there are two distinct cases. Assume for simplicity that the three points being interpolated are $x(-1)$, $x(0)$, and $x(1)$. For $Q = 3$ a second-order polynomial is fit exactly to these three values. Now the question is: Which interpolated values are obtained from the polynomial, the ones between samples $x(-1)$ and $x(0)$ (case 1), or those between $x(0)$ and $x(1)$ (case 2)? Distinctly different results are obtained in the two cases. For $Q = 4$, a third-order polynomial is fit to the four samples $x(-1)$, $x(0)$, $x(1)$, and $x(2)$, and three interpolation regions (cases) are possible. Case 1 is interpolation between $x(-1)$ and $x(0)$; case 2 is interpolation between $x(0)$ and $x(1)$; case 3 is interpolation between $x(1)$ and $x(2)$. In the general case of interpolating between Q samples with a $(Q - 1)$st-degree polynomial, there are $(Q - 1)$ regions in which the fitted polynomial can be used. Each such case corresponds to a uniquely different interpolation filter.

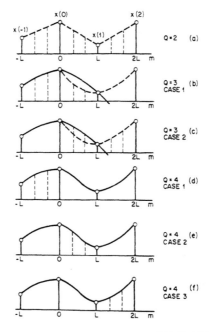

Figure 4.22 Illustration of the interpolation regions of a Q-point interpolator, for $Q = 2$, 3, and 4.

In this section we consider only the case of Lagrangian interpolation for values $Q = 2R$, and for a symmetric interpolation region. This set of cases leads to filter designs, $h(m)$, that are symmetric, and can be analyzed simply using polyphase filters. All other cases have been discussed by Schafer and Rabiner [4.23] and will not be discussed here.

For a $Q = 2R$ point Lagrange interpolator, each polyphase filter $p_\rho(n)$ is a $2R$ point filter. We assume that $p_\rho(n)$ is defined for $n = -R$, $-R + 1, ..., 0, ..., R - 1$, and that the resulting filter, $h(m)$, is defined for $-RL \leqslant m \leqslant RL$ for an L-point interpolator. As in the case of linear interpolators, we use the block diagram of Fig. 4.2(b) to solve for $y(m)$ at $m = nL + \rho$ as

$$y(nL + \rho) = y_\rho(nL + \rho) = p_\rho(-R)x(n + R) + p_\rho(-R + 1)x(n + R - 1)$$

$$+ ... + p_\rho(0)x(n) + ... + p_\rho(R - 1)x(n - R + 1) \qquad (4.144)$$

Based on the Lagrangian polynomial fit to the samples $x(n + R)$, ..., $x(n - R + 1)$, we get [4.24]

$$y(nL + \rho) = A^{2R}_{-R}\left[\frac{\rho}{L}\right]x(n + R) + A^{2R}_{-R+1}\left[\frac{\rho}{L}\right]x(n + R - 1)$$

$$+ ... + A^{2R}_{0}\left[\frac{\rho}{L}\right]x(n) + ... + A^{2R}_{R}\left[\frac{\rho}{L}\right]x(n - R + 1) \quad (4.145)$$

where $A^{2R}_\lambda(\rho/L)$ is the Lagrangian function [4.24]

$$A^{2R}_\lambda\left(\frac{\rho}{L}\right) = \frac{(-1)^{\lambda+R} \prod\limits_{i=1}^{2R} (\rho/L + R - i)}{(\lambda + R)!(R - 1 - \lambda)!(\rho/L - \lambda)} \qquad (4.146)$$

which is obtained by fitting a polynomial of degree $(2R - 1)$ to the set of equally spaced samples $x(n + R)$, ..., $x(n - R + 1)$. From Eqs. (4.145) and (4.146) we can solve directly for $p_\rho(\lambda)$, $\lambda = -R, ..., R - 1$, as

$$p_\rho(\lambda) = A^{2R}_\lambda\left[\frac{\rho}{L}\right], \quad \lambda = -R, ..., R - 1; \rho = 0, 1, ..., L - 1 \qquad (4.147)$$

and $h(m)$ can then be obtained by combining the interpolated $p_\rho(\lambda)$'s via Eq. (4.15). However, unlike the linear interpolation filter, no simple closed-form expression exists for $h(m)$ for the $2R$-point Lagrange interpolation filter.

To compare the Lagrangian interpolation filters in terms of their spectral properties, Fig. 4.23 shows a series of plots of the linear and log magnitude frequency responses of Lagrange interpolation filters for $L = 5$ and values of R from 1 (linear interpolation) to 4. It is readily seen that as R increases, the

Lagrange interpolation filter begins to approximate the ideal lowpass interpolation filters by a filter with a monotonic passband (no ripples), and a series of stopbands with multiple zeros at the frequencies $\omega = 2\pi r/L$, $r = 1, 2, ..., L - 1$. Oetken [4.19] was able to show that the R-point Lagrange interpolation filter is maximally flat at $\omega = 0$ (leading to the monotonic passband). Furthermore, based on the maximally flat property, Oetken [4.19] showed that the polyphase filters $p_\rho(n)$ of the Lagrange interpolator satisfied a Vandermonde set of equations of the form

$$\sum_{\lambda=-R}^{R-1} p_\rho(\lambda)\lambda^u = \left[-\frac{\rho}{L}\right]^u, \quad u = 0, 1, ..., 2R - 1; \rho = 0, 1, ..., L - 1 \quad (4.148)$$

providing an alternative procedure to Eq. (4.146) for solving for the coefficients of the polyphase filters of the Lagrange interpolator.

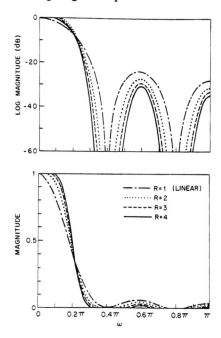

Figure 4.23 Frequency-response characteristics (on log and linear magnitude scales) of Lagrange interpolators for $R = 1, 2, 3,$ and 4.

4.3.13 Discussion

This concludes our discussion of FIR design techniques for the filters for decimators and interpolators. As we have shown, there are many techniques for designing such filters, and they are all based on slightly different criteria. Some techniques are convenient because of their simplicity, some because they optimize a specific error criterion, and others are of interest strictly from a historical point of view. As in most filter design problems, it is up to the user to decide which of a set of alternative solutions to the filter design problem is most applicable to the problem at hand.

4.4 FILTER DESIGN PROCEDURES FOR IIR DECIMATORS AND INTERPOLATORS

In the preceding section we spent considerable effort on outlining FIR design methods for general- and special-purpose decimation and interpolation filters. In this section we consider, more briefly, design techniques for IIR filters for decimation and interpolation. As discussed previously, a *major* distinction between the FIR and IIR systems is that FIR designs can achieve *exactly* linear phase responses (i.e., constant group delay) and thus the envelopes of the time waveforms of the signals being decimated or interpolated are preserved. On the contrary, with IIR designs, exact linear phase responses are not possible and thus waveform envelopes are not preserved. Thus this class of designs can only be used in applications where linear phase response is *not* essential. For this reason we will not go into as much detail in discussing these techniques. For more information on this subject and some related FIR designs the reader is referred to Refs. 4.25 to 4.31.

Before considering specific IIR design techniques, we first need to modify the ideal response characteristics for decimators and interpolators as discussed in Section 4.2 to allow for the arbitrary phase characteristics of the IIR designs. Also we need to briefly recall, from Chapter 3, the constraints imposed on the IIR system functions due to the practical structures and implementations of these systems.

4.4.1 Ideal Characteristics and Practical Realizations of IIR Decimators and Interpolators

If we relax the constraints on phase, the ideal characteristic for the prototype filter $h(k)$ (see Section 4.2 and Fig. 4.1) can be specified (for the M-to-1 decimator) as

$$\widetilde{H}_{\text{IIR}}(e^{j\omega'}) = \begin{cases} e^{j\phi(\omega')}, & |\omega'| \leqslant \dfrac{\pi}{M} \\ 0, & \text{otherwise} \end{cases} \tag{4.149}$$

where $\phi(\omega')$ is allowed to be an arbitrary function of ω'. It follows from Section 4.2.3 that the ideal polyphase filter responses can be specified as

$$\widetilde{P}_\rho(e^{j\omega}) = e^{j\omega\rho/M} e^{j\phi(\omega/M)} \tag{4.150}$$

A similar pair of ideal responses for IIR interpolators can be defined by replacing M by L and scaling the right side of Eq. (4.149) by L.

In Section 3.4, three basic approaches were discussed for realizing IIR decimators or interpolators. The first approach was that of a direct application of *conventional* IIR design methods to the prototype structure of decimators or interpolators as represented by Fig. 4.1. In this approach, well-known methods for designing and implementing time-invariant IIR filters can be applied to this problem [4.1-4.3]. The transfer function in this case has the form

$$H(z) = \frac{N(z)}{D(z)} = \frac{\sum\limits_{r=0}^{N-1} b_r z^{-r}}{1 - \sum\limits_{r=1}^{D} a_r z^{-r}} \qquad (4.151)$$

where $D = N - 1$ is the order of the filter (i.e., the order of the numerator and denominator polynomials $N(z)$ and $D(z)$ respectively). Design techniques that can be applied directly to give the b_r and a_r coefficients are summarized briefly in Section 4.4.2.

A modification of this approach is to realize the IIR filter in a direct or transposed direct-form structure and to commute the FIR part of the structure to the low-sampling-rate side for greater computational efficiency, as illustrated by the example in Fig. 3.35. A difficulty with this modification, however, is that the direct-form IIR structure is known to have undesirable round-off noise and coefficient sensitivity properties for high-order, sharp-cutoff designs.

Another approach to the realization of IIR systems is to restrict the system function to the form

$$H(z) = \frac{\sum\limits_{r=0}^{N-1} \hat{b}_r z^{-r}}{1 - \sum\limits_{r=1}^{R} \hat{c}_r z^{-rM}} = \frac{\hat{N}(z)}{\hat{D}(z)} \qquad (4.152)$$

where $\hat{N}(z)$ is an $(N - 1)$th-order numerator polynomial, $\hat{D}(z)$ is a polynomial of order R in z^M, and M is the decimation factor (for interpolators replace M by L). In this case the recursive part of the structure can be commuted to the low-sampling-rate side of the decimator (or interpolator) for greater computational efficiency, as discussed in Section 3.4. Figure 3.40(a) illustrates this structure for the decimator and Fig. 3.40(b) shows a similar structure for the interpolator, where the FIR decimators or interpolators in these structures can be implemented with any of the methods discussed in Section 3.3. More specific examples of such IIR structures are given in Figs. 3.38, 3.39, and 3.41.

The form of the system function in Eq. (4.152) is no longer the conventional one, and therefore special design techniques are required. Section 4.4.3 covers one (nonoptimal) approach to the design of filters with this form of system function. These designs are based on the conversion of conventional IIR designs in the form of Eq. (4.151) to the form of Eq. (4.152) through the application of a polynomial factorization relation.

Section 4.4.4 discusses a design technique that is applied directly to system functions of Eq. (4.152). It results in IIR filters that are optimal in the minimax sense (in frequency).

4.4.2 Conventional IIR Filter Designs

The design of conventional IIR filters is a subject which has received a great deal of attention and is well documented in the literature [4.1-4.3]. Hence we will not go into these methods in detail, but only point out a few relatively widely used methods and discuss their general characteristics. These designs are:

1. The Butterworth approximation
2. The Bessel approximation
3. The Chebyshev approximation
4. The elliptic approximation

Butterworth filters have the property that they are all-pole filters whose magnitude (frequency) response is maximally flat at $\omega' = 0$. For a D-pole lowpass filter this means that the first $2D - 1$ derivatives are zero at $\omega' = 0$. The magnitude response also has the property that it is a monotonically decreasing function of ω' (for lowpass filters). An example showing the log magnitude response and the group delay of an $D = 11$-pole, 500-Hz (10 kHz sampling rate) Butterworth lowpass filter is given in Fig. 4.24(a).

Bessel filters have the property that they are all-pole filters whose group delay is maximally flat at $\omega' = 0$ (for lowpass designs). Their step responses typically exhibit extremely low (less than 1%) overshoot. Both the impulse response and the magnitude response tend toward Gaussian responses as the filter order is increased. Thus they have a significant magnitude deviation in the passband. An example showing the log magnitude response and the group delay of an $D = 10$ pole, 500-Hz Bessel lowpass filter is given in Fig. 4.24(b)

Chebyshev filters have the property that over a prescribed band of frequencies the peak magnitude of the approximation error is minimized. If it is minimized over the passband (type I designs), the magnitude response is equiripple in the passband and monotonic in the stopband. If it is minimized over the stopband (type II designs), the magnitude response is monotonic in the passband and equiripple in the stopband. The group delay response of Chebyshev filters exhibits significant group delay dispersion in the passband for both types of designs. Examples of the log magnitude responses and group delays of $D = 6$-pole, 500-Hz

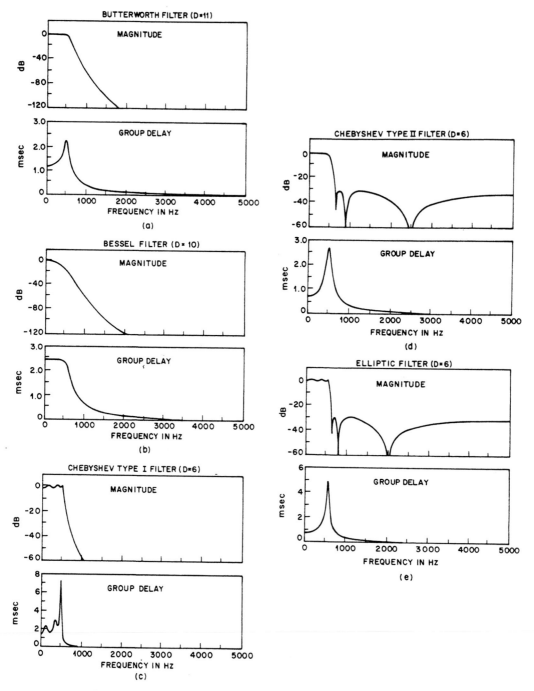

Figure 4.24 Log magnitude and group delay responses of; (a) $D = 11$, Butterworth lowpass filter; (b) $D = 10$, Bessel lowpass filter; (c) $D = 6$, Chebyshev type I lowpass filter; (b) $D = 6$, Chebyshev type II lowpass filter; (e) $D = 6$, elliptic lowpass filter.

Chebyshev lowpass filters are given in Fig. 4.24(c) (type I) and Fig. 4.24(d) (type II).

Elliptic filters have the property that these magnitude responses are equiripple in both the passband and stopband. It can be shown that equiripple filters are optimum in the sense that for a given-order filter, and for given ripple specifications, no other filter achieves a faster transition between the passband and stopband. The group delay dispersion in the passband of elliptic filters is extremely large. However, because of this optimality property, these filters are often used in many practical applications. An example of the log magnitude response and group delay of an $D = 6$-pole, 500-Hz elliptic filter is given in Fig. 4.24(e).

Design charts for comparing the IIR filter designs above (for lowpass filters) and conventional FIR filter designs are given in Refs. 4.1 and 4.3.

4.4.3 Special IIR Designs Based on the Transformation of Conventional Designs

The conventional filter designs lead to system functions of the form of Eq. (4.151). On the other hand, it was shown previously that system functions of the form of Eq. (4.152) are often more desirable for FIR decimators or interpolators. Bellanger, Bonnerot, and Coudreuse [4.25] have proposed a method of transforming system functions of the form of Eq. (4.151) to the form of Eq. (4.152) in a nonoptimal way. Their technique allows all the conventional designs, discussed in the preceding section, to be applied to any of the decimator or interpolator structures.

This approach is based on the polynomial factorization relation

$$z^M - p_r^M = (z - p_r)(z^{M-1} + p_r z^{M-2} + \cdots + p_r^{M-1}) \tag{4.153}$$

To apply this relation to the design problem described above, we need to express Eq. (4.151) in the factored form

$$H(z) = \frac{N(z)}{D(z)} = A z^{D-N-1} \frac{\prod\limits_{r=1}^{N-1} (z - z_r)}{\prod\limits_{r=1}^{D} (z - p_r)} \tag{4.154}$$

where z_r and p_r are the zeros and poles of the system function $H(z)$, respectively, and A is the gain factor.

By solving Eq. (4.153) for the quantity $z - p_r$ and substituting it into Eq. (4.154) for each value of r, we get

$$H(z) = \frac{\hat{N}(z)}{\hat{D}(z)}$$

$$= Az^{D-N-1} \frac{\displaystyle\prod_{r=1}^{N-1} (z - z_r) \left[\prod_{r=1}^{D} (z^{M-1} + p_r z^{M-2} + \dots + p_r^{M-1}) \right]}{\displaystyle\prod_{r=1}^{D} (z^M - p_r^M)} \qquad (4.155)$$

This system function is now in the form of Eq. (4.152), as desired, and it has the same magnitude and phase response as $H(z)$ in Eq. (4.154).

In the process of making this transformation, the order of the numerator has been increased from that of an $(N-1)$th-order polynomial to an $\hat{N} - 1 = [N - 1 + D(M - 1)]$th-order polynomial. This increase in the numerator order increases the amount of computation required for the FIR part of the structure. However, this increased computation is offset by the reduction in computation which can now be gained by commuting the IIR part of the structure to the low-sampling-rate side as discussed in Section 3.4. The net result is that this approach leads to a reduction in the required overall computation rate compared to the straightforward approach of using the conventional IIR design in the prototype model of Fig. 4.1 [4.25]. A different but similar form of transformation has been developed by Vary [4.26].

4.4.4 A Direct Design Procedure for Equiripple IIR Filters for Decimation and Interpolation

Although the technique described above, based on transforming conventional designs, leads to system functions of the form of Eq. (4.152) (and more efficient implementations), these designs are no longer optimal even when the filters that we start with are optimal designs [4.27]. This is because of the increased order of the numerator, which is a consequence of the transformation technique. The additional numerator terms do not contribute in the most efficient way toward meeting the required design criteria.

To avoid this problem, Martinez and Parks [4.27, 4.28] have proposed a design technique that leads to optimized designs (in the Chebyshev sense) for the class of system functions represented by Eq. (4.152). In principle this design procedure is similar to that for the FIR equiripple designs discussed in Sections 4.3.2 and 4.3.3. An error tolerance criterion of the form of Fig. 4.10 or 4.12 is defined. A multiple exchange algorithm similar to the Remez algorithm is then applied to the design. Since the IIR design involves a design of both a numerator and denominator polynomial, this modified algorithm is achieved by working iteratively with the numerator and denominator separately in an alternate fashion (a number of other, related, design methods have also been proposed [4.29] to [4.31]). The details of these design procedures are too involved to discuss here; however, we will discuss a number of properties and characteristics of these designs.

The primary difference between these (optimum) designs and the transformed designs is in the number of passband zeros that are located on the unit circle. For example, if we start with a conventional elliptic design of order $N - 1 = D$, there are $N - 1$ zeros in the frequency response (all located in the stopband, on the unit circle). In the transformed design there are also $N - 1$ zeros in the stopband since the system function remains unchanged; however, the numerator order has now been increased to $\hat{N} - 1 = N - 1 + D(M-1) = DM$ (where M is the decimation ratio). If we design an optimum filter with this same numerator order, $\hat{N} - 1$, using the Martinez and Parks approach, it will be found that there are $\hat{N} - 1$ zeros on the unit circle in the stopband, where they are more effective in reducing the stopband ripple. Alternatively, this improved stopband performance can be traded, to some extent, for a reduced passband ripple or a narrower transition band.

Figure 4.25(a) illustrates this case for a second-order elliptic filter ($N - 1 = 2$). If we transform this design to a design for an $M = 3$ decimator according to the method introduced in the previous section, the resulting filter has a numerator order of $\hat{N} - 1 = 6$. The resulting filter has the same frequency response as in Fig. 4.25(a). Alternatively, if we design an optimal filter according to the Martinez-Parks method with the same numerator and denominator orders as the transformed design, the frequency response in Fig. 4.25(b) will result.

Figure 4.26(a) helps to give a better idea of where the pole and zero locations occur in these designs [with system functions of the form of Eq. (4.152) for the case $M = 5$, $\hat{N} - 1 = 36$, and $\hat{D} = 6$]. Note that there are three complex pole pairs that are used to give the passband behavior and these pole patterns are repeated $M = 5$ times (equispaced) on the unit circle. In the stopband the effects of these poles are attenuated by closely spaced zeros on the unit circle [see Fig. 4.26(b)]. At first look this pole-zero plot seems to indicate that there may not be any advantage in using these filters because of the pole repetition pattern. But it

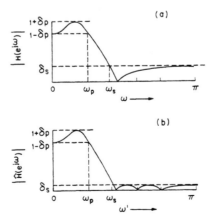

Figure 4.25 (a) Illustration of the magnitude response of an elliptic filter of order $D = N - 1 = 2$; (b) illustration of an optimum (in the Chebyshev sense) IIR filter with numerator order $\hat{N} - 1 = N - 1 + D(M - 1) = 6$, $M = 3$ and denominator order $D = 2$.

has been shown [4.27] that these filters are considerably more efficient in terms of computation designs discussed previously for a wide range of specifications. Figure 4.26(c) shows the group delay for the design above. The group delays for this class of designs are similar to those of elliptic filters.

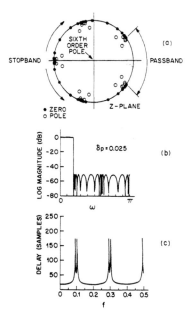

Figure 4.26 (a) Pole-zero location; (b) log magnitude response; and (c) group delay for an optimal design for $M = 5$, $\hat{N} - 1 = 36$, and $\hat{D} = 6$.

4.5 COMPARISONS OF IIR AND FIR DESIGNS OF INTERPOLATORS AND DECIMATORS

We have already obtained some indications as to the comparisons of various IIR design methods. In this section we consider another example that will help to place in perspective the computation rates required of the various FIR and IIR techniques discussed so far (for the case of a Chebyshev criterion in frequency).

The design parameters for a 5-to-1 decimator are:

$$\delta_p = 0.025 \quad = \text{passband ripple}$$
$$\delta_s = 0.0031 \quad = \text{stopband ripple}$$
$$\omega_s = 0.2\pi \quad = \text{stopband edge}$$
$$\Delta F = \frac{\omega_s - \omega_p}{2\pi} \quad = \text{transition band (design variable)}$$
$$M = 5 \quad = \text{decimation ratio}$$

where ΔF is considered as a variable parameter in this example, and the input sampling rate is normalized to 1 Hz. Note that in this comparison we are assuming that the phase response is not important.

Figure 4.27 shows a plot of the required multiplication rates as a function of ΔF for the designs [4.27]:

1. An equiripple FIR design that takes advantage of symmetry (Section 4.3.2)
2. An elliptic design in a prototype structure as in Fig. 3.35 (Section 4.4.2)
3. A transformed elliptic design (Section 4.4.3)
4. Optimized IIR designs (Section 4.4.4)

As seen from this figure, the optimum designs of Section 4.4.4 are most efficient in terms of multiplication rates and the elliptic and transformed elliptic designs are considerably less efficient. The FIR design is less efficient as ΔF becomes small. However, as ΔF becomes large, the efficiency of the FIR design approaches that of the optimum equiripple design. Again, if linear phase is of concern, the FIR designs is the only one that can meet this constraint.

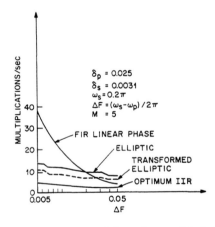

Figure 4.27 Multiplication rate vs ΔF for various equiripple FIR and IIR single-stage decimator designs.

In Chapter 5 we consider further comparisons of the designs listed above and include multistage designs which will again change some of this picture. In particular we will see that designs involving very small values of ΔF are often associated with decimators or interpolators involving very large changes in sampling rates. For these designs considerably more efficiency is obtained by using multistage designs instead of the single-stage designs discussed here. In such cases the FIR design can be nearly as efficient as the optimal IIR designs in terms of the multiplication rate.

As pointed out in the beginning of this chapter, multiplication rate is only one measure of complexity and in a practical application numerous other factors must also be considered. Thus there is simply no universal answer to the question of what approach (FIR, IIR, single-stage, multistage, equiripple design, mean-square-error design, etc.) is best, and each application must be evaluated according to its own practical requirements.

REFERENCES

4.1   L. R. Rabiner and B. Gold, *Theory and Application of Digital Signal Processing*, Englewood Clifs, N.J.:   Prentice-Hall, Inc., 1975.

4.2   A. V. Oppenheim and R. W. Schafer, *Digital Signal Processing*, Englewood Cliffs, N.J.:   Prentice-Hall, Inc., 1975.

4.3   L. R. Rabiner, J. F. Kaiser, O. Herrmann, and M. T. Dolan, "Some Comparisons between FIR and IIR Digital Filters," *Bell Syst. Tech. J.*, Vol. 53, No. 2, pp. 305-331, February 1974.

4.4   J. F. Kaiser, "Nonrecursive Digital Filter Design Using the I_0 - Sinh Window Function," Proc. IEEE Int. Symp. Circuits Syst., April 1974, pp. 20-23, (Also reprinted in *Digital Signal Processing, II*, New York: IEEE Press, pp. 123-126, 1975.

4.5   A. W. Crooke and J. W. Craig, "Digital Filters for Sample-Rate Reduction," *IEEE Trans. Audio Electroacoust.*, Vol. AU-20, No. 4, pp. 308-315, October 1972.

4.6   T. W. Parks and J. H. McClellan, "Chebyshev Approximation for Nonrecursive Digital Filters with Linear Phase," *IEEE Trans. Circuit Theory*, Vol. CT-19, No. 2, pp. 189-194, March 1972.

4.7   E. Hofstetter, A. V. Oppenheim, and J. Siegel, "A New Technique for the Design of Nonrecursive Digital Filters," *Proc. 5th Annual Princeton Conf. Inf. Sci. Syst.*, pp. 64-72, 1971.

4.8   L. R. Rabiner, "The Design of Finite Impulse Response Digital Filters Using Linear Programming Techniques," *Bell Sys. Tech. J.*, Vol. 51, No. 6, pp. 1177-1198, July-August 1972.

4.9   J. H. McClellan, T. W. Parks, and L. R. Rabiner, "A Computer Program for Designing Optimum FIR Linear Phase Digital Filters," *IEEE Trans. Audio Electroacoust.*, Vol. AU-21, No. 6, pp. 506-526, December 1973.

4.10 D. W. Rorabacher, "Efficient FIR Filter Design for Sample Rate Reduction or Interpolation," *Proc. Int. Symp. Circuits Syst.*, April 1975.

4.11 M. G. Bellanger, J. L. Daguet, and G. P. Lepagnol, "Interpolation, Extrapolation, and Reduction of Computation Speed in Digital Filters," *IEEE Trans. Acoust. Speech Signal Process.*, Vol. ASSP-22, No. 4, pp. 231-235, August 1974.

4.12 M. G. Bellanger, "Computation Rate and Storage Estimation in Multirate Digital Filtering with Half-Band Filters," *IEEE Trans. Acoust. Speech Signal Process.*, Vol. ASSP-25, No. 4, pp. 344-346, August 1977.

4.13 G. Oetken and H. W. Schuessler, "On the Design of Digital Filters for Interpolation," *Arch. Elek. Ubertrag.*, Vol. 27, pp. 471-476, 1973.

4.14 G. Oetken, T. W. Parks and H. W. Schuessler, "New Results in the Design of Digital Interpolators," *IEEE Trans. Acoust. Speech Signal Process.*, Vol. ASSP-23, No. 3, pp. 301-309, June 1975. (See also G. Oetken, T. W. Parks, and H. W. Schüssler, "A Program for Digital Interpolator Design," in *Programs for Digital Signal Processing*, New York: pp. 8.1-1 to 8.1-6, 1979).

4.15 C. H. Lu and S. C. Gupta, "Optimal Design of Multirate Digital Filters with Application to Interpolation," *IEEE Trans. Circuits Syst.*, Vol. CAS-26, No. 3, pp. 160-165, March 1979.

4.16 J. Durbin, "Efficient Estimation of Parameters in Moving Average Models," *Biometrika*, Vol. 46, Pts. 1 and 2, pp. 306-316, 1959.

4.17 J. Durbin, "The Fitting of Time Series Models," *Rev. Inst. Int. Stat.*, Vol. 28, No. 3, pp. 233-243, 1960.

4.18 J. P. Burg, "Maximum Entropy Spectral Analysis," Ph.D. dissertation, Dept. Biophysics, Stanford Univ., Stanford, Calif., 1975.

4.19 G. Oetken, "A New Approach for the Design of Digital Interpolating Filters," *IEEE Trans. Acoust. Speech Signal Process.*, Vol. ASSP-27, No. 6, pp. 637-643, December 1979.

4.20 T. W. Parks and D. P. Kolba, "Interpolation Minimizing Maximum Normalized Error for Band-Limited Signals," *IEEE Trans. Acoust. Speech Signal Process.*, Vol. ASSP-26, No. 4, pp. 381-384, August 1978.

4.21 O. Herrmann, L. R. Rabiner, and D. S. Chan, "Practical Design Rules for Optimum Finite Impulse Response Lowpass Digital Filters," *Bell Syst. Tech. J.*, Vol. 52, No. 6, pp. 769-799, July-August 1973.

4.22 A. Polydoros and E. N. Protonotarios, "Digital Interpolation of Stochastic Signals," *Proc. Zurich Semin. Comm.*, pp. 349-367, March 1978.

4.23 R. W. Schafer and L. R. Rabiner, "A Digital Signal Processing Approach to Interpolation," *Proc. IEEE*, Vol. 61, No. 6, pp. 692-702, June 1973.

4.24 R. W. Hamming, *Numerical Methods for Scientists and Engineers*, New York: McGraw Hill, 1962.

4.25 M. G. Bellanger, G. Bonnerot, and M. Coudreuse, "Digital Filtering by Polyphase Network: Application to Sample-Rate Alteration and Filter Banks," *IEEE Trans. Acoust. Speech Signal Process.*, Vol. ASSP-24, pp. 109-114, April 1976.

4.26 P. Vary, "On the Design of Digital Filter Banks Based on a Modified Principle of Polyphase," *Arch. Elek. Ubertrag.*, Vol. AEU 33, pp. 293-300, July/August 1979.

4.27 H. G. Martinez and T. W. Parks, "A Class of Infinite-Duration Impulse Response Digital Filters for Sampling Rate Reduction," *IEEE Trans. Acoust. Speech Signal Process.*, ASSP-27, No. 2, pp. 154-162, April 1979.

4.28 H. G. Martinez and T. W. Parks, "Design of Recursive Digital Filters with Optimum Magnitude and Attenuation Poles on the Unit Circle," *IEEE Trans. Acoust. Speech Signal Process.*, Vol. ASSP-26, No. 2, pp. 150-156, April 1978.

4.29 H. Gockler, "Design of Recursive Polyphase Networks with Optimum Magnitude and Minimum Phase," *Signal Processing*, Vol. 3, No. 4, pp. 365-376, October 1981.

4.30 D. Pelloni, "Synthese von Digitalen Polyphasennetzwerken," AGEN-Mitt., Vol. 24, pp. 3-11, December 1977.

4.31 T. Saramaki, "A New Class of Linear-Phase FIR Filters for Decimation, Interpolation, and Narrow-Band Filtering," *Proc. IEEE Int. Symp. Circuits Syst.*, pp. 808-811, Rome, Italy, May 10-12, 1982.

5

Multistage Implementations
of Sampling Rate Conversion

5.0 INTRODUCTION

We have shown several times throughout this book that a general conceptual model for changing the sampling rate of a signal by the rational factor L/M is of the type shown in Fig. 5.1. It was shown in Chapter 2 that the processing involved in implementing this model could be viewed as a two-stage system: first interpolating the sequence by a factor of L, followed by a stage of decimation by a factor of M. We showed in Chapter 3 that the computational load in implementing this system (i.e., the digital filtering operations) could be efficiently performed at the lowest sampling rate of the system and in Chapter 4 we discussed filter designs for this single-stage approach. In this chapter we consider cascaded (multistage) implementations of these sampling rate conversion systems for even greater efficiencies in some cases. We will concentrate primarily on integer conversion ratios; however, the concepts can be readily applied to rational ratios as well, particularly at the last stage of conversion.

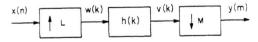

Figure 5.1 General structure for sampling rate conversion by a ratio of L/M.

The concept of using a series of stages to implement a sampling rate conversion system can be applied to the case of integer interpolators and decimators [5.1-5.4], as shown in Fig. 5.2. Consider first a system for decimating a signal by an integer factor of M, as shown in Fig. 5.2(a). We denote the original sampling frequency of the input signal, $x(n)$, as F_0, and the final sampling frequency of the decimated signal $y(m)$ as F_0/M. If the decimation ratio M can be factored into the product

$$M = \prod_{i=1}^{I} M_i \qquad (5.1)$$

where each M_i, $i = 1, 2, ..., I$, is an integer, we can express this network in the form shown in Fig. 5.2(b) (see Section 3.1.2). This structure, by itself, does not provide any inherent advantage over the structure of Fig. 5.2(a). However, if we modify the structure by introducing a lowpass filter between *each* of the sampling rate compressors, we obtain the structure of Fig. 5.2(c). This structure has the property that the sampling rate reduction occurs in a series of I stages, where each stage (shown within dashed boxes) is an *independent* decimation stage. Note that the intermediate sampling frequencies at the output of the stages are denoted as F_1, F_2, ..., F_I for stages $i = 1, 2, ..., I$, respectively.

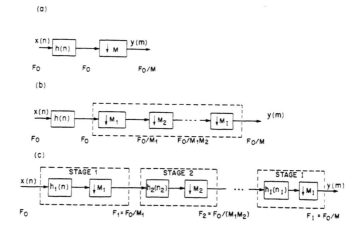

Figure 5.2 Steps in constructing a multistage decimator for decimation by a factor M.

Similarly, for an I-stage 1-to-L interpolator, if the overall interpolation rate, L, can be factored into the product

$$L = \prod_{i=1}^{I} L_i \qquad (5.2)$$

then the general single-stage interpolator structure of Fig. 5.3(a) can be converted into the multistage structure of Fig. 5.3(c). Again each of the stages within the structure of Fig. 5.3(b) is an *independent* interpolation stage. The reader should note that the stages of the interpolator structure in Fig. 5.3(c) are numbered in *reverse* order (i.e., the output stage is stage 1, and the input stage is stage *I*). This is often done for notational convenience to emphasize the transpose relationship between multistage decimators and interpolators (see Section 3.1.3). As will be seen later, an *I*-stage interpolator can be designed by designing an *I*-stage decimator and taking its transpose. Since the transposition operation reverses the direction of all signal paths in the structure, the output stage of the multistage decimator becomes the input stage of the multistage interpolator. Thus it is convenient, notationally, to number the interpolator stages in reverse order, as shown in Fig. 5.3. For the same reason the input sampling frequency of the multistage interpolator is labeled F_I and the output sampling frequency is labeled F_0. The input and output sampling frequencies for stage i of the multistage interpolator are F_i and F_{i-1}, respectively, as seen in Fig. 5.3. Because of this transpose relationship between multistage decimators and interpolators we will generally discuss only decimator designs in this chapter, with the understanding that the same operations can be applied to interpolators by the process of transposition.

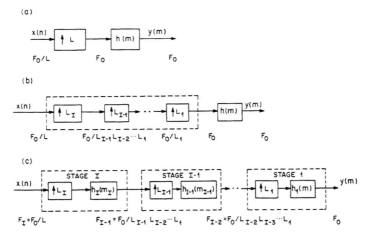

Figure 5.3 Steps in constructing a multistage interpolator for interpolation by a factor L.

The obvious question that arises from the discussion above is why consider such multistage structures. At first glance it would appear as if we are greatly increasing the overall computation (since we have inserted filters between each stage) of the structure. This, however, is precisely the opposite of what occurs in practice. The reasons for considering multistage structures, of the types shown in Figs. 5.2(c) and 5.3(c), are:

1. Significantly reduced computation to implement the system
2. Reduced storage in the system
3. Simplified filter design problem (i.e., wider normalized transition bands allowed at each stage)
4. Reduced finite-word-length effects (i.e., round-off noise and coefficient sensitivity) in the implementations of the digital filters

These structures however are not without some drawbacks. They include:

1. Increased control structure required to implement a multistage process
2. Difficulty in choosing the appropriate values of I of Eq. (5.1) or (5.2) and the best factors L_i (or M_i)

It is the purpose of this chapter to show why and how a multistage implementation of a sampling rate conversion system can be (and generally is) more efficient than the standard single-stage structure for the following cases:

1. $L \gg 1$ ($M = 1$) case 1
2. $M \gg 1$ ($L = 1$) case 2
3. $L/M \approx 1$ but $L \gg 1, M \gg 1$ case 3

Cases 1 and 2 are high-order interpolation and decimation systems, and case 3 is when a slight change in sampling rate is required (e.g., $L/M = 80/69$).

5.1 COMPUTATIONAL EFFICIENCY OF A TWO-STAGE STRUCTURE—A DESIGN EXAMPLE

Since the motivation for considering multistage implementations of sampling rate conversion systems is the potential reduction in computation, it is worthwhile presenting a simple design example that illustrates the manner in which the computational efficiency is achieved.

The design example is one in which a signal, $x(n)$, with sampling rate of 10,000 Hz, is to be decimated by a factor of $M = 100$ to give the signal $y(m)$ at a 100-Hz rate. Figure 5.4(a) shows the standard, single-stage, decimation network which implements the desired process. It is assumed that the passband of the signal is from 0 to 45 Hz, and that the band from 45 to 50 Hz is a transition band. We also assume that a passband ripple of 0.01 and a stopband ripple of 0.001 are required. Hence the specifications of the required lowpass filter are as shown in Fig. 5.4(b). We assume, for simplicity, that the design formula for FIR equiripple designs

$$N \approx \frac{D_\infty(\delta_p, \delta_s)}{\Delta F / F} \tag{5.3}$$

can be used to give the number of taps, N, of a symmetric FIR filter with maximum passband ripple δ_p, maximum stopband ripple δ_s, transition width ΔF, and sampling frequency F (an explanation of this form of the equation is given in Section 5.3.1). For the lowpass filter of Fig. 5.4(b) we have

$$\Delta F = 50 - 45 = 5 \text{ Hz}$$
$$F = 10,000 \text{ Hz}$$
$$\delta_p = 0.01$$
$$\delta_s = 0.001$$
$$D_\infty(\delta_p, \delta_s) = 2.54$$

giving, from Eq. (5.3), $N \approx 5080$. Let the overall computation, in multiplications per second (MPS) necessary to implement this system be denoted by R_T^* where the * denotes *multiplication* rate. [Later we will let R_T^+ denote total computation in additions per second (APS)]. Then it can be shown that

$$R_T^* \approx \frac{NF}{2M} = \frac{(5080)(10,000)}{2(100)} = 254,000 \quad \text{MPS}$$

that is, a total of 254,000 multiplications per second is required to implement the system of Fig. 5.4(a) [assuming an efficient decimator structure which takes advantage of symmetry of $h(k)$].

Consider now the two-stage implementation shown in Fig. 5.4(c). The first

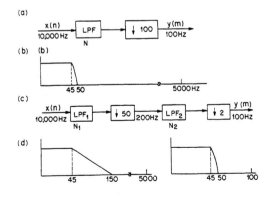

Figure 5.4 Simple example of a one-stage and a two-stage network for decimation by a factor of 100-to-1.

stage decimates the signal by a factor of 50,[1] and the second stage decimates the (already decimated) signal by a factor of 2, giving a total decimation factor of 100. The resulting filter specifications are illustrated in Fig. 5.4(d). For the first stage the passband of the N_1 tap filter, LPF_1, is from 0 to 45 Hz, but the transition band extends from 45 Hz to 150 Hz. Since the sampling rate at the output of the first stage is 200 Hz, the residual signal energy from 100 to 150 Hz gets aliased back into the range 50 to 100 Hz after decimation by the factor of 50. This aliased signal then gets removed in the second stage. For the second stage the passband of the N_2 tap filter, LPF_2, extends from 0 to 45 Hz and the transition band extends from 45 to 50 Hz. One other change in the filter specifications occurs because we are using a two-stage filtering operation. The passband ripple specification of each stage of the two-stage structure is reduced to $\delta_p/2$ (since each stage can theoretically add passband ripple to each preceding stage). The stopband ripple specification does not change since the cascade of two lowpass filters only reduces the stopband ripples. Hence the $D_\infty(\delta_p, \delta_s)$ function in the filter design equation becomes $D_\infty(\delta_p/2, \delta_s)$ for the filters in the two-stage implementation. Since the function $D_\infty(\delta_p, \delta_s)$ is relatively insensitive to factors of 2 in δ_p or δ_s, only slight changes occur in its value (from 2.54 to 2.76) due to this factor. For the specific example of Fig. 5.4(c) we get (for the first stage)

$$N_1 = \frac{2.76}{(150 - 45)/10,000} = 263 \quad \text{taps}$$

$$R_1 = \frac{N_1 F}{M_1} = \frac{(263)(10,000)}{50} = 52,600 \quad \text{MPS (not including symmetry)}$$

For the second stage we get

$$N_2 = \frac{2.76}{5/200} = 110.4 \quad \text{taps}$$

$$R_2 = \frac{(110.4)(200)}{2} = 11,000 \quad \text{MPS (not including symmetry)}$$

The total computation in MPS for the two-stage implementation, *assuming* the use of symmetry, is then

$$R_T^* = \frac{R_1}{2} + \frac{R_2}{2} = \frac{52,600 + 11,000}{2} = 31,800 \quad \text{MPS}$$

Thus a reduction in computation of almost 8 to 1 is achieved in the two-stage decimator over a single-stage decimation for this design example.

---

[1] We explain later in this chapter how the individual decimation factors of the stages are obtained.

It is easy to see where the reduction in computation comes from for the multistage decimator structure by examining Eq. (5.3). We see that the required filter orders are directly proportional to $D_\infty(\delta_p, \delta_s)$ and F, and inversely proportional to ΔF, the filter transition width. For the early stages of a multistage decimator, although the sampling rates are large, equivalently the transition widths are very large, thereby leading to relatively small values of filter length N. For the last stages of a multistage decimator, the transition width becomes small, but so does the sampling rate and the combination again leads to relatively small values of required filter lengths. We see from the analysis above that computation is kept low in each stage of the overall multistage structure.

As discussed earlier, a two-stage interpolator with a 1-to-100 change in sampling rate can be designed by transposing the decimator design of Fig. 5.4(c). The first stage (stage 2) increases the sampling rate by a factor of $L_2 = 2$ (i.e., from 100 Hz to 200 Hz) and the second stage (stage 1) increases the sampling rate by the factor $L_1 = 50$ (from 200 Hz to 10,000 Hz). As in the decimator, it can be shown that this two-stage design is more efficient in computation than a single-stage interpolator by a factor of approximately 8.

The simple examples presented above are by no means a complete picture of the capabilities and sophistication that can be found in multistage structures for sampling rate conversion. They are merely intended to show why such structures are of fundamental importance for many practical systems in which sampling rate conversion is required. In the next section we set up a formal structure for dealing with multistage sampling rate conversion networks, and show how it can be used in a variety of implementations.

5.2 TERMINOLOGY AND FILTER REQUIREMENTS FOR MULTISTAGE DESIGNS

There are a large number of parameters and trade-offs involved in the design of multistage decimator and interpolator systems, including the choice of the number of stages I, the decimation or interpolation ratios M_i or L_i at each stage, the filter requirements, and the actual filter designs at each stage. Several general design procedures have been suggested for choosing these parameters and they are discussed in more detail in Sections 5.3 to 5.6. In this section we define the terminology used to specify the filter requirements for each stage of a multistage system.

5.2.1 Overall Filter Requirements

Consider the I-stage decimator of Fig. 5.2(c), where the total decimation rate of the system is M. The sampling rate (frequency) at the output of the ith stage $(i = 1, 2, ..., I)$ is

$$F_i = \frac{F_{i-1}}{M_i}, \quad i = 1, 2, ..., I \tag{5.4}$$

with initial input sampling frequency F_0, and final output sampling frequency F_I, where

$$F_I = \frac{F_0}{\prod_{i=1}^{I} M_i} = \frac{F_0}{M} \tag{5.5}$$

Any of the structures discussed in Chapter 3 and any of the filter designs in Chapter 4 can be used for each stage of the network. Using the ideas developed in Chapter 4, we define the useful frequency range of the output signal $y(m)$ as

$$0 \leqslant f \leqslant F_p \qquad \text{passband} \tag{5.6a}$$

$$F_p \leqslant f \leqslant F_s \leqslant \frac{F_I}{2} \quad \text{transition band} \tag{5.6b}$$

where f, F_p, and F_s are analog frequencies (i.e., they are not normalized to any sampling rate). The filter specifications of Eq. (5.6) are illustrated in Fig. 5.5(a). Because of the large number of sampling rates involved in multistage systems, we will generally refer to frequencies in this chapter in analog terms rather than in digital terms, to avoid confusion.

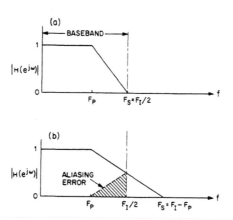

Figure 5.5 (a) Filter requirements such that no aliasing occurs in the baseband of $y(m)$; (b) alternative design which allows aliasing above F_p.

Note that in the specification above we have restricted the final stopband edge, F_s, to be less than or equal to $F_I/2$. [In Fig. 5.5(a) we have drawn it so that $F_s = F_I/2$.] This protects the entire baseband (from 0 to $F_I/2$) of $y(m)$ from aliasing. In *each* stage of processing this baseband must be protected since once any significant aliasing occurs, it cannot be removed. Any other frequency band can be aliased in an early stage since subsequent processing will remove any signal components from the band, as seen in the simple design example of Section 5.1.

In some applications it may be acceptable to allow some aliasing to occur in the upper region of the transition band of $y(m)$ above F_p. In this case we can choose $F_I/2 \leqslant F_s \leqslant F_I - F_p$ as shown in Fig. 5.5(b). (Hence we have drawn the case where $F_s = F_I - F_p$.) This permits the use of a filter with a wider transition band (up to twice as wide) in the final stage, and therefore the filter order is reduced to about half that required when no aliasing can occur. In most of our discussions, however, we assume that the entire baseband of the signal $y(m)$ is to be protected and therefore the condition in Eq. (5.6b) and Fig. 5.5(a) applies.

In the case of an I-stage interpolator [see Fig. 5.3(c)], the sampling rate F_{i-1} at the output of the ith stage is

$$F_{i-1} = F_i L_i, \quad i = I, I - 1, ..., 2, 1 \qquad (5.7a)$$

and

$$F_0 = F_I \prod_{i=1}^{I} L_i = LF_I \qquad (5.7b)$$

where F_I is the initial input sampling frequency and F_0 is the final output sampling frequency. [Note that Eqs. (5.7a) and (5.7b) are the same as Eqs. (5.4) and (5.5), respectively, with M replaced by L.] The frequency bands of interest for the interpolator filters are

$$0 \leqslant f \leqslant F_p \qquad \text{passband} \qquad (5.8a)$$

$$F_p \leqslant f \leqslant F_s \leqslant \frac{F_I}{2} \quad \text{transition band} \qquad (5.8b)$$

and any harmonic images above $F_I/2$ must be removed by the filtering process.

The overall filter requirements above impose specific restrictions on the filter specifications for individual stages in the multistage designs. These restrictions are affected by the choice of F_p and F_s and the frequencies F_i, $i = 0, 1, 2, ..., I$. For convenience we define a dimensionless transition bandwidth measure Δf as

$$\Delta f = \frac{F_s - F_p}{F_s} \qquad (5.9)$$

This definition will be used throughout this chapter. As Δf approaches 0, the width of the transition band becomes small, and as Δf approaches 1.0, the width of the transition band becomes large (see Fig. 5.5). Hence Δf is a measure of the complexity of the filters needed to ensure that no aliasing (decimator) or imaging (interpolator) occurs in the multistage design (particularly in stage I of the design).

5.2.2 Lowpass Filter Requirements for Individual Stages

If we now examine the lowpass filter frequency regions for the individual stages of the I-stage decimator of Fig. 5.6(a) we get results of the type shown in Figs. 5.6(b)-(d). For the first stage the passband is defined from

$$0 \leqslant f \leqslant F_p \quad \text{stage 1 passband} \tag{5.10a}$$

and the transition band is from

$$F_p \leqslant f \leqslant F_1 - F_s \quad \text{stage 1 transition band} \tag{5.10b}$$

Signal energy in the transition band in this stage will alias back upon itself (after the decimation by M_1) only from $f = F_s$ up to $f = F_1/2$; hence the baseband 0 to F_s is protected against aliasing. The stopband of the stage 1 lowpass filter is from

$$F_1 - F_s \leqslant f \leqslant \frac{F_0}{2} \quad \text{stage 1 stopband} \tag{5.10c}$$

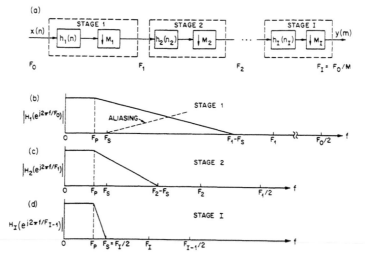

Figure 5.6 Signal processing operations and filter specifications for an I-stage decimator.

For the second stage of the system, the passband and stopband regions of the lowpass filter are

$$0 \leqslant f \leqslant F_p \quad \text{stage 2 passband} \qquad (5.11a)$$

$$F_2 - F_s \leqslant f \leqslant \frac{F_1}{2} \quad \text{stage 2 stopband} \qquad (5.11b)$$

as shown in Fig. 5.6(c). Again the band from $0 \leqslant f \leqslant F_s$ is protected from aliasing in the decimation stage.

For the ith stage of the system, the lowpass filter passband and stopband regions are

$$0 \leqslant f \leqslant F_p \qquad \text{stage } i \text{ passband} \qquad (5.12a)$$

$$F_i - F_s \leqslant f \leqslant \frac{F_{i-1}}{2} \quad \text{stage } i \text{ stopband} \qquad (5.12b)$$

It is readily seen from Eqs. (5.4), (5.6), and (5.12) and Fig. 5.6(d) that, for the last stage, the transition band of the lowpass filter is the same as the transition band of the one-stage implementation filter. However, the sampling rate of the system is substantially reduced in most cases.

Figure 5.7 illustrates the lowpass filter passband and transition band regions for the case of an I-stage interpolator. At any stage i ($i=I, I-1, ..., 2, 1$), the input sampling rate is F_i and the output sampling rate is F_{i-1} and they are related according to Eq. (5.7a). The process of sampling rate increase by L_i introduces harmonic images of the baseband spectrum centered at multiples of F_i. To remove

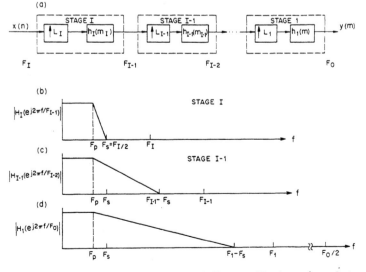

Figure 5.7 Signal processing operations and filter specifications of an I-stage interpolator.

these images and retain the baseband signal, the filter $h_i(m_i)$ must filter the signal over the bands

$$0 \leqslant f \leqslant F_p \qquad \text{stage } i \text{ passband} \qquad (5.13a)$$

$$F_i - F_s \leqslant f \leqslant \frac{F_{i-1}}{2} \qquad \text{stage } i \text{ stopband} \qquad (5.13b)$$

As discussed previously, these filter bands are identical to those for the decimator [Eqs. (5.12)] because of the transpose relationship between these two systems and the way in which we numbered the stages.

5.2.3 Filter Specifications for Individual Stages That Include Don't-Care Bands

In the preceding section we specified the filter frequency bands of the individual stages in terms of lowpass filter requirements (i.e., passband, transition band, and stopband regions). These lowpass characteristics are often used in practice because simple, straightforward design techniques are readily available for lowpass designs. These filter characteristics are, however, more restrictive than necessary in early stages of multistage decimators (or final stages of multistage interpolators). In these stages the frequency band 0 to F_s often represents a relatively small percentage of the output sampling rate of that stage (or the input sampling rate of an interpolator stage) (i.e., $F_s \ll F_i$). When this condition occurs, the stopband region can be divided into a set of true stopband and ϕ (don't-care) regions.

To illustrate this point, consider the filtering requirements for the ith stage of an I-stage interpolator (where $F_s \ll F_i$), as shown in Fig. 5.8(a). Figure 5.8(b) shows the spectrum of the input signal $x_i(n)$ and Fig. 5.8(c) shows the spectrum of the signal $w_i(m)$ after sampling rate expansion by the factor L_i. The role of the filter $h_i(m)$ is to remove the harmonic images of the baseband in $w_i(m)$ located at F_i, $2F_i$, $3F_i$, ..., as seen by the characteristic in Fig. 5.8(d). In between these images the filter response can be unconstrained and these regions are called don't-care or ϕ bands.

As F_s becomes much smaller than F_i, the widths of the ϕ bands become wider and the required stopband regions become narrower. In the limit as $F_s \rightarrow 0$, the required filter response approaches that of a "comb filter," as will be discussed later. Also, recall from Section 4.3.3 that by taking advantage of the ϕ bands, the order of the required filters can be reduced significantly when F_s becomes small compared to F_i. In the case of minimum mean-square-error filter designs, the ϕ bands are automatically taken advantage of in the design (see Chapter 4). We discuss the effects of incorporating the ϕ bands in the filter design later in this chapter.

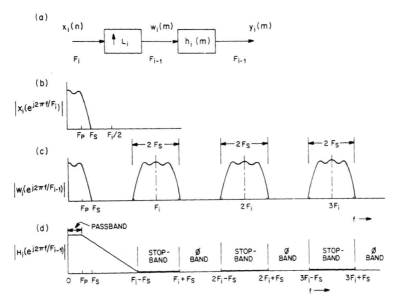

Figure 5.8 General multiple stopband specifications for stage i of an I-stage interpolator.

5.2.4 Passband and Stopband Tolerance Requirements

Up to now we have been concerned solely with the regions of definition of the individual filter frequency bands. Another consideration in the design equations is the magnitude specifications on the filter response in each of the frequency bands. If we denote the canonic one-stage lowpass filter response as $H_{\text{LP}}(e^{j2\pi f/F})$, we have

$$1 - \delta_p \leqslant |H_{\text{LP}}(e^{j2\pi f/F})| \leqslant 1 + \delta_p \quad f \in \text{passband} \qquad (5.14\text{a})$$

$$0 \leqslant |H_{\text{LP}}(e^{j2\pi f/F})| \leqslant \delta_s \qquad f \in \text{stopband} \qquad (5.14\text{b})$$

where δ_p is the maximum deviation from 1.0 in the passband, δ_s is the maximum deviation from 0.0 in the stopband, and F is the sampling rate of the filter.[2] For an I-stage implementation, the overall passband magnitude response is

$$H_0(e^{j2\pi f/F_0}) = \prod_{i=1}^{I} H_i(e^{j2\pi f/F_{i-1}}) \quad f \in \text{passband} \qquad (5.15)$$

where $H_i(e^{j2\pi f/F_{i-1}})$ is the response of the ith lowpass filter, and $H_0(e^{j2\pi f/F_0})$ is the

---

[2] Recall also that for interpolators an additional scale factor of L_i is associated with the filters. For convenience we will ignore this factor in this discussion.

overall response. Equation (5.15) is valid in the passband since each filter response is well defined for the common passband range. If each lowpass filter individually satisfied Eq. (5.14a) in the passband, and if the responses lined up precisely in frequency, we would have

$$(1 - \delta_p)^I \leqslant |H_0(e^{j2\pi f/F_0})| \leqslant (1 + \delta_p)^I \quad f \in \text{passband} \qquad (5.16a)$$

or (for small δ_p)

$$1 - I\delta_p \leqslant |H_0(e^{j2\pi f/F_0})| \leqslant 1 + I\delta_p \quad f \in \text{passband} \qquad (5.16b)$$

Hence to compensate for this "growth" of the passband ripple with the number of stages, the design specification of the ith stage for each lowpass filter must be modified to

$$1 - \frac{\delta_p}{I} \leqslant |H_i(e^{j2\pi f/F_{i-1}})| \leqslant 1 + \frac{\delta_p}{I} \quad f \in \text{passband} \qquad (5.17)$$

The constraint of Eq. (5.17) guarantees that the desired passband ripple constraints [Eq. (5.14a)] are met by the composite filter characteristic of the I-stage network.

For the stopband no changes need be made in the ripple specification [Eq. (5.14b)] since δ_s is generally small and the cascade of stages reduces the stopband ripple when the frequency bands of two stages overlap.

5.2.5 Design Considerations

Given the network structure of Fig. 5.2c for the I-stage decimator [or Fig. 5.3(c) for the interpolator], we will be interested in the specification of the following:

1. The number of stages, I, to realize an overall decimation factor of M
2. The choice of decimation factors, M_i, $i = 1, 2, ..., I$, that are appropriate for the chosen implementation
3. The types of digital filters used in each stage of the structure (e.g., FIR versus IIR designs, equiripple versus mean-square-error versus specialized designs)
4. The structure used to implement the filter chosen for each stage
5. The required filter order (impulse response duration) in each stage of the structure
6. The resulting amount of computation and storage required for each stage, and for the overall structure

As in most signal processing design problems, there are a number of factors that influence each choice, and it is not a simple matter to make any choice over all others. However, it is the purpose of this chapter to highlight the kinds of trade offs that can be made and to illustrate the effects of some of these trade-offs in practical design examples.

In Sections 5.3 to 5.5 we discuss three distinct approaches to the design of FIR multistage structures that take these considerations into account. Each approach is based on a somewhat different design philosophy and has slightly different advantages and disadvantages. In practice, however design trade-offs are often highly applications dependent, and combinations of these methods may give the best results. In Section 5.6 we consider, briefly, multistage IIR designs, and in Section 5.7 we discuss a practical method of implementation of multistage systems. Although we refer only to decimator designs in the following discussion, the reader is again reminded that the same techniques apply equally to interpolators through the process of transposition.

5.3 MULTISTAGE FIR DESIGNS BASED ON AN OPTIMIZATION PROCEDURE

The first approach that we consider for designing a multistage decimator is based on setting up the problem as an optimization problem [5.3-5.5]. Given that the filter parameters F_0, M, F_p, F_s, δ_p and δ_s are specified, the objective is to optimize the efficiency of the design (i.e., minimize computation, storage, etc.) by varying the remaining free parameters M_i, $i = 1, 2, ..., I$, and the total number of stages I. The design procedure is performed by finding the most efficient solution (i.e., the best choice of values M_i) for each value of I, $I = 1, 2, 3, ...$, and then selecting that value of I that gives the best overall solution.

The objective function to be minimized is an analytical measure of the efficiency of the design. One meaningful measure of this is the total number of multiplications per second (MPS) that must be performed. If this value is denoted as R_T^*, and R_i is the number of MPS for stage i, then[3]

$$R_T^* = \sum_{i=1}^{I} R_i \qquad (5.18)$$

where Eq. (5.18) does not include the use of symmetry of $h(k)$. Letting N_i denote the FIR filter length required for stage i, R_i can be expressed in the form (using the efficient structures in Chapter 3)

$$R_i = N_i F_i \qquad (5.19)$$

If the FIR filters are symmetrical, the total number of multiplications can be reduced by a factor of 2, or if they contain regularly spaced, zero-valued

---

[3] We use the notation R_T^* to denote total computation in terms of multiplications per second (MPS), and R_T^+ to denote total computation in terms of additions per second (APS).

coefficients (as in half-band filters), these properties can be utilized in the design, as will be seen later. For our purposes we assume that Eq. (5.19) applies.

Another measure of design complexity is the amount of memory required for coefficient and data storage. Since this quantity is approximately proportional to the sum of the filter lengths in each of the stages, a second objective criterion that we will consider is total storage, N_T, defined as

$$N_T = \sum_{i=1}^{I} N_i \tag{5.20}$$

In the optimization procedure we then wish to solve for

$$\widetilde{R}_T^* = \min_{I,\{M_i\}} (R_T^*) \tag{5.21a}$$

or

$$\widetilde{N}_T = \min_{I,\{M_i\}} (N_T) \tag{5.21b}$$

where, for convenience, the ratios M_i, $i = 1, 2, ..., I$, are allowed to be continuous variables. Once these ideal values of M_i are found, the nearest appropriate choice of integer values can be chosen such that Eq. (5.1) is satisfied.

To formulate the problem we need first to express the required filter lengths N_i for each stage i in terms of the passband and stopband requirements discussed above. This is discussed in Section 5.3.1. We then derive expressions for the objective function R_T^* and N_T in Sections 5.3.2 and 5.3.3, respectively, and give analytic solutions for the case of $I = 2$. In Section 5.3.4 we discuss a computer-aided optimization procedure for obtaining solutions when I is greater than 2 and discuss the results of these simulations. Finally, in Section 5.3.5 we give a design example. Later, in Sections 5.6 and 5.7, we consider the problem of multistage designs with IIR filters and some specific implementation issues.

5.3.1 Analytic Expressions for the Required Filter Order for Each Stage of a Multistage Design

As seen from the discussion above, in order to formulate the optimization procedure it is necessary to express the filter order in each stage in terms of the filter requirements for that stage. Analytic expressions for many of the filter designs discussed in Chapter 4 are not available. This is especially the case for filters with the general multiple stopband specifications discussed in Section 5.2.3.

However, for the case of equiripple designs and the lowpass characteristics discussed in Section 5.2.2, a good approximate design formula is available and therefore our optimization will be based on this particular class of designs. The design equation is (see Chapter 4)

$$N \simeq \frac{D_\infty(\delta_p, \delta_s)}{\Delta F/F} - \frac{f(\delta_p, \delta_s)\Delta F}{F} \qquad (5.22a)$$

where $D_\infty(\delta_p, \delta_s)$ and $f(\delta_p, \delta_s)$ are as defined in Chapter 4, ΔF is the transition bandwidth of the filter (expressed in analog terms), and F is the sampling rate of the filter.

In the particular application that we are considering, the second term in Eq. (5.22a) is generally insignificant compared to the first term, especially when ΔF is much smaller than F [5.4, 5.5]. Therefore, for convenience, it is generally sufficient to simplify Eq. (5.22a) to the form

$$N \simeq \frac{D_\infty(\delta_p, \delta_s)}{\Delta F/F} \qquad (5.22b)$$

Applying the lowpass filter requirements in Sections 5.2.2 and 5.2.4 to Eq. (5.22b) then gives the desired filter length N_i for stage i in the form

$$N_i \simeq \frac{D_\infty(\delta_p/I, \delta_s)F_{i-1}}{F_i - F_s - F_p}, \quad i = 1, 2, ..., I \qquad (5.23)$$

5.3.2 Design Criteria Based on Multiplication Rate

By substituting Eq. (5.23) into Eqs. (5.18) and (5.19), the overall computation rate of the multistage system can be expressed as

$$R_T^* = D_\infty\left[\frac{\delta_p}{I}, \delta_s\right] \sum_{i=1}^{I} \frac{F_{i-1}F_i}{F_i - F_s - F_p} \qquad (5.24)$$

By careful application of Eqs. (5.1), (5.4), (5.5), (5.6), and (5.9) and assuming that

$$F_s = \frac{F_I}{2} \qquad (5.25)$$

it can be shown [5.4] that Eq. (5.24) can be expressed in the form

$$R_T^* = D_\infty\left[\frac{\delta_p}{I}, \delta_s\right] \cdot F_0 \cdot S \quad \text{MPS} \qquad (5.26a)$$

where

$$S = S(\Delta f, M, I; M_1, M_2, ..., M_{I-1})$$

$$= \frac{2}{\Delta f \prod_{i=1}^{I-1} M_i} + \sum_{i=1}^{I-1} \frac{M_i}{\left[\prod_{j=1}^{i} M_j\right]\left[1 - \frac{2 - \Delta f}{2M}\left[\prod_{j=1}^{i} M_j\right]\right]} \qquad (5.26b)$$

This form of the expression for R_T^* is particularly revealing because it expresses R_T^* only in terms of the specified design parameters δ_p, δ_s, F_0, Δf, M, and I [note that F_p and F_s are absorbed in Δf in Eq. (5.9)] and in terms of the $I - 1$ *independent* design variables $M_1, M_2, ..., M_{I-1}$. Thus the expression for R_T^* is in the proper form for an algorithmic optimization procedure, as will be seen shortly. It is important to recognize that R_T^*, in this form, is not a function of the final decimation ratio M_I. This is because once the values $M_1, M_2, ..., M_{I-1}$ and M are specified, M_I can be determined [from Eq. (5.1)]. Thus M_I is a dependent variable, not an independent variable.

We can now proceed to minimize the objective function R_T^* as a function of the independent variables. For $I = 1$ there is no flexibility in the design since $M_1 = M$. For $I = 2$ we can minimize R_T^* as a function of M_1 by taking the derivative of Eq. (5.26) with respect to M_1 and setting it to zero. This yields a quadratic function in M_1 and leads to two solutions, only one of which is valid and is of the form [5.4]

$$M_{1_{opt}} = \frac{2M[1 - \sqrt{M\Delta f/(2 - \Delta f)}\ ]}{2 - \Delta f(M + 1)}, \qquad I = 2$$

$$= \frac{2M}{(2 - \Delta f) + \sqrt{2M\Delta f - M(\Delta f)^2}}, \quad I = 2 \qquad (5.27a)$$

$$M_{2_{opt}} = \frac{M}{M_{1_{opt}}}, \qquad\qquad\qquad I = 2 \qquad (5.27b)$$

For I greater than 2, this analytic approach is not possible and we must resort to more powerful computer-aided optimization techniques [5.4]. The results of this approach, including design charts, are discussed more fully in Section 5.3.4.

5.3.3 Design Criteria Based on Storage Requirements

Following a similar procedure to the one described above, we can generate designs that minimize the objective function N_T rather than R_T^*. By substituting Eq. (5.23) into Eq. (5.20), we get

$$N_T = D_\infty\left(\frac{\delta_p}{I}, \delta_s\right) \sum_{i=1}^{I} \frac{F_{i-1}}{F_i - F_s - F_p} \qquad (5.28)$$

and with the aid of Eqs. (5.1), (5.4), (5.5), (5.9), and (5.25) we can express N_T in the form [5.5]

$$N_T = D_\infty\left(\frac{\delta_p}{I}, \delta_s\right) T \qquad (5.29a)$$

where

$$T = T(\Delta f, M, I; M_1, M_2, ..., M_{I-1})$$

$$= \frac{2M}{\Delta f \displaystyle\prod_{i=1}^{I-1} M_i} + \sum_{i=1}^{I-1} \frac{M_i}{1 - [(2 - \Delta f)/2M]\displaystyle\prod_{j=1}^{i} M_j} \qquad (5.29b)$$

Again, this expression for N_T is the proper form for an optimization procedure since N_T is expressed only in terms of the essential design parameters δ_p, δ_s, F_0, Δf, M, and I, and in terms of the $I - 1$ *independent* design variables M_1, M_2, ..., M_{I-1}.

For the case $I = 2$, an analytic solution can again be obtained by taking the derivative of N_T with respect to M_1 and setting it to zero. This leads to

$$M_{1_\text{opt}} = \frac{2M}{(2 - \Delta f) + \sqrt{2M\Delta f}}, \quad I = 2 \qquad (5.30a)$$

$$M_{2_\text{opt}} = \frac{M}{M_{1_\text{opt}}}, \qquad\qquad I = 2 \qquad (5.30b)$$

For I greater than 2, a computer optimization routine must be used to find the minimum of N_T with respect to the independent variables, as discussed in the next section.

5.3.4 Design Curves Based on Computer-Aided Optimization

In the procedure described above for minimizing the objective functions R_T^* or N_T, several interesting observations can be made. From Eq. (5.26) it is seen that R_T^* is basically a product of three factors, F_0, $D_\infty(\delta_p/I, \delta_s)$ (which is a function of δ_p, δ_s, and I), and S (which is a function of Δf, M, I, and the independent design parameters M_i, $i = 1, 2, ..., I - 1$). Similarly, from Eqs. (5.29) it is seen that N_T

is a product of two factors $D_\infty(\delta_p/I, \delta_s)$ and T (which is a function of Δf, M, I, and M_i, $i = 1, 2, ..., I - 1$). This factorization property makes the problem much simpler to deal with and to grasp intuitively. For example, we can observe the effects that δ_p and δ_s have on R_T^* or N_T by examining the behavior of $D_\infty(\delta_p/I, \delta_s)$. We can also see from the curves of $D_\infty(\delta_p, \delta_s)$ in Chapter 4 (Fig. 4.11) that $D_\infty(\delta_p/I, \delta_s)$ is a very weak function of I and thus this factor has very little effect on the choice of I in minimizing N_T or R_T^* as a function of I. For example, in going from a one-stage ($I = 1$) deign to a two-stage ($I = 2$) design, $D_\infty(\delta_p/I, \delta_s)$ typically changes by a factor of approximately 10%.

In contrast, the factors S in R_T^* or T in N_T are very strong functions of I, as well as the design parameters Δf and M, and they are independent of the parameters δ_p and δ_s. We can therefore minimize R_T^* or N_T as a function of M_i, $i = 1, 2, ..., I - 1$ for each I by minimizing only the respective factors S or T. These solutions can then be plotted as a function of the design variables Δf and M.

Figure 5.9 illustrates the basic algorithmic approach for minimizing the functions S or T with respect to M_i, $i = 1, 2, ..., I - 1$, with an optimization procedure such as the Hooke and Jeeves algorithm [5.4-5.7]. An initial guess for the M_i values is made first to start the procedure. The algorithm then systematically searches the space of M_i values in an iterative manner until the minimum of S (or T) is found. The Hooke and Jeeves algorithm is well suited to this problem. It does not require the evaluation of derivatives and it does not generally get trapped in local minima [5.4].

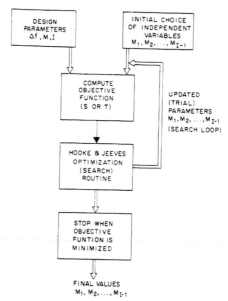

Figure 5.9 Minimization of S (or T) with the Hooke and Jeeves optimization routine.

Design curves based on this optimization procedure for minimum R_T^* designs can be found in Ref. 5.4 and similar curves for minimum N_T designs can be found in Ref. 5.5. It turns out that designs based on minimizing N_T also result in designs that are essentially minimized for R_T^* as well. This occurs because minima, for the function S (or R_T^*) are relatively broad minima whereas minima for the function T (or N_T) are slightly narrower. Therefore, the designs for minimized N_T are preferred since they cover both cases. The fact that essentially the same ideal values of M_i minimize both functions R_T^* and N_T can be seen for the case $I = 2$ by comparing the analytical solutions in Eqs. (5.27) and (5.30). The only difference in these two expressions is the $(\Delta f)^2$ term in Eq. (5.27a), which rapidly becomes negligible as Δf becomes small.

Figures 5.10 to 5.13 illustrate four examples of design curves based on the optimization procedure described above for the cases $\Delta f = 0.2, 0.1, 0.05,$ and 0.02 (i.e., transition bandwidths of 20, 10, 5, and 2% of the baseband, respectively). Figures 5.10(a) to 5.13(a) show plots of minimized values of S (called $\widetilde{S}$) in Eq. (5.26) as a function of M for each value of I. Figures 5.10(b) to 5.13(b) show similar minimized values of T (called $\widetilde{T}$) in Eqs. (5.29). Finally, Figs. 5.10(c) to 5.13(c) show ideal values of M_i, $i = 1, 2, ..., I$ as a function of M for each I, which result in the foregoing minimized values of S and T. Several important properties of this design procedure can be seen from these figures and the discussion above. They include:

1. Both the minimization of computation rate (proportional to $\widetilde{S}$) and storage (proportional to $\widetilde{T}$) can be achieved with essentially the same choice of ratios M_i, $i = 1, 2, ..., I$, which minimize T for a given I. (This result was discussed above.)

2. In terms of minimizing the computation rate [see Eq. 5.26(a) and the curves for minimum S in Figs. 5.10(a) to 5.13(a)], most of the gain in efficiency in multistage designs is achieved when going from a one-stage to a two-stage design. Only small additional gains are achieved when going from two-stage to three- or four-stage designs.

3. For minimizing the storage requirements, significant reductions in storage can still be achieved in going from two to three or four stages for large M, although the largest decrease is obtained in going from one to two stages.

4. The gain in efficiency (either in computation or storage) in going from one to I stages increases dramatically as Δf gets smaller and as M gets larger. For example, a computational reduction of about 80 to 1 and a reduction in storage requirements of about 450 to 1 can be achieved by using a three-stage structure instead of a one-stage structure in implementing a 1000-to-1 decimator (or interpolator) with a normalized transition width of $\Delta f = 0.02$ [see Figs. 5.13(a) and (b)].

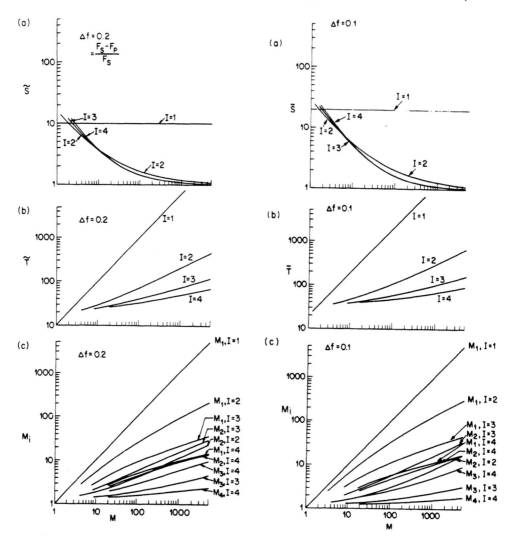

Figure 5.10 Design curves for multistage FIR decimators (or interpolators) with minimized values of S and T and ideal values of M_i, $i = 1, 2, ..., I$, and $\Delta f = 0.2$.

Figure 5.11 Design curves for multistage FIR decimators or interpolators for $\Delta f = 0.1$.

5.   The decimation ratios of an I-stage minimized design follow the relation

$$M_1 > M_2 > \cdots > M_I$$

as anticipated from the discussion early in this chapter.

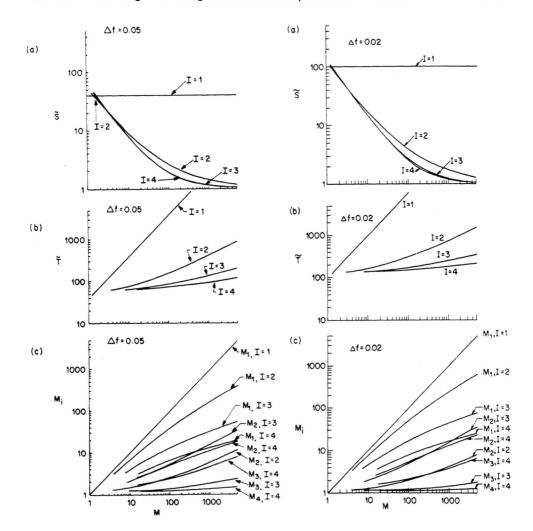

Figure 5.12 Design curves for multistage FIR decimators or interpolators for $\Delta f = 0.05$.

Figure 5.13 Design curves for multistage FIR decimators or interpolators for $\Delta f = 0.02$.

6. The required computation rate or storage requirements of a minimized I-stage design are relatively insensitive to small changes in the M_i values for each stage. Thus the nearest integer values for M_i, $i = 1, 2, ..., I$, can generally be used with little loss in efficiency.

7. The design curves above can also be applied to the design of decimators or interpolators for a minimized computation rate (but not storage) in which the overall conversion factor is a ratio of integers M/L or where some of the stages have conversion ratios of the form M_i/L_i, where M_i and L_i are

are integers [5.4]. In this case the structure in Fig. 5.1 is used in those stages and the nearest ratios of integers are still obtained from the design curves of Figs. 5.10(c) to 5.13(c).

8. The choices of ideal M_i values in Figs. 5.10(c) to 5.13(c) are valid for any class of filter designs in which the length of the filter is inversely proportional to the width of the transition band (i.e., $N \propto 1/\Delta F$). However, they cannot account for specific kinds of filter designs such as half-band designs, comb filter designs, or designs that take advantage of don't-care bands. These issues are discussed in the next sections.

9. The design curves are unchanged if we assume that all the filters N_i, $i = 1, 2, ..., I$, are symmetrical and that this symmetry is used to reduce the total number of multiplies. In this case the total number of multiplications/sec is approximately $R_T^*/2$; however, the total number of additions/sec, R_T^+, is still equal to R_T^*.

5.3.5 Application of the Design Curves and Practical Considerations

The design curves in Figs. 5.10 to 5.13 are intended only to be used as a guideline in selecting practical integer (or ratio of integer) values of M_i, $i = 1, 2, ..., I$, and the number of stages I. Once practical values of M_i are selected, they are applied to Eqs. (5.4) [or (5.7)], (5.12), (5.13) or the multiple stopband conditions in Fig. 5.8(d) to obtain the cutoff frequency requirements for each stage. The equiripple filter design techniques discussed in Chapter 4 can then be applied to design the appropriate filters, and any of the structures in Chapter 3 can be used for their implementations. Alternatively, any other filter designs that can be implemented more efficiently than the equiripple designs and still meet the same frequency requirements can be applied to any of the stages for additional efficiency. To illustrate the application of the ideas of this section, we present a simple design example.

Design Example 1

Consider the design of an I-stage structure to decimate a signal by a factor of $M = 64$, with overall filter specifications: $\delta_p = 0.01$, $\delta_s = 0.001$, $F_p = 0.45$ Hz, and $F_s = 0.5$ Hz. The input sampling frequency is $F_0 = 64$ Hz, and the output sampling rate is $F_I = F_0/64 = 1.0$ Hz.

Solution. For an $I = 1$-stage implementation we get

$$D_\infty(\delta_p, \delta_s) = D_\infty(0.01, 0.001) = 2.54$$

$$N_T = N_1 = \frac{D_\infty(\delta_p, \delta_s)}{(F_s - F_p)/F_0} = 3251 \quad \text{taps}$$

$$R_T^* = R_T^+ = R_1 = N_1 F_1 = (3251)(1) = 3251 \quad \text{MPS}$$

If we assume that the symmetry of the filter is used in the implementation, we then get

$$R_T^+ = R_1 = 3251 \quad \text{APS}$$

$$R_T^* = \frac{R_1}{2} = 1625 \quad \text{MPS}$$

where we have denoted the number of additions per second as APS. For an $I = 2$ stage design, we can determine the ideal choices of M_1 and M_2 from Eqs. (5.30). Alternatively, noting that $\Delta f = (0.5 - 0.45)/0.5 = 0.1$, we may use the appropriate design curves for $\Delta f = 0.1$ in Fig. 5.11(c). Using either method we get the ideal values $\widetilde{M}_1 \simeq 23.4$ and $\widetilde{M}_2 \simeq 2.74$. The nearest choice of integer values such that $M_1 M_2 = 64$ is either $(M_1 = 32, M_2 = 2)$ or $(M_1 = 16, M_2 = 4)$. We will use the second choice in this example. This leads to the design parameters:

$$D_\infty \left[\frac{\delta_p}{2}, \delta_s \right] = 2.76$$

$$N_1 = \frac{D_\infty(\delta_p/2, \delta_s) F_0}{F_1 - F_s - F_p} = 58 \quad \text{taps}$$

$$N_2 = \frac{D_\infty(\delta_p/2, \delta_s) F_1}{F_2 - F_s - F_p} = 221 \quad \text{taps}$$

$$N_T = N_1 + N_2 = 279 \quad \text{taps}$$

$$R_1 = N_1 F_1 = (58)(4) = 232 \quad \text{MPS}$$

$$R_2 = N_2 F_2 = (221)(1) = 221 \quad \text{MPS}$$

If symmetry is used in the implementation, we get for the two-stage design,

$$R_T^+ = R_1 + R_2 = 453 \quad \text{APS}$$

$$R_T^* = \frac{R_1 + R_2}{2} = 227 \quad \text{MPS}$$

Thus the two-stage design is more efficient than the one-stage design by a factor of about 7.2 in computation rate and a factor of about 11.7 in storage requirements. This is in agreement with that predicted by the curves for S and T, respectively, in Fig. 5.11. Also, we can see from these curves that a slightly greater efficiency can be achieved by going to an $I = 3$-stage design.

Following the same procedure for three-stage design, we get

$$M_1 = 8, \ M_2 = 4, \ M_3 = 2$$

$$N_1 = 26, \ R_1 = 208$$

$$N_2 = 22, \ R_2 = 44$$

$$N_3 = 115, \ R_3 = 115$$

Therefore, N_T is

$$N_T = N_1 + N_2 + N_3 = 163$$

and, assuming the use of symmetry,

$$R_T^+ = R_1 + R_2 + R_3 = 367 \quad \text{APS}$$

$$R_T^* = \frac{R_1 + R_2 + R_3}{2} = 184 \quad \text{MPS}$$

Thus we see that a three-stage design achieves a reduction in storage of a factor of 19.9 and a reduction in computation of a factor of 8.9 over a one-stage design.

By using the multistopband designs discussed in Chapter 4 for early stages in the two- or three-stage design above it is possible to reduce the order of the filters in those stages. For example, from Fig. 4.13 it can be estimated that for the three-stage design, N_1 can be reduced by a factor of about 22% or $N_1 \simeq 20$, and N_2 can be reduced by a factor of about 10% or $N_2 \simeq 20$. This gives final values of $N_T = 155$, $R_T^+ = 315$ (APS), and $R_T^* = 157$ (MPS) and savings factors of 21 and 10.4, respectively, for storage and computation over that of a one-stage design.

5.4 MULTISTAGE STRUCTURES BASED ON HALF-BAND FIR FILTERS

A second philosophy in the design of multistage structures for decimators and interpolators is based on the use of half-band filters [5.1, 5.8, 5.9]. These filters are based on the symmetrical FIR designs discussed in Chapter 4 and they have the property that approximately half of the filter coefficients are exactly zero. Hence the number of multiplications in implementing such filters is half of that needed for symmetrical FIR designs and one-fourth of that needed for arbitrary FIR designs. The half-band filter is appropriate only for sampling rate changes of 2 to 1. Hence the design approach in an M-to-1 (I-stage) half-band decimator structure is to reduce the sampling rate by factors of 2 in each of the first $I - 1$ stages, that is,

$$M_1 = M_2 = \ldots = M_{I-1} = 2 \qquad (5.31a)$$

The final stage then has an integer reduction factor of

$$M_I = \frac{M}{2^{I-1}} \qquad (5.31b)$$

or more generally it may have a rational factor

$$\frac{M_I}{L_I} = \frac{M}{2^{I-1}} \qquad (5.31c)$$

in which case it is implemented by the structure in Fig. 5.1. This strategy works best when M is a power of 2; however, it can be generally applied to arbitrary values of M with, perhaps, some difficulty. For example, a conversion ratio of $M = 14$ can be factored into the form

$$M = 14 = 2\cdot2\cdot2\cdot2\cdot\frac{7}{8}$$

where the final stage has a conversion ratio of $M_I/L_I = 7/8$ and requires a high-order filter to achieve this conversion (in practice a two-stage, $M_1 = 7$ and $M_2 = 2$, design might be more appropriate for this particular problem).

As discussed in Chapter 4, and illustrated in Fig. 5.14(a), the half-band filter has the following symmetry properties:

$$\delta_p = \delta_s = \delta \qquad (5.32a)$$

$$0 \leqslant f \leqslant F_c \quad \text{passband} \qquad (5.32b)$$

$$\frac{F}{2} - F_c \leqslant f \leqslant \frac{F}{2} \quad \text{stopband} \qquad (5.32c)$$

where F is the sampling rate and F_c is the passband cutoff frequency of the half-band filter. These symmetry properties impose certain restrictions on the use of half-band designs in multistage decimators (or interpolators) where the final transition band (F_p to F_s) is to be protected from aliasing. These restrictions and a design example are discussed in Section 5.4.1. Alternatively, the half-band symmetry properties naturally conform to the design of multistage decimators (or interpolators) in which M is a power of 2 and aliasing is permitted in the final transition band (F_p to F_s). This special class of structures is discussed in Section 5.4.2.

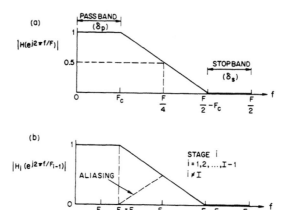

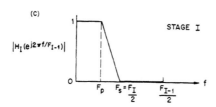

Figure 5.14 (a) Symmetry constraints of half-band filters; (b) filter requirements for half-band filters in stage i ($i=1, 2, ..., I-1$) of an I-stage decimator; (c) filter requirements for stage I.

5.4.1 Half-Band Designs with No Aliasing in the Final Transition Band

Consider the design of an I-stage decimator (or interpolator) with a passband from 0 to F_p, and a stopband at $F_s = F_I/2$, where F_I is the final sampling rate. Because of the symmetry properties of the half-band filters, the cutoff frequency requirements F_c for the filters in stages $i = 1, 2, ..., I - 1$ must be altered slightly to accommodate these constraints. Figure 5.14(b) illustrates this modification, where

$$F_c = F_s \tag{5.33a}$$

$$0 \leqslant f \leqslant F_c \quad \text{passband} \tag{5.33b}$$

$$F_i - F_c \leqslant f \leqslant F_i \quad \text{stopband}, \quad i = 1, 2, ..., I - 1 \tag{5.33c}$$

Note that the choice of F_p does not enter into these constraints.

Figure 5.14(c) shows the filter requirements for stage I (assuming that

$M_I = 2$), which is unchanged from that discussed in Section 5.2.2. Since this is not a symmetrical frequency response about $f = F_I/2$, this final filter cannot be a half-band design. The consequences of this result are illustrated in the following example.

Design Example 2

Consider the design of a six-stage structure to decimate a signal by a factor of $M = 2^6 = 64$, with overall filter specifications $\delta_p = 0.01$, $\delta_s = 0.001$, $F_p = 0.45$ Hz, $F_s = 0.5$ Hz, and where the frequency range from F_p to F_s is to be protected from aliasing. The input sampling frequency is $F_0 = 64$ Hz and the output sampling frequency is 1.0 Hz (this is identical to Design Example 1).

Solution.   The half-band tolerance constraints, δ, for stages $i = 1, 2, ..., I - 1$ must satisfy

$$\delta = \min\left[\frac{\delta_p}{6}, \delta_s \right] = 0.001$$

Therefore,

$$D_\infty(0.001, 0.001) = 3.25$$

From the filter constraints in Eq. (5.33) and the discussion in Section 5.3.1, we get for the filter order

$$N_i = \frac{D_\infty(0.001, 0.001)}{(F_i - 2F_c)/2F_i}$$

Alternatively, the design curves for half-band designs in Chapter 4 (Fig. 4.16) can be used. The filter orders are approximately (based on the next-largest odd-order filter from the estimate above) $N_1 = 7$, $N_2 = 7$, $N_3 = 9$, $N_4 = 9$, and $N_5 = 13$. Note that because these are half-band designs, approximately half of the coefficients in these filters are zero.

For the final stage ($I = 6$) a half-band filter cannot be used, so we will assume that an equiripple FIR design is used. Thus

$$D_\infty\left[\frac{\delta_p}{6}, \delta_s \right] = 3.10$$

and

$$N_6 = \frac{D_\infty(\delta_p/6, \delta_s)}{(F_s - F_p)/F_5} = 124$$

The total storage requirement, N_T, for this design is then

$$N_T \simeq N_6 + \sum_{i=1}^{5} 0.5N_i \simeq 147$$

where the factor of 0.5 in the expression above accounts for the fact that approximately one-half of the coefficients in the half-band filters are zero-valued. The total computation rate is approximately

$$R_T^+ = N_6 F_6 + \sum_{i=1}^{5} \frac{1}{2} N_i F_i = 359 \quad \text{APS}$$

and assuming the use of symmetry,

$$R_T^* = \frac{R_T^+}{2} = 180 \quad \text{MPS}$$

Thus it is seen that the savings factors (with regard to storage and computation) for this six-stage half-band example over that of a one-stage approach (see Design Example 1) are similar to those for the three-stage optimized design in the preceding section. The question as to whether it is more efficient to use a three-stage design rather than a six-stage design depends on the particular application. More will be said about this in Section 5.7.

5.4.2 Half-Band Designs for Power-of-2 Conversion Ratios and Aliasing in the Final Transition Band

As stated earlier, the symmetry properties of the half-band filters naturally conform to the design of I-stage decimators or interpolators in which the conversion ratios are

$$M_i = 2, \quad i = 1, 2, ..., I \tag{5.34a}$$

such that

$$M = 2^I \tag{5.34b}$$

and where aliasing is permitted in the final transition band (F_p to $F_I/2$) [5.8]. The cutoff frequency requirements for the filters in each stage i (including stage I) are then (see Fig. 5.15)

$$F_c = F_p \tag{5.35a}$$

$$0 \leqslant f \leqslant F_p \quad \text{passband} \tag{5.35b}$$

$$F_i - F_p \leqslant f \leqslant F_i \quad \text{stopband}, \quad i = 1, 2, ..., I \tag{5.35c}$$

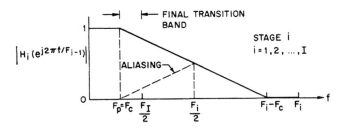

Figure 5.15 Filter requirements for stage i, $i = 1, 2, ..., I$, of an I-stage half-band decimator with $M = 2^I$ and aliasing permitted in the final transition band.

Based on these cutoff frequency requirements the filter order for the ith stage is

$$N_i \simeq \frac{D_\infty(\delta, \delta)}{(F_i - 2F_p)/F_{i-1}} \qquad (5.36)$$

Noting that

$$F_i = \frac{F_0}{2^i} = \frac{MF_I}{2^i} \qquad (5.37)$$

and defining a normalized measure of the final passband bandwidth as

$$k = \frac{F_p}{F_I/2} \qquad (5.38)$$

Eq. (5.36) can be expressed in the form

$$N_i = \frac{2D_\infty(\delta, \delta)}{1 - k/2^{I-i}} \qquad (5.39)$$

Since approximately half of the coefficients in these filters are zero valued, the total storage requirement for this design is

$$N_T \simeq \frac{1}{2} \sum_{i=1}^{I} N_i$$

$$\simeq D_\infty(\delta, \delta) \sum_{i=1}^{I} \left[1 - \frac{k}{2^{I-i}} \right]^{-1} \qquad (5.40)$$

Similarly, because of the zero-valued coefficients, the computation rate (in additions/sec) for this class of structures is

$$R_T^+ \simeq \frac{1}{2} \sum_{i=1}^{I} N_i F_i \qquad (5.41)$$

Applying Eqs. (5.39) and (5.40) gives

$$R_T^+ = D_\infty(\delta, \delta) \cdot F_0 \cdot S_h(k, I) \quad \text{APS} \qquad (5.42a)$$

where

$$S_h(k, I) = \sum_{i=1}^{I} \frac{1}{2^i} \left[1 - \frac{k}{2^{I-i}} \right]^{-1} \qquad (5.42b)$$

and, assuming the use of symmetry, the computation rate in multiplications/sec is

$$R_T^* = \frac{R_T^+}{2} \quad \text{MPS} \qquad (5.43)$$

Figure 5.16 shows plots of $S_h(k, I)$ as a function of k and with I as a parameter. It can be seen that $S_h(k, I) \approx 1$ except for small values of I and for values of k near 1. Thus, under these conditions, R_T^+ can be further approximated as

$$R_T^+ = 2R_T^* \approx D_\infty(\delta, \delta) F_0 \qquad (5.44)$$

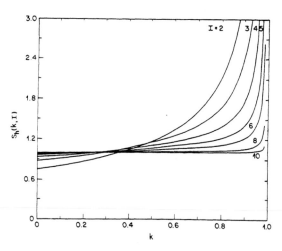

Figure 5.16 Plot of $S_h(k, I)$ as a function of k and the number of stages I.

It is of interest to compare this computation rate to that required for a single-stage (nonhalf-band) design which allows aliasing in the transition band. From Fig. 5.5 and the discussion above it can be seen that the filter order for a single-stage design is

$$N_1 \Big|_{\substack{\text{single} \\ \text{stage}}} \simeq \frac{D_\infty(\delta_p, \delta_s)}{(F_I - 2F_p)/F_0} \tag{5.45a}$$

$$\simeq \frac{M D_\infty(\delta_p, \delta_s)}{1} - k \tag{5.45b}$$

and the multiplication rate (assuming symmetry) is

$$R_T^* \Big|_{\substack{\text{single} \\ \text{stage}}} \cong \frac{N_1 F_0}{2M} \tag{5.46a}$$

$$\simeq \frac{D_\infty(\delta_p, \delta_s) \cdot F_0}{2(1 - k)} \tag{5.46b}$$

Therefore, the gain in computational efficiency of an I-stage half-band structure [Eq. (5.44)] over that of a single-stage M-to-1 structure is

$$GAIN \triangleq \frac{R_T^* \Big|_{\substack{\text{single} \\ \text{stage}}}}{R_T^* \Big|_{\substack{I-\text{stage} \\ \text{half-band}}}} \tag{5.47a}$$

$$= \frac{D_\infty(\delta_p, \delta_s)}{D_\infty(\delta, \delta)} \frac{1}{1 - k} \tag{5.47b}$$

$$\approx \frac{1}{1 - k} \tag{5.47c}$$

It is seen from this expression that the gain is approximately inversely proportional to the normalized width of the final transition band. The following example illustrates the differences when we can allow aliasing in the final stage of a multistage decimator.

Design Example 3

Consider the design of a six-stage half-band structure with aliasing in the final transition band and compare it with a similar one-stage design. The design parameters are $\delta_p = 0.01$, $\delta_s = 0.001$, $F_p = 0.45$ Hz, $F_0 = 64$ Hz, and $M = 2^6 = 64$.

Solution.   For a single-stage design we get

$$D_\infty(\delta_p, \delta_s) = 2.54$$

$$k = \frac{0.45}{0.5} = 0.9$$

and

$$N_T = N_1 \simeq \frac{MD_\infty(\delta_p, \delta_s)}{1-k} = 1625 \quad \text{taps}$$

$$R_T^+ = N_1 F_1 = 1625 \quad \text{APS}$$

Using symmetry gives

$$R_T^* = \frac{R_T^+}{2} = 813 \quad \text{MPS}$$

For a six-stage half-band design we get

$$\delta = \min\left[\frac{\delta_p}{6}, \delta_s\right] = 0.001$$

$$D_\infty(\delta, \delta) = 3.25$$

From Eq. (5.39) the filter orders are approximately (based on the next-largest odd order from the estimate above) $N_1 = 7$, $N_2 = 7$, $N_3 = 9$, $N_4 = 9$, $N_5 = 13$, and $N_6 = 65$. Therefore, the storage requirements are

$$N_T = \frac{1}{2} \sum_{i=1}^{6} N_i = 55$$

and the computational requirements (using symmetry) are

$$R_T^+ = \frac{1}{2} \sum_{i=1}^{6} N_i F_i = 268 \quad \text{APS}$$

$$R_T^* = \frac{R_T^+}{2} = 134 \quad \text{MPS}$$

In comparing this design to that of the one-stage design it is seen that the six-stage design is more efficient in computation by a factor of 6.1.

Comparing these designs, which allow aliasing in the final band, to those in

Design Examples 1 and 2, which do not allow aliasing, it is also seen that it is primarily the final stage that gets changed in the designs.

5.5 MULTISTAGE FIR DESIGNS BASED ON A SPECIFIC FAMILY OF HALF-BAND FILTER DESIGNS AND COMB FILTERS

A third design approach incorporates the use of comb filters, where possible, in the initial stages of multistage decimators (or the final stages of multistage interpolators) followed by half-band designs and equiripple designs [5.10, 5.11]. The idea here is to use a large number of stages to implement a large change in sampling rates, and to use extremely simple linear-phase FIR filters when possible. The types of FIR filters that are used in these cases are generally comb filters and half-band filters. Thus in this section we first discuss the characteristics of comb filters and then we discuss a design procedure for multistage decimators (or interpolators) based on this approach.

5.5.1 Comb Filter Characteristics

The class of FIR linear-phase filters called *comb filters* are characterized by the impulse response

$$h(n) = \begin{cases} 1, & 0 \leqslant n \leqslant N-1 \\ 0, & \text{elsewhere} \end{cases} \qquad (5.48)$$

where N is the number of taps in the filter. It can be shown that the Fourier transform of the impulse response of Eq. (5.48) is

$$H(e^{j\omega}) = \frac{\sin(\omega N/2)}{\sin(\omega/2)} e^{-j(N-1)\omega/2} \qquad (5.49a)$$

and the magnitude of the frequency response is therefore

$$|H(e^{j\omega})| = \left| \frac{\sin(\omega N/2)}{\sin(\omega/2)} \right| \qquad (5.49b)$$

Figure 5.17 illustrates a simple, one-stage interpolator using an $N = 5$-comb filter (where F_i is the input sampling rate and $5F_i$ is the output sampling rate). A block diagram of the interpolator is shown in Fig. 5.17(a); Fig. 5.17(b) shows the impulse response of the comb filter, and Fig. 5.17(c) shows the magnitude frequency response (on a log scale). Figure 5.18 shows the impulse and frequency responses for an $N = 10$ comb filter.

It is seen from Figs. 5.17 and 5.18, as well as from Eqs. (5.49), that these

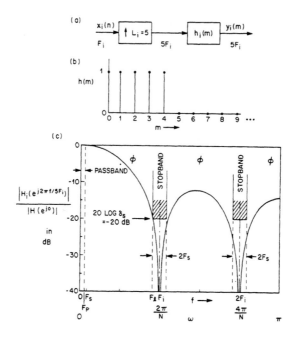

Figure 5.17 Example of an $N = 5$-comb filter used for a 1-to-5 interpolator stage.

comb filters have zeros in their frequency response at frequencies

$$\omega = \frac{2\pi k}{N}, \quad k = 1, 2, ..., N - 1 \tag{5.50}$$

Examination of the frequency-response characteristic shows that it is of the same form as that required for the multiple stopband specifications for the ith stage of a multistage decimator or interpolator as discussed in Section 5.2.3 (see Fig. 5.8) when we choose

$$N = N_i = L_i \quad \text{interpolator} \tag{5.51a}$$

or

$$N = N_i = M_i \quad \text{decimator} \tag{5.51b}$$

It is also the same form as the multiple stopband filter discussed in Chapter 4. To emphasize this point, the ideal multiple stopband characteristics for an $L_i = 5$ and an $L_i = 10$ interpolator have been superimposed on the frequency responses in Figs. 5.17(c) and 5.18(b), respectively, where it is assumed that the stopband tolerances, 20 log δ_s, are -20 dB and -30 dB, respectively.

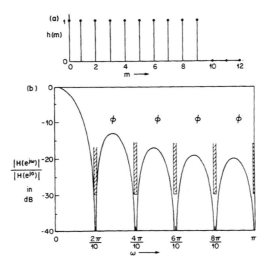

Figure 5.18 Comb filter responses for $N = 10$ design.

Because of the choice of N in Eqs. (5.51) and the fact that all the coefficient values in the comb filter are unity, these comb filters are extremely simple to implement since they do not require any multiplications. In the case of an $N = M_i$-to-1 decimator, they can be implemented by a simple sum-and-dump procedure (and scaling by $1/N$) in which the input signal is blocked into groups of M_i samples. Each group of M_i samples produces a single output sample. In the case of a 1-to-L_i interpolator, the process is even simpler. Each input sample is simply repeated L_i times to produce a signal whose sampling rate is L_i times that of the input sampling rate (i.e., it is a digital sample-and-hold).

It can also be seen from Figs. 5.17(c) and 5.18(b) that comb filters work only when the required widths of the stopbands, $2F_s$, are small compared to the sampling rate F_i, and therefore these types of filters clearly have a limited range of applications (i.e., in the initial stages of decimators or the final stages of interpolators when $2F_s \ll F_i$).

The required design equations for this class of comb filters can be obtained by observing that the most critical stopband tolerance requirement is reached at the lower edge of the first stopband [see Fig. 5.17(c)], that is, at the frequency

$$F_l = F_i - F_s \tag{5.52}$$

or equivalently at the frequency

$$\omega_l = \frac{2\pi F_l}{F_{i-1}} = 2\pi \frac{(1 - F_s/F_i)}{L_i} \tag{5.53}$$

where $F_{i-1} = L_i F_i$ is the output sampling rate of stage i according to the notation defined in Fig. 5.7. At this frequency we require that the stopband tolerance δ_s be met or exceeded, that is

$$\delta_s \geqslant \frac{|H(e^{j\omega_l})|}{|H(e^{j0})|}$$

$$\geqslant \frac{1}{L_i} \left| \frac{\sin [\pi(1 - F_s/F_i)]}{\sin [(\pi/L_i)(1 - F_s/F_i)]} \right|$$

$$\geqslant \left| \frac{\sin (\pi F_s/F_i)}{L_i \sin [(\pi/L_i)(1 - F_s/F_i)]} \right| \tag{5.54a}$$

For $F_s \ll F_i$ this becomes

$$\delta_s \geqslant \left| \frac{\pi F_s}{L_i F_i \sin (\pi/L_i)} \right| \tag{5.54b}$$

and for $L_i \gg 1$ it simplifies to the condition

$$\delta_s \geqslant \frac{F_s}{F_i} \tag{5.54c}$$

Similarly, at the passband edge we require that the passband ripple δ_p/I (for an I-stage design) be met or exceeded, that is

$$\frac{\delta_p}{I} \geqslant 1 - \frac{|H(e^{j\omega_p})|}{|H(e^{j0})|} \tag{5.55a}$$

where

$$\omega_p = \frac{2\pi F_p}{F_{i-1}} = \frac{2\pi F_p}{L_i F_i} \tag{5.55b}$$

Applying Eqs. (5.49b) and (5.51) gives

$$\frac{\delta_p}{I} \geqslant 1 - \frac{1}{L_i} \left| \frac{\sin (\pi F_p/F_i)}{\sin (\pi F_p/L_i F_i)} \right| \tag{5.56a}$$

and for $F_p \ll F_i$ and $L_i \gg 1$ we can apply the approximations $\sin(x) \simeq x - x^3/6$ to the numerator and $\sin(x) \simeq x$ to the denominator to get

$$\frac{\delta_p}{I} \geq 1 - \frac{1}{L_i}\left[\frac{(\pi F_p/F_i) - (\pi^3 F_p^3/6F_i^3)}{(\pi F_p/L_i F_i)}\right] \qquad (5.56b)$$

This expression simplifies to the condition

$$\frac{\delta_p}{I} \geq \frac{1}{6}\left(\frac{\pi F_p}{F_i}\right)^2 \qquad (5.56c)$$

Equations (5.54c) and (5.56c) can therefore be used to determine quickly whether a comb filter design can be applied to a particular stage of a multistage decimator or interpolator, given the parameters I, F_p, F_s, F_i, δ_p, and δ_s. If the inequalities are satisfied, the comb filter can be used, and if they are not satisfied, it cannot be used. Generally, the stopband requirement of Eq. (5.54c) will be the most critical one.

5.5.2 A Design Procedure Using a Specific Class of Filters

We will now discuss a design procedure by Goodman and Carey [5.10] for multistage decimators and interpolators based on the use of a specific family of half-band filter designs and comb filters. Table 5.1 gives the values of the filter coefficients $h(n)$ for a set of nine specially designed symmetric FIR digital filters centered at $n = 0$ (only one symmetric half of the coefficient values are given).

TABLE 5.1   COEFFICIENTS OF SPECIALIZED FIR FILTERS

Filter	N	$h(0)$	$h(1)$	$h(3)$	$h(5)$	$h(7)$	$h(9)$
$F1^*$	N	1	1				
$F2$	3	2	1				
$F3$	7	16	9	-1			
$F4$	7	32	19	-3			
$F5$	11	256	150	-25	3		
$F6$	11	346	208	-44	9		
$F7$	11	512	302	-53	7		
$F8$	15	802	490	-116	33	-6	
$F9$	19	8192	5042	-1277	429	-116	18

*$F1$ is an N-tap comb filter

The first filter, denoted as $F1$ in the table, is an N-tap comb filter of the type discussed in the preceding section. Filters $F2$ through $F9$ are specialized half-band filter designs (see Section 5.4) with impulse response durations of from 3 to 19 samples. The magnitude responses of filters $F2$, $F3$, and $F5$ (normalized to a dc gain of 1.0) are plotted in Fig. 5.19(a). These three filters all have monotone passband responses. The passband magnitude responses (normalized to a dc gain of 1.0) of filters $F4$ and $F6$ through $F9$ (in the passband) are plotted in Figs. 5.19(b) and 5.19(c). Each of these filters is seen to have two or more ripples in the passband. Before discussing how these nine filters can be used in a multistage decimator, three key points must be made, namely:

1. The comb filter $F1$ can be used for decimation or interpolation by any integer factor (as long as its filtering characteristics meet specifications); however, half-band filters $F2$ through $F9$ can be used only in stages that decimate by a factor of 2.
2. Since filters $F1$, $F2$, $F3$, and $F5$ have monotonic passband response, some compensation of their falloff is generally required by later-stage filters to guarantee that the overall filtering characteristic of the multistage structure meets specifications. Thus the applicability of filters $F1$, $F2$, $F3$, and $F5$ is determined primarily by given stopband ripple specifications; for the other filters, both passband and stopband ripple specifications are important.
3. For all nine filters of Table 5.1, the filter coefficients are expressible with a small number of bits. In fact, many of the filter coefficients are integer powers of 2. Hence these filters can be implemented (on special-purpose hardware) even more efficiently than a general half-band filter.

For a given set of design specifications of a multistage structure, the way in which the specific filters (i.e., of Table 5.1) are chosen is as follows. From the given filter specifications (i.e., $\{F_p, F_s, \delta_p, \delta_s\}$) we define a level ripple factor D as

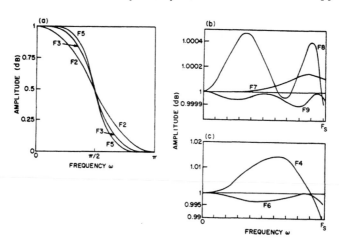

Figure 5.19 (a) Monotonic frequency responses of half-band filters F_2, F_3, and F_5; (b) and (c) passband responses of filters F_4, F_6 through F_9.

$$D = -20 \log(\delta_s) \quad \text{for filters } F1, F2, F3, F5 \quad \text{(monotone response)} \qquad (5.57a)$$

or

$$D = -20 \log \left[\min \left(\frac{\delta_p}{I}, \delta_s \right) \right] \quad \text{for filters } F4, F6\text{-}F9 \quad \text{(nonmonotone response)}$$
$$(5.57b)$$

We next define an "oversampling" frequency ratio at the input to stage i, O_i, as

$$O_i = \frac{F_{i-1}}{F_I} = M \left(\frac{F_{i-1}}{F_0} \right) \qquad (5.58a)$$

At the output of stage i, the oversampling ratio is

$$O_{i+1} = M \left(\frac{F_i}{F_0} \right) \qquad (5.58b)$$

For conventional FIR filter length designs we would calculate the required filter based on the ripple specification of Eq. (5.57) and the required transition width. Since the idea in this implementation is to use one of the fixed set of filters (whenever possible), the procedure here is to determine the range of D and O_i such that each of the nine filters of Table 5.1 is applicable. This range is plotted in Fig. 5.20(a), which shows boundaries, in the (O_i, D) plane, for which each of the nine filters meets specifications. The way the data of Fig. 5.20(a) are used is as follows. For stage 1, the grid point (O_1, D) is located. If the boundary line of filter $F1$ lies to the left of the grid point, filter $F1$ is used in the first stage, and the decimation rate of this stage, M_1, is the largest integer such that the grid point $(O_1/M_1, D)$ also lies to the right of the $F1$ boundary line. For the second, and subsequent stages, decimation ratios of 2 are used, and the filter required is the filter whose boundary line lies to the left of the grid point $(O_i/2, D)$ for stage i. This procedure can be used until the next-to-last stage. A more general FIR filter is generally required to meet overall filter specifications for the last stage, using one of the design techniques discussed in Chapter 4.

We will now illustrate the use of specialized filters in a multistage decimator.

Design Example 4

To illustrate use of the procedure described above, again consider the design of a specialized filter structure to decimate a signal by a factor of $M = 64$, with filter specifications $\delta_p = 0.01$, $\delta_s = 0.001$, $F_p = 0.45$ Hz, $F_s = 0.5$ Hz, $F_0 = 64$ Hz, and $F_I = 1$ Hz. (This is identical to Design Examples 1 and 2.)

Solution.   We first calculate the ripple factor D as

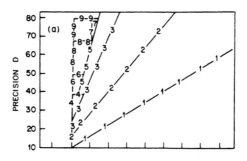

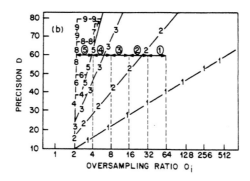

Figure 5.20 (a) Design chart for specific filter family; (b) design example using the chart.

$$D = - 20 \log (0.001) = - 60 \text{ dB} \qquad (F1, F2, F3, F5)$$

$$D = - 20 \log (\min (0.01/6, 0.001)) = - 60 \text{ dB} \quad (F4, F6\text{-}F9)$$

where we have used six as the largest possible number of stages in the design. The oversampling index O_1 for the first stage is 64, showing that an $F2$ filter is required, as shown in Fig. 5.20(b). Note that the comb filter (filter $F1$) cannot be used in this design. Since the $F2$ filter is a half-band filter, a 2-to-1 sampling rate reduction is used in the first stage and the oversampling index O_2 at the input to the second stage is 32. For stages 2 and 3, filter $F3$ can be used as O_i goes from 32 to 16 (stage 2), and 16 to 8 (stage 3). For stage 4 (as O_i goes from 8 to 4) an $F5$ filter is required, whereas for stage 5 (as O_i goes from 4 to 2) an $F8$ filter is required. For the final stage, a 124-point FIR filter (the same used in Design Example 2, stage 6 of the three-stage design in Section 5.4.1) is required.

The total storage requirements and computational requirements (using symmetry and excluding multiplication by −1, 1, or 2) of the resulting six-stage design are shown in Table 5.2.

TABLE 5.2 FILTERS AND COMPUTATION RATES FOR DESIGN EXAMPLE 4

Stage i	Filter	Number of Storage Locations	Number of Multiplications per Second (MPS)	Number of Additions per Second (APS)
1	F_2	3	0	64
2	F_3	5	32	64
3	F_3	5	16	32
4	F_5	7	16	24
5	F_8	9	10	16
6	$N_6 = 124$	124	62	124
Total		153	136	324

From this example it can be seen that there is only a slight savings in the multiplication rate in this design over that of Design Examples 1 and 2, and the number of additions/sec and storage requirements are similar. Some extra manipulations of the data to avoid multiplications by coefficients of −1 and 2 and to do data normalization in the structure are required. As mentioned previously, the principal advantage of this type of design is that the digital filter coefficients can be represented by a small number of bits, and specialized hardware structures can exploit this result to reduce the computation of the structure. A disadvantage of this approach, as in the half-band approaches discussed in Design Examples 2 and 3, is that generally a larger number of stages are required for the implementation. As discussed previously, the large number of stages add extra requirements on the control structure in the implementation which also must be considered in the design. This issue will be taken up in more detail in Section 5.7.

5.6 MULTISTAGE DECIMATORS AND INTERPOLATORS BASED ON IIR FILTER DESIGNS

Until now we have considered the use of FIR filters only in the implementation of decimators and interpolators. One reason for this is that FIR filters can be designed to have exactly linear phase. However, if linear phase is not an important consideration, then IIR filters can be considered as an attractive alternative. In this section it is shown that an optimization procedure similar to the one discussed in Section 5.3 can be used to design IIR multistage decimators and interpolators when conventional equiripple designs are used. In this discussion we consider only the case of integer decimation or interpolation stages. It is fairly straightforward to extend the results to noninteger stages.

To formulate the problem it is useful to reexamine the design of decimator and interpolator stages with integral ratios of sampling rates. As shown in Fig. 5.2(a), a decimator with a decimation rate of M can be implemented by filtering the input signal with a lowpass filter (in this case an IIR filter) and then decreasing the sampling rate by selecting one out of every M samples of the output of the filter. Unlike the FIR implementation, however, the output of the IIR filter (or all states of its recursive part) must be computed for *all values of n* prior to decimating by M, unless the filter design satisfies the denominator requirements discussed in Chapters 3 and 4. In a multistage decimator or interpolator design, the specifications on the lowpass filters are the same as those discussed in Section 5.2. As in the FIR case, transposition also applies in the case of IIR designs of decimators and interpolators, and therefore all design formulas and curves for multistage IIR decimators also apply to multistage IIR interpolators.

With the concepts above in mind we can now formulate the design relations for the multistage IIR decimator (or interpolator) [5.5]. We assume that the IIR filters are *elliptic designs* [5.12] (see Section 4.4.2). This choice of filters is made over that of the specific designs discussed in Section 4.4.4 because the order of the elliptic filters can be conveniently expressed in analytic form in order to define the objective function for the optimization procedure. For the designs in Section 4.4.4, such analytic expressions are not presently available.

Given the filter requirements for stage i, (see Section 5.2) the required elliptic filter order for that stage can be shown to be of the form [5.12]

$$N_i = A\left(\frac{\delta_p}{I}, \delta_s\right) B_i(M, \Delta f, I; M_1, M_2, ..., M_{I-1}) \qquad (5.59)$$

where $A(\cdot)$ is a function of the ripples and $B_i(\cdot)$ is a function of the cutoff frequencies and decimation ratios. The function $A(\cdot)$ has the form

$$A = A\left(\frac{\delta_p}{I}, \delta_s\right)$$

$$= \frac{K(\sqrt{1 - r^2})}{K(r)} \qquad (5.60)$$

where

$$r = \frac{2\delta_s \sqrt{\delta_p/I}}{(1 - \delta_p/I)\sqrt{(1 - \delta_p/I)^2 - \delta_s^2}} \qquad (5.61)$$

$K(\cdot)$ is the complete elliptic integral of the first kind, δ_p is passband ripple tolerance, δ_s is the stopband ripple tolerance, and I is the total number of stages.

The function $B_i(\cdot)$ has the form

$$B_i = B_i(M, \Delta f, I; M_1, M_2, ..., M_{I-1})$$

$$= \frac{K(t_i)}{K(\sqrt{1 - t_i^2})} \qquad (5.62)$$

where

$$t_i = \text{transition ratio of the filter for the } i\text{th stage}$$

$$= \frac{\tan (\pi F_p / F_{i-1})}{\tan [\pi (F_i - F_s) / F_{i-1}]} \qquad (5.63\text{a})$$

and using the previous terminology

$$F_i = \frac{F_0}{\prod_{j-1}^{i} M_j} \qquad (5.63\text{b})$$

$$\Delta f = \frac{F_s - F_p}{F_s} \qquad (5.63\text{c})$$

$$F_s = \frac{F_I}{2} \qquad (5.63\text{d})$$

$$M_I = \frac{M}{\prod_{i-1}^{I-1} M_i} \qquad (5.63\text{e})$$

With the expression above for filter order, the problem of minimizing a multistage IIR decimator or interpolator design in terms of its multiplication rate can readily be formulated. The multiplication rate for a single decimator stage i is approximately proportional to its IIR filter order N_i times its (input) sampling rate F_{i-1}. For an I-stage design, the total multiplication rate of the decimator (or interpolator) R_T^* can then be expressed in the form

$$R_T^* \approx \sum_{i-1}^{I} G' N_i F_{i-1}$$

$$\approx G' F_0 \sum_{i-1}^{I} \frac{N_i M_i}{\prod_{j-1}^{i} M_j} \qquad (5.64)$$

where G' is a proportionality factor that is dependent on the method of implementation of the IIR filter structure. For example, in a conventional cascade structure, three multiplications are required for the implementation of a second-order section, and therefore for this structure G' is approximately $3/2$ (and we must add one multiplication for the gain constant).

With the aid of Eq. (5.59) the expression above can be written in the form

$$R_T^* \approx G' \cdot A\left(\frac{\delta_p}{I}, \delta_s\right) F_0 \widetilde{S} \tag{5.65a}$$

where

$$\widetilde{S} = \widetilde{S}(M, \Delta f, I; M_1, M_2, ..., M_{I-1})$$

$$= \sum_{i=1}^{I} \frac{B_i M_i}{\prod\limits_{j=1}^{i} M_j}\Bigg|_{M_I = M} \Bigg/ \sum_{j=1}^{I-1} M_j \tag{5.65b}$$

The reader should note the similarity of this expression to that of S in Eq. (5.26) for the FIR case (but not to be confused with $\widetilde{S}$ in Figs. 5.10 to 5.13). It is seen that R_T^* can be expressed as a product of a function of the ripples, the initial sampling frequency, and a function of the cutoff frequencies and decimation ratios. In a similar manner, R_T^* can be minimized by minimizing $\widetilde{S}$ as a function of the decimation (interpolation) ratios for each value of I and then choosing the value of I that minimizes the product. For convenience, the function $A(\delta_p/I, \delta_s)$ is tabulated in Table 5.3.

The minimization of $\widetilde{S}$ can again be performed by a routine such as the Hooke and Jeeves algorithm. This optimization is not a trivial one. The evaluation of the elliptic functions is extremely sensitive numerically [5.5]. Great care must be used in controlling the range of parameter values to avoid both arithmetic overflow and round-off error, and also in constraining the optimization search to be within the region of realizable solutions.

Using the Hooke and Jeeves method, the minimization problem above was solved. Plots of minimized values of $\widetilde{S}$ are given in Figs. 5.21(a) to 5.24(a) for final transition bandwidths (Δf) of 0.2, 0.1, 0.05, and 0.02. The corresponding optimum decimation ratios and $B_i(\cdot)$ values are given in Figs. 5.21(b) to 5.24(b) and 5.21(c) to 5.24(c), respectively. These curves and Table 5.3 (which gives values for the function A as δ_p and δ_s vary) can be used in a manner similar to that of the FIR design curves as a guide toward selecting practical integer values of decimation ratios for decimator (or interpolator) designs.

Several interesting observations can be made from these curves and tables. As in the FIR case, the function of the ripples $A(\delta_p/I, \delta_s)$ (see Table 5.3) is a weakly varying function of I, and again most of the interesting observations can be made from curves of the $\widetilde{S}$ function. From plots of $\widetilde{S}$ [Figs. 5.21(a) to 5.24(a))]

TABLE 5.3 $A(\delta_p/I, \delta_s)$ FOR SEVERAL
VALUES OF δ_p, δ_s, AND I

δ_p	δ_s	$I = 1$	$I = 2$	$I = 3$	$I = 4$
0.100	0.0100	4.10	4.32	4.45	4.55
0.100	0.0050	4.54	4.77	4.90	4.99
0.100	0.0010	5.57	5.79	5.92	6.01
0.100	0.0005	6.01	6.23	6.36	6.45
0.100	0.0001	7.03	7.26	7.39	7.48
0.050	0.0100	4.32	4.55	4.68	4.77
0.050	0.0050	4.77	4.99	5.12	5.21
0.050	0.0010	5.79	6.01	6.14	6.23
0.050	0.0005	6.23	6.45	6.58	6.67
0.050	0.0001	7.26	7.48	7.61	7.70
0.010	0.0100	4.84	5.06	5.19	5.28
0.010	0.0050	5.28	5.50	5.63	5.72
0.010	0.0010	6.30	6.53	6.65	6.75
0.010	0.0005	6.75	6.97	7.10	7.19
0.010	0.0001	7.77	7.99	8.12	8.21
0.005	0.0100	5.06	5.28	5.41	5.50
0.005	0.0050	5.50	5.72	5.85	5.94
0.005	0.0010	6.53	6.75	6.88	6.97
0.005	0.0005	6.97	7.19	7.32	7.41
0.005	0.0001	7.99	8.21	8.34	8.43
0.001	0.0100	5.57	5.79	5.92	6.01
0.001	0.0050	6.01	6.23	6.36	6.45
0.001	0.0010	7.04	7.26	7.39	7.48
0.001	0.0005	7.48	7.70	7.83	7.92
0.001	0.0001	8.50	8.72	8.85	8.94

we observe that improvements in efficiency of approximately two- or threefold are possible for moderate values of M (20-50) by using a multistage design and gains of up to about eightfold are possible for large values of M and small values of Δf. Although these gains are not as striking as those for FIR designs (which can be orders of magnitude), they do represent modest improvements. Another conclusion that can be drawn from these curves is that little, if any, improvement in efficiency can be gained by using a three-stage IIR design over a two-stage IIR design, and therefore two stages, at most, are sufficient for most purposes.

The curves in Figs. 5.21(c) to 5.24(c) represent B_i values for the optimized designs. By noting from Eq. (5.59) that $N_i = AB_i$, it is clear that the B_i values represent the normalized (e.g., N_i/A) filter orders. We can observe from these curves that the order of the final stage of a two- or three-stage design is essentially equal to that of a one-stage design. This occurs because the orders of the IIR

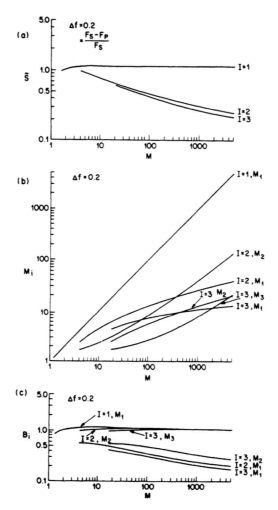

Figure 5.21 Design curves for multistage IIR decimators (or interpolators) with minimized values of $\widetilde{S}$ and ideal values of M_i, $i = 1, 2, ..., I$ and $\Delta f = 0.2$.

filters are determined by the ratio F_p/F_s. As the sampling frequency is reduced it is clear that the transition ratio t_i [see Eq. 5.63(a)] of the final stage I remains essentially unchanged (except for a slight warping due to the bilinear transformation), and therefore N_I and B_I remain essentially unchanged as I is increased. The filter orders for the earlier stages of a multistage design are of course lower, as seen in Figs. 5.21(c) to 5.24(c).

It can also be seen that the sum of the filter orders for a multistage design (i.e., the total required storage), will always be greater than that of a one-stage

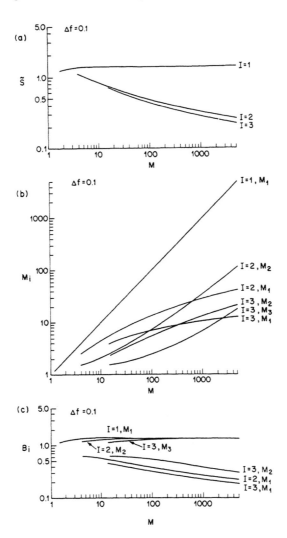

Figure 5.22 Design curves for IIR multistage decimators or interpolators for $\Delta f = 0.1$.

design. Therefore, a multistage IIR decimator or interpolator is always less efficient, in terms of storage, than a single-stage IIR design. Thus, unlike the FIR design, where both computation and storage could be reduced, the multistage IIR design represents a trade-off between computation and storage.

Design Example 5

To illustrate the use of the IIR design tables and curves, we will choose a decimator with the specifications $M = 100$, $F_0 = 100$ Hz, $F_I = 1$ Hz, $F_p = 0.475$ Hz,

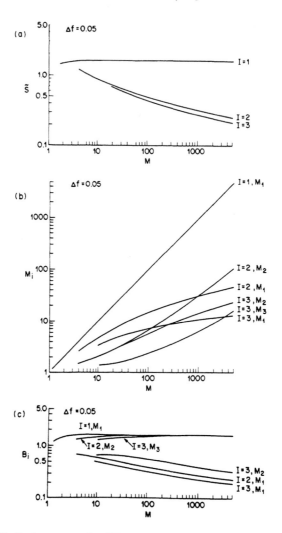

Figure 5.23 Design curves for IIR multistage decimators or interpolators for $\Delta f = 0.05$.

$F_s = 0.5$ Hz, $\delta_p = 0.001$, and $\delta_s = 0.0001$, giving $\Delta f = (0.5 - 0.475)/0.5 = 0.05$. For a one-stage IIR design it can be seen from Table 5.3 that $A = 8.50$ and from Fig. 5.23(c) that $B_1 = 1.61$. Therefore, the theoretical filter order is 13.7 [from Eq. (5.59)] and an actual value of 14 must be used. For a cascade implementation this will require a multiplication rate of $(1 + 14 \times 3/2)F_0$, or 2200 MPS. For a two-stage IIR design it is seen from Fig. 5.23(b) that the optimum (theoretical) decimation ratios are $M_1 = 14.5$ and $M_2 = 6.9$. From Table 5.3 and Fig. 5.23(c) we find that $A = 8.72$, $B_1 = 0.39$, and $B_2 = 1.6$, giving theoretical filter orders of $N_1 \simeq 3.4$ and $N_2 \simeq 13.9$. One practical choice of decimation ratios is $M_1 = 20$ and

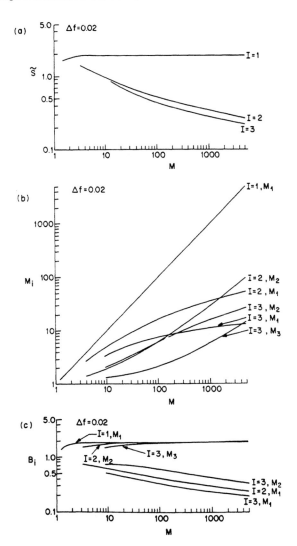

Figure 5.24 Design curves for IIR multistage dcimators or interpolators for $\Delta f = 0.02$.

$M_2 = 5$. This leads to theoretical filter orders of $N_1 = 3.8$ and $N_2 = 13.8$ (these values cannot be obtained from Fig. 5.23(c); see instead the tables in [5.12]). Actual filter orders for this design are, therefore, 4 and 14 for N_1 and N_2, respectively.

Another practical choice of decimation ratios might be $M_1 = 10$ and $M_2 = 10$. This results in theoretical filter orders of $N_1 = 3.1$ and $N_2 = 13.97$ and actual values of 4 and 14, respectively. With careful design, a third-order filter might be substituted for N_1, however.

The results for these three IIR designs together with corresponding FIR designs for

TABLE 5.4   COMPARISONS FOR AN $M = 100$ DECIMATOR

	One-Stage IIR	Two-Stage IIR	Two-Stage IIR	Two-Stage FIR	Three-Stage FIR
Decimation Ratios	$M_1 = 100$	$M_1 = 20$ $M_2 = 5$	$M_1 = 10$ $M_2 = 10$	$M_1 = 50$ $M_2 = 2$	$M_1 = 10$ $M_2 = 5$ $M_3 = 2$
Filter Lengths (taps)	$N_1 = 14$	$N_1 = 4$ $N_2 = 14$	$N_1 = 4$ $N_2 = 14$	$N_1 = 423$ $N_2 = 347$	$N_1 = 38$ $N_2 = 38$ $N_3 = 356$
Computation (MPS)	2200	810	920	598	406
Computational Savings over One-Stage IIR Design	1.0	2.7	2.4	3.7	5.4
Storage	14	18	18	798	436

the $M = 100$ decimation are given in Table 5.4. As seen in this table, a good choice
for the IIR design might be the two-stage approach with $M_1 = 20$ and $M_2 = 5$. This
results in a savings in computation of 2.7 over that of a one-stage design. It is
achieved at the cost of four extra storage locations and the added complexity of a
two-stage design.

In order to compare the IIR designs to FIR designs, a two-stage and a three-stage
FIR decimator design are also included in Table 5.4. The design of the two-stage
FIR decimator was taken from [5.13] and the three-stage design is taken from [5.5].
From this comparison we can observe that the FIR designs are more efficient in terms
of computation than the IIR designs; however, they require considerably more storage
for both data and coefficients.

Greater efficiency can be expected in multistage IIR decimator and
interplator designs when the more efficient IIR filter designs in Section 4.4.4 are
used. An example in Ref. 5.14, for an $M = 20$ decimator, shows that this type of
multistage IIR design is slightly more efficient than a comparable FIR design, but
that it requires more coefficient and data storage than an IIR design based on
conventional IIR filters as discussed above. A difficulty with this approach is that
no simple analytic expression for the filter order exists for these more efficient IIR
filter designs. Therefore it is difficult to establish a systematic design procedure for
multistage designs incorporating these filters.

5.7 CONSIDERATIONS IN THE IMPLEMENTATION OF MULTISTAGE DECIMATORS AND INTERPOLATORS

In the preceding sections we have presented three methods for implementing
multistage FIR digital systems for decimation or interpolation and one approach

using IIR filters. As in most real-world situations, there is no simple answer to the question as to when to use each of these structures. In the design examples of these sections, we have shown that each structure has special properties that can be exploited to minimize computation, storage, and cost. For example, the half-band structure is especially suitable for sampling rate changes of powers of 2, and in a pipeline architecture where the repetitive stages with decimation factors of 2 are easily realized. The structure based on the optimization procedure in Section 5.3 is especially suitable for sampling rate changes involving any product of factors, and for architectures that favor a small number of stages (e.g., software or array processor implementations). The design based on a specific family of comb and half-band filters is most suitable for designs with large ripple specifications, and for architectures where the cost of multiplication greatly exceeds the cost of shifting or adding. Finally, the IIR designs are useful where coefficient and storage requirements must be kept at a minimum and where a nonlinear phase response is acceptable.

In addition to these fairly straightforward differences between the different multistage structures, several other factors must be considered for implementing any of the proposed structures. One important factor is the cost of control of a hardware or software implementation of the structure. For many practical systems the cost (computation) of setting up the control structure of a multistage, multirate digital system may exceed the cost (computation) of the process being simulated. To illustrate some of the considerations, a simple software implementation of a three-stage FIR decimator and a three-stage FIR interpolator is described next [5.15].

We recall from Chapter 2 that the basic computation for implementing a sampling rate conversion by a factor of M/L is

$$y(m) = \sum_{n=-\infty}^{\infty} g_m(n)x\left[\lfloor \frac{mM}{L} \rfloor - n\right] \tag{5.66}$$

where

$$g_m(n) = h(nL + mM \oplus L), \text{ all } m \text{ and all } n \tag{5.67}$$

and where $\lfloor u \rfloor$ denotes the largest integer less than or equal to u. Thus for each stage, if the impulse response coefficients of the FIR filter, $h(n)$, are stored as the scrambled coefficient sets $g_m(n)$ of Eq. (5.67), we get $y(m)$ as a convolution of x and g_m (see Section 3.3.4).

If we now consider an implementation of a three-stage decimator with decimation ratios of M_1, M_2, M_3 and with $M = M_1 M_2 M_3$, we get a block diagram of the type shown in Fig. 5.25(a), with control structure of the type shown in Fig. 5.25(b). The decimator has three data buffers, labeled S1, S2, and S3 in Fig. 5.25(a). These buffers are used for storage of internal data for each stage. The size of the ith buffer, N_i', is given as

$$N_i' = M_i \left\lceil \frac{N_i}{M_i} \right\rceil \geq N_i, \quad i = 1, 2, 3 \tag{5.68}$$

where $\lceil u \rceil$ denotes the smallest integer greater than or equal to u, and N_i is the length of the FIR filter for stage i. Each data buffer is partitioned into blocks of size M_i. A total of

$$Q_i = \left\lceil \frac{N_i}{M_i} \right\rceil \tag{5.69}$$

blocks are used at stage i. Three additional buffers hold the filter coefficients of each stage.[4]

The operation of the three-stage decimator structure is illustrated in the control structure of Fig. 5.25(b). Initially, M_1 samples of the input signal are transferred from the main input data buffer to S1 (i.e., S1 is updated by M_1 new samples). One output sample is then computed via Eq. (5.66), and stored in data buffer S2. This process is repeated M_2 times until data buffer S2 has M_2 samples, at which point one output is computed for stage 2 of the decimator, and stored in S3. This process repeats M_3 times until M_3 samples are stored in S3, at which point one output sample is computed for stage 3 of the decimator, which is the final output of the three-stage structure. At this point the decimator cycle is completed; $M = M_1 M_2 M_3$ samples have been read from the main input buffer, and one output sample has been computed. The process can now be repeated on each consecutive input block of M samples until all processing is completed.

Figure 5.26 shows a similar data flow structure and control structure for a three-stage 1-to-L interpolator where

$$L = L_1 L_2 L_3 \tag{5.70a}$$

$$Q_i = \left\lceil \frac{N_i}{L_i} \right\rceil, \quad i = 1,2,3 \tag{5.70b}$$

The interpolator has five data buffers associated with it, three (S1, S2, S3) of lengths Q_1, Q_2, and Q_3 samples, respectively, and two (T2, T3) of lengths L_2 and L_3 samples, respectively. In addition, each stage i has a buffer for N_2' "scrambled" coefficients which are partitioned into L_i blocks of Q_i each (see Section 3.3.4), where

$$N_i' = Q_i L_i, \quad i = 1, 2, 3 \tag{5.70c}$$

In operation, one input sample of the signal $x(n)$ is stored in S3 and L_3

[4] Clearly, storage and computation reductions can be obtained based on symmetries in filter impulse responses. We neglect such effects here.

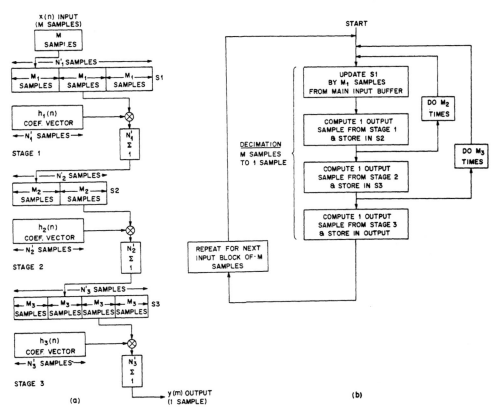

Figure 5.25 (a) Data flow structure; and (b) control structure for a three-stage FIR decimator.

output samples are computed from stage 3 of the interpolator and temporarily stored in T3. One sample from T3 is then stored in S2 and L_2 output samples are computed from stage 2 and stored in T2. S1 is then updated by one sample from T2 and L_1 outputs are computed and stored in the main output buffer. This is repeated until all L_2 samples in T2 are removed. T2 is then refilled by updating S2 with one sample from T3 and computing L_2 more samples. Upon completion of the interpolator cycle one input was transferred into the interpolator and L output samples are computed and stored in the main output data buffer. The process can then be repeated for the next input sample.

The purpose of the examples above is to illustrate the complexity of the control structure, even for a three-stage structure, and the interaction between the signal processing computation and the data flow within the system. The processes are described as serial processes and represent a straightforward approach for a software implementation. If high-speed and/or special-purpose hardware is used, the structures in Figs. 5.25 and 5.26 are also particularly attractive as they lend

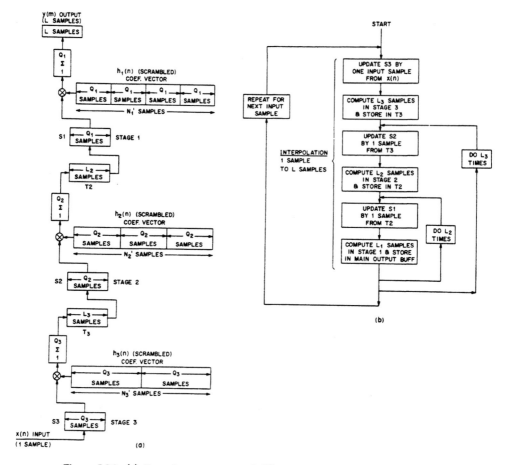

Figure 5.26 (a) Data flow structure and (b) control structure for a three-stage FIR interpolator.

themselves easily (with slight modification) to various degrees of parallel processing and pipelining. For example, in the interpolation stages all L_i outputs of a stage i can be computed in parallel [5.16]. Similar degrees of parallelism are possible in the implementation of the decimator stages. Pipelining is also an attractive possibility with these structures since they are essentially completely feedforward structures [5.17]. Each stage can be separately implemented in hardware. In this case it may be more attractive to choose the designs such that the amount of computation is equally divided among the stages rather than minimizing the total computation or total storage.

In Section 7.7.3 we will see another example of a multistage decimator and interpolator structure based on the use of polyphase filter structures in each stage. It will be shown that this is a particularly effective approach to the realization of a digital filter bank in which the computation for the different filter bank channels is shared in the various stages of the multistage structure and distributed in time for effective control of the real-time processing.

5.8 SUMMARY

In this chapter we have shown how a significant reduction in computation can be achieved by realizing a decimator (or interpolator) with a large decimation (interpolation) ratio in stages. The larger the overall decimation or interpolation ratio, the higher the savings that can be achieved (over that of a single-stage design). Depending on the detailed specifications of the system, several different structures can be utilized to achieve the sampling rate reduction. It was also shown that, under some conditions, reductions in computation and/or storage could be achieved by using multiple stopband filters or comb filters (instead of conventional lowpass designs), and IIR filters instead of FIR linear phase designs.

Although we have concentrated primarily on multistage implementations of decimators, it should be clear that all the results of this chapter apply equally to multistage interpolators (because of duality and transposition), and to multistage realizations of networks making small rational changes in sampling rate (i.e., L/M conversion) with large values of either L or M.

REFERENCES

5.1   M. G. Bellanger, J. L. Daguet, and G. P. Lepagnol, "Interpolation, Extrapolation, and Reduction of Computation Speed in Filters," *IEEE Trans. Acoust. Speech Signal Process.*, Vol. ASSP-22, No. 4, pp. 231-235, August 1974.

5.2   G. A. Nelson, L. L. Pfeifer, and R. C. Wood, "High-Speed Octave Band Digital Filtering," *IEEE Trans. Audio Electroacoust.*, Vol. AU-20, No. 1, pp. 58-65, March 1972.

5.3   R. R. Shively, "On Multistage FIR Filters with Decimation," *IEEE Trans. Acoust. Speech Signal Process.*, Vol. ASSP-23, No. 4, pp. 353-357, August 1975.

5.4   R. E. Crochiere and L. R. Rabiner, "Optimum FIR Digital Filter Implementations for Decimation, Interpolation, and Narrow-Band Filtering," *IEEE Trans. Acoust. Speech Signal Process.*, Vol. ASSP-23, No. 5, pp. 444-456, October 1975.

5.5   R. E. Crochiere and L. R. Rabiner, "Further Considerations in the Design of Decimators and Interpolators," *IEEE Trans. Acoust. Speech Signal Process.*, Vol. ASSP-24, No. 4, pp. 296-311, August 1976.

5.6   R. Hooke and T. A. Jeeves, "Direct Search Solution of Numerical and Statistical Problems," *J. Assoc. Comput. Mach.*, Vol. 8, No. 4, pp. 212-229, April 1961.

5.7   J. L. Kuester and J. H. Mize, *Optimization Techniques with Fortran.* New York: McGraw-Hill, 1973.

5.8   M. G. Bellanger, "Computation Rate and Storage Estimation in Multirate Digital Filtering with Halfband Filters," *IEEE Trans. Acoust. Speech Signal Process.*, Vol. ASSP-25, No. 4, pp. 344-346, August 1977.

5.9   D. W. Rorabacher, "Efficient FIR Filter Design for Sample-Rate Reduction and Interpolation," *Proc. IEEE Int. Symp. Circuits Syst.*, pp. 396-399, April 1975.

5.10   D. J. Goodman and M. J. Carey, "Nine Digital Filters for Decimation and Interpolation," *IEEE Trans Acoust. Speech Signal Process.*, Vol. ASSP-25, No. 2, pp. 121-126, April 1977.

5.11   E. B. Hogenauer, "An Economical Class of Digital Filters for Decimation and Interpolation," *IEEE Trans. Acoust. Speech Signal Process.*, Vol. ASSP-29, No. 2, pp. 155-162, April 1981.

5.12   L. R. Rabiner, J. F. Kaiser, O. Herrmann, and M. T. Dolan, "Some Comparisons between FIR and IIR Digital Filters," *Bell Syst. Tech. J.*, Vol. 53, No. 2, pp. 305-331, February 1974.

5.13   L. R. Rabiner and R. E. Crochiere, "A Novel Implementation for Narrow-Band FIR Digital Filters," *Trans. Acoust. Speech Signal Process.*, Vol. ASSP-23, No. 5, pp. 457-464, October 1975.

5.14   H. G. Martinez and T. W. Parks, "A Class of Infinite-Duration Impulse Response Digital Filters for Sampling Rate Reduction," *IEEE Trans. Acoust. Speech Signal Process.*, Vol. ASSP-27, No. 2, pp. 154-162, April 1979.

5.15   R. E. Crochiere and L. R. Rabiner, "A Program for Multistage Decimation, Interpolation, and Narrow-Band Filtering," in *Programs for Digital Signal Processing.* New York: IEEE Press, 1979, pp. 8.3-1 to 8.3-14.

5.16   H. Urkowitz, "Parallel Realization of Digital Interpolation Filters for Increasing the Sampling Rate," *IEEE Trans. Circuits Syst.*, Vol. CAS-22, pp. 146-154, February 1975.

5.17   R. E. Crochiere and A. V. Oppenheim, "Analysis of Linear Digital Networks," *Proc. IEEE*, Vol. 63, No. 4, pp. 581-595, April 1975.

6

Multirate Implementations of Basic Signal Processing Operations

6.0 INTRODUCTION

In Chapters 1 to 5 we have been primarily concerned with the theory, design and implementation of multirate systems for sampling rate conversion. We have shown how to implement sampling rate conversions digitally (Chapter 2), how to realize the systems in efficient, single-stage structures (Chapter 3), how to design digital filters suitable for sampling rate conversion (Chapter 4), and how to use more than one stage to achieve large computational efficiencies for sampling rate conversion structures (Chapter 5). In this and the following chapter, we concentrate on some basic signal processing operations: lowpass and bandpass filtering, phase shifting (fractional sample delays), Hilbert transformation, spectral analysis, and filter bank implementation, and show how the concepts of multirate signal processing, as developed in Chapters 2 to 5, can be used to derive efficient realizations of these operations. Our goal here is to discuss both the theoretical and implementational aspects of each of these realizations, and to relate them to the conventional (linear, shift-invariant, single-rate) methods used to implement these operations.

We begin this chapter with the two most basic signal processing operations: lowpass (or equivalently highpass) filtering [6.1, 6.2] and bandpass (or equivalently bandstop) filtering [6.3]. We show that in the case where the filter bandwidth is small relative to the Nyquist rate, multirate techniques can lead to significant reductions in computation. We then proceed to a discussion of a fractional sample phase shifter [6.4] in which the set of M polyphase filters are shown to yield a set of delays of value $\Delta_k = k/M$, $k = 0, 1, ..., M - 1$ [i.e., fractions of a sample from

0 to $(M - 1)/M$ in steps of $1/M$ samples]. Thus, for M sufficiently large, any fractional delay can be approximately achieved (relative to some fixed delay). We next show how a standard Hilbert transformer can be implemented via a modulation filtering network using simple half-band lowpass filters. This section provides some new and insightful interpretations of the Hilbert transform relations [6.5-6.8]. This is followed by a brief discussion of narrow-band, high-resolution spectral analysis where we show how computation can be substantially reduced by multirate techniques [6.9]. We conclude with a discussion of sampling rate conversion between digital systems with incommensurate clocks or sampling rates.

6.1 MULTIRATE IMPLEMENTATION OF LOWPASS FILTERS

In Chapter 2 we introduced the idea of cascading a 1-to-L interpolator with an M-to-1 decimator (Fig. 2.17) and showed that the resulting structure implemented a sampling rate conversion by a factor of L/M. Consider now cascading an M-to-1 decimator with a 1-to-M interpolator, as shown in Fig. 6.1(a). Intuitively, we see that the overall system relating the output $y(n)$ to the input $x(n)$ acts like a lowpass filter [due to $h_1(n)$ and $h_2(n)$] with aliasing (due to the decimation) and imaging (due to the interpolation). Whenever aliasing and imaging are negligible, the system in Fig. 6.1(a) acts like a well-behaved digital filter. We now formally show that this is the case.

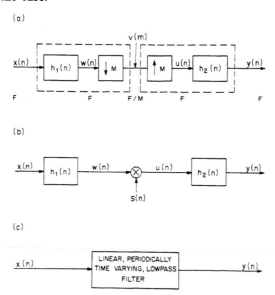

Figure 6.1 (a) Block diagram of a cascaded decimator/interpolator structure for multirate implementation of a lowpass filter; (b) the equivalent single rate modulation structure; (c) the equivalent single rate time-varying filter.

The filters $h_1(n)$ and $h_2(n)$ are conventional filters (i.e., at the sampling frequency F) and we define their z-transforms as $H_1(z)$ and $H_2(z)$. If the input to $h_1(n)$ is $x(n)$, with z-transform $X(z)$, then, using the results of Chapter 2, we have

$$W(z) = X(z)H_1(z) \tag{6.1}$$

and

$$V(z) = \frac{1}{M} \sum_{l=0}^{M-1} W(z^{1/M} e^{-j(2\pi l/M)}) \tag{6.2}$$

for the decimation stage, and

$$U(z) = V(z^M) \tag{6.3}$$

$$Y(z) = U(z)H_2(z) \tag{6.4}$$

for the interpolation stage. Combining Eqs. (6.1) to (6.4) and evaluating the result on the unit circle gives

$$Y(e^{j\omega}) = \frac{H_2(e^{j\omega})}{M} \sum_{l=0}^{M-1} X(e^{j(\omega - (2\pi l/M))}) H_1(e^{j(\omega - (2\pi l/M))}) \tag{6.5}$$

or, by taking the inverse Fourier transform, we get

$$y(n) = \frac{h_2(n)}{M} * \sum_{l=0}^{M-1} \left[x(n)e^{j(2\pi ln/M)} \right] * \left[h_1(n)e^{j(2\pi ln/M)} \right] \tag{6.6}$$

where $*$ denotes linear convolution. Performing the inner convolution gives

$$y(n) = \frac{h_2(n)}{M} * \sum_{l=0}^{M-1} \left[\sum_m x(m)e^{j(2\pi lm/M)} h_1(n-m)e^{j(2\pi l(n-m)/M)} \right] \tag{6.7}$$

and interchanging the summations over l and m gives

$$y(n) = h_2(n) * \left[\sum_m x(m)h_1(n-m) \sum_{l=0}^{M-1} \frac{e^{j(2\pi ln/M)}}{M} \right] \tag{6.8}$$

$$= h_2(n) * [\sum_m x(m)h_1(n-m)s(n)] \tag{6.9}$$

where

$$s(n) = \begin{cases} 1, & n = 0, \pm M, \pm 2M, ... \\ 0, & \text{otherwise} \end{cases} \qquad (6.10)$$

The overall "equivalent" system of Eq. (6.9) is shown in Fig. 6.1(b), where the modulator samples the convolution of $x(n)$ and $h_1(n)$, and the output is then filtered by $h_2(n)$. The operations of Fig. 6.1(b) can also be anticipated by the discussion lending up to Fig. 3.10(b), where it was shown that a cascade of a sampling rate compressor (by M-to-1) and a sampling rate expander (by 1-to-M) is equivalent to modulation by a periodic impulse train.

We now consider the properties of the system of Fig. 6.1. From Eq. (6.9) we can see that the system is linear. By considering $x(n) = u_0(n)$ (an impulse), and $x(n) = u_0(n - 1)$, a delayed impulse, we see that for $x(n) = u_0(n)$ we get

$$y(n) = h_2(n) * [s(n)h_1(n)] \qquad (6.11a)$$

$$= h_2(n) * \hat{p}_0(n) \qquad (6.11b)$$

where $\hat{p}_0(n)$ is the expanded zeroth polyphase filter of $h_1(n)$ as discussed in Section 4.2.1. Alternatively, for $x(n) = u_0(n - 1)$ we get

$$y(n) = h_2(n) * [s(n)h_1(n - 1)] \qquad (6.12a)$$

$$= h_2(n) * \hat{p}_1(n) \qquad (6.12b)$$

where $\hat{p}_1(n)$ is the expanded first polyphase filter of $h_1(n)$. In general, for $x(n) = u_0(n - m)$ we get

$$y(n) = h_2(n) * [s(n)h_1(n - m)] \qquad (6.13a)$$

$$= h_2(n) * \hat{p}_{m \oplus M}(n) \qquad (6.13b)$$

where $\hat{p}_{m \oplus M}$ is the expanded $(m \oplus M)$th polyphase filter of $h_1(n)$. Thus we see that the system of Fig. 6.1 is periodically time-varying. Since as shown in Chapter 3 all the polyphase filters of $h_1(n)$ have the same magnitude response (but different delays), we can readily see that the equivalent overall frequency response magnitude of the system of Fig. 6.1 is

$$|H_{\text{EQ}}(e^{j\omega})| = |H_2(e^{j\omega})||\hat{P}_0(e^{j\omega})| \qquad (6.14)$$

which, for the band $|\omega| \leqslant \pi/M$, is equal to (assuming no aliasing)

$$H_{\text{EQ}}(e^{j\omega}) = H_2(e^{j\omega})H_1(e^{j\omega}) \qquad (6.15)$$

Thus in the case where $H_1(e^{j\omega})$ and $H_2(e^{j\omega})$ are lowpass filters, with bandwidth less than or equal to π/M, and with sufficient out-of-band rejection to avoid aliasing (for the decimation stage) and imaging (for the interpolation stage), the equivalent structure is that of a *lowpass filter*.

Now that we have shown that the system of Fig. 6.1(a) actually does perform lowpass filtering of $x(n)$, the important question is how the computation of this system [in multiplies per second (MPS)] compares to that of a conventional FIR lowpass filter. To answer this question consider a polyphase implementation of this system as illustrated in Fig. 6.2, in which we have combined a counterclockwise commutator model for the decimator with the counterclockwise model for the interpolator. We assume that both $h_1(n)$ and $h_2(n)$ are N-point FIR lowpass filters, where $N = QM$ (i.e., N is a multiple of the decimator rate M), and that $p_\rho(m)$ and $q_\rho(m)$ are the ρth polyphase filters of $h_1(n)$ and $h_2(n)$. We also assume that for the direct, single-rate lowpass filter, an N-point FIR filter is required. For an input sampling rate of F samples per second we require that

$$R_0^* = FN \quad \text{MPS} \tag{6.16}$$

to implement the N-point FIR filter (assuming that we don't use any symmetry in the filter to reduce computation). For the polyphase structure of Fig. 6.2 we require that

$$R_1^* = \frac{F}{M} \cdot N + \frac{F}{M} \cdot N = \frac{2FN}{M} \quad \text{MPS} \tag{6.17}$$

giving a ratio, G_1, of computation as

$$G_1 = \frac{R_0^*}{R_1^*} = \frac{M}{2} \tag{6.18}$$

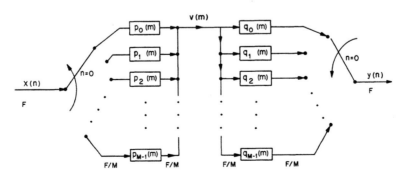

Figure 6.2 Polyphase implementation of a multirate lowpass filter.

Clearly, for $M > 2$ we get a reduction in computation for the multirate lowpass filter over the conventional FIR filter implementation.

If we define the lowpass bandwidth ω_Δ as

$$\omega_\Delta = 2\pi F_s/F \qquad (6.19)$$

where F_s is the (analog) stopband edge of a conventional lowpass filter and F is its sampling rate, then we get that

$$M = \lfloor \frac{F}{2F_s} \rfloor = \lfloor \frac{\pi}{\omega_\Delta} \rfloor \qquad (6.20)$$

where $\lfloor u \rfloor$ is the largest integer less than or equal to u, and we then get for the computational improvement ratio

$$G_1 = \lfloor \frac{\pi}{2\omega_\Delta} \rfloor \qquad (6.21)$$

showing that as $\omega_\Delta \to 0$ (narrow-band lowpass filters), the potential computational gain tends to infinity! Thus the use of a multirate structure to realize a narrow band lowpass filter is well worth considering. The price paid for this increased efficiency in computation is that the total delay of the system is doubled [i.e., it is the sum of the group delays of $h_1(n)$ and $h_2(n)$]. This is easily seen from Fig. 6.1(b).

6.1.1 Design Characteristics of the Lowpass Filters

We assume that the lowpass filters, $h_1(n)$ and $h_2(n)$, of Fig. 6.1(a) are N-point FIR equiripple designs with passband ripple $\widetilde{\delta}_p$ and stopband ripple $\widetilde{\delta}_s$. Also assume that the desired lowpass filter characteristic, $H_{EQ}(e^{j\omega})$, of the overall system has specifications

$$1 - \delta_p \leqslant |H_{EQ}(e^{j\omega})| \leqslant 1 + \delta_p, \ 0 \leqslant |\omega| \leqslant \omega_p < \frac{\pi}{M} \qquad (6.22a)$$

$$|H_{EQ}(e^{j\omega})| \leqslant \delta_s, \qquad |\omega| \geqslant \omega_s = \frac{\pi}{M} \qquad (6.22b)$$

then the question that remains is how to choose $\widetilde{\delta}_p$ and $\widetilde{\delta}_s$ (in terms of δ_p and δ_s) to guarantee that the multirate structure meets the specifications of Eq. (6.22).

To see how to choose ripple specifications for $H_1(e^{j\omega})$ and $H_2(e^{j\omega})$, consider the sequence of plots of Fig. 6.3 which are the transforms of the signals represented in Fig. 6.1 respectively. [We assume $|X(e^{j\omega})| = 1$ for this example.] Figure 6.3(a) shows the magnitude response of $W(e^{j\omega})$, which is seen to be identical to that of $|H_1(e^{j\omega})|$ with passband ripple $\tilde{\delta}_p$ and stopband ripple $\tilde{\delta}_s$. Following decimation the magnitude response of the signal $V(e^{j\omega'})$ [or equivalently of any of the polyphase filters $p_\rho(m)$] is as shown in Fig. 6.3(b). If $\tilde{\delta}_s$ is sufficiently small, we can assume that no aliasing occurs in the signal $V(e^{j\omega'})$. Following sample rate increase, the magnitude spectrum of $U(e^{j\omega})$ is as shown in Fig 6.3(c). Finally, application of $H_2(e^{j\omega})$ gives the signal $Y(e^{j\omega})$, whose magnitude spectrum is shown in Fig. 6.3(d). From this figure we see that the equivalent filter response is

$$1 - 2\tilde{\delta}_p \leqslant |H_{EQ}(e^{j\omega})| \leqslant 1 + 2\tilde{\delta}_p, \ 0 \leqslant |\omega| \leqslant \omega_p < \frac{\pi}{M} \qquad (6.23a)$$

$$|H_{EQ}(e^{j\omega})| \leqslant \tilde{\delta}_s, \qquad\qquad |\omega| > \frac{\pi}{M} \qquad (6.23b)$$

It is also seen from Fig. 6.3(d) that for the imaged transition regions the magnitude response is not equiripple, but instead the response in the image transition regions is strictly less than $\tilde{\delta}_s$; however, this property cannot be conveniently used in the filter design procedure.

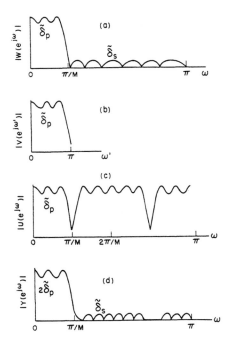

Figure 6.3 Illustration of the spectra of the signals in the diagram of Fig. 6.1.

Based on Eqs. (6.22) and (6.23) we can now see that the specifications on

$H_1(e^{j\omega})$ and $H_2(e^{j\omega})$ are

$$\widetilde{\delta}_p \leqslant \frac{\delta_p}{2} \tag{6.24a}$$

$$\widetilde{\delta}_s \leqslant \delta_s \tag{6.24b}$$

to guarantee that the resulting lowpass filter, $|H_{EQ}(e^{j\omega})|$, meets all the required filtering specifications. Furthermore, we require δ_s be sufficiently small to ensure that no aliasing occurs in the decimation stage and no imaging occurs in the interpolation stage (e.g., $\delta_s < 0.001$ is sufficient for most applications).

By way of example Fig. 6.4(a) and (b) show the "impulse" and "frequency" responses of a multirate lowpass filter with design specifications $F = 1$, $f_p = \omega_p/2\pi = 0.025$, $f_s = \omega_s/2\pi = 0.05$, $\delta_p = 0.01$, and $\delta_s = 0.001$ which was realized by a structure of the type shown in Fig. 6.1(a) (or Fig. 6.2) with $M = 10$. The individual filter lengths required for $h_1(n)$ and $h_2(n)$ were 121 points each. (The standard fixed rate system required a 110-point filter because of the different specification on δ_p.) Examination of Fig. 6.4(a) shows that the equivalent filter impulse response is that of a 241-point filter $[h_1(n)s(n)] * h_2(n)$, and the frequency response shows the characteristic gaps due to the imaging of the decimated signal transition band.

6.1.2 Multistage Implementations of the Lowpass Filter Structure

In Chapter 5 we showed that, in general, a decimator (or an interpolator) with a sufficiently large value of M (or L) could be realized more efficiently in a multistage structure than in a single-stage structure. Clearly, this same line of reasoning can be applied to the multirate lowpass filter system of Fig. 6.1(a), leading to structures with even greater efficiencies than that given in Eq. (6.18). This situation is depicted in Figure 6.5. Fig. 6.5(a) illustrates the same one-stage decimator/interpolator realization as in Fig. 6.1 (i.e., one stage of decimation followed by one stage of interpolation). Alternatively, Fig. 6.5(b) illustrates an equivalent I-stage design involving I stages of decimation followed by I stages of interpolation.

To see the potential reductions in computation in going from a one-stage to an I-stage design for both the decimator and the interpolator, the curves of Figs. 5.10 to 5.13 can be used. (Equivalently the curves in Figs. 5.21 through 5.24 can be used for IIR designs.) If we denote the reduction in computation for an I-stage design as compared to a one-stage design as

$$G_{I/1} = \frac{\widetilde{S}_{(I-1)}}{\widetilde{S}_{(I)}} \tag{6.25}$$

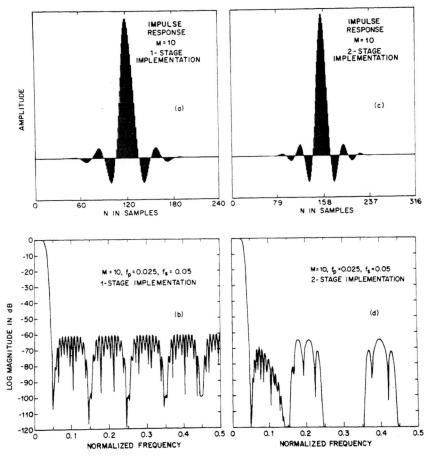

Figure 6.4 Impulse response and log magnitude frequency responses for a single and two-stage implementation of a simple lowpass filter.

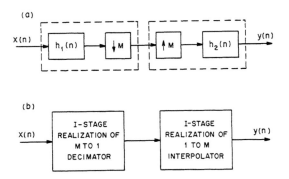

Figure 6.5 Block diagram of a multistage realization of a multirate lowpass filter structure.

where we read off values of $\widetilde{S}$ from the curves of Figs. 5.10 to 5.13, we see that the total reduction in computation for an I-stage design, compared to a direct-form, fixed-rate implementation, is

$$G_I = G_1 \cdot G_{I/1} \qquad (6.26a)$$

$$= \frac{M}{2} \frac{\widetilde{S}_{(I-1)}}{\widetilde{S}_{(I)}} \qquad (6.26b)$$

where G_1 is obtained from Eq. (6.18). Slightly greater gains can be achieved using the multiple stopband designs discussed in Chapters 4 and 5.

By way of example, consider again the lowpass filter with $F = 1.0$, $f_p = 0.025$, $f_s = 0.05$, $\delta_p = 0.01$, and $\delta_s = 0.001$. We can use a value of $M = 10$, giving $G_1 = M/2 = 5$, and from design curves of the type given in Chapter 5 we get $G_{I/1} = 2$ for $I = 2$, giving a potential speedup of about 10 to 1 over a conventional lowpass filter implementation. The actual filters for a two-stage realization of the lowpass filter were designed and the entire multirate, multistage lowpass filter was implemented. The resulting impulse and frequency responses of the two-stage lowpass filter are shown in Fig. 6.4(c) and (d), respectively [note that the time scales in Fig. 6.4(a) and (c) are different]. The remarkable similarity of the impulse response of the two-stage structure to that of the one-stage structure [Fig. 6.4(a)] is clearly seen. Due to the different imaging characteristics of the two-stage structure, the stopband frequency response of Fig. 6.4(d) is markedly different from the stopband frequency response of the one-stage structure of Fig. 6.4(b). However, the peak stopband ripple is still δ_s, indicating that no simple changes can be made to the specifications of the lowpass filters in each individual stage of the design.

6.1.3 Some Comments on the Resulting Lowpass Filters

It should be obvious to the reader from the discussion above and the material presented in Chapter 5 that the use of multistage implementations makes the filter design and implementation significantly more efficient and less sensitive than that required for single-stage structures. For example, the 317-point impulse response of Fig. 6.4(c) was obtained from the convolution of an $N = 25$-point filter, at one rate, with an $N = 27$-point filter at a lower rate. Direct design of such a 317-point filter would be affected by the sensitivities of almost any known design method for FIR filters.

It can also be seen that the implementations of the individual filters in the multistage structure leads to significantly less round-off noise than for the single-stage structure since the round-off noise is directly related to filter size. Additionally, round-off noise generated in one stage of a multistage structure is often partially filtered out by a subsequent stage. Hence the multistage structure has side advantages other than the simple reduction of computation. (A disadvantage is that the system is not strictly time-invariant and care must be observed in the design to minimize effects of aliasing and imaging.)

Although we have concentrated our attention on the use of FIR filters in each stage of the multirate, multistage implementation of lowpass filters, it should be clear that we could equally well use IIR filters. The trade-offs between FIR and IIR filters, as discussed in Chapter 4 for single-stage designs and in Chapter 5 for multistage designs, apply directly to the lowpass filter implementation problem. Hence we will not discuss these issues separately here.

6.1.4 Design Example Comparing Direct and Multistage Implementations of a Lowpass Filter

To illustrate more explicitly the concepts discussed in Section 6.1, we conclude this section with a design example that illustrates the gains in efficiency attainable using a multirate, multistage implementation of a narrow-band lowpass filter, over that of a standard, direct-form implementation.

Consider the design of a narrow-band lowpass filter with specifications $F = 1.0$, $f_p = \omega_p/2\pi = 0.00475$, $f_s = \omega_s/2\pi = 0.005$, $\delta_p = 0.001$, and $\delta_s = 0.0001$. Clearly, this filter is a very narrow band lowpass filter with very tight ripple specifications.

For the single-rate, standard, direct-form implementation, the *estimate* of filter order, based on the standard equiripple design formula (see Chapter 4), is

$$N_0 = \frac{D_\infty(0.001, 0.0001)}{0.00025} - f(0.001, 0.0001)(0.00025) + 1$$

$$= 15{,}590$$

that is, an extremely high order FIR filter is required to meet these filter specifications. Such a filter would never be designed in practice, and even if it could be designed, the implementation would yield excessive round-off noise and therefore would not be useful. However, for theoretical purposes, we postulate such a filter, and compute its multiplication rate (using symmetry of the impulse response) as

$$R_0^* = \frac{N_0}{2} = 7795 \quad \text{MPS}$$

For a multirate implementation, a decimation ratio of $M = 100 = 0.5/0.005$ can be used. For a one-stage implementation of the decimator and interpolator, the (estimated) required filter order (see Chapter 4) is

$$N_1 = \frac{D_\infty(0.001/2, \, 0.0001)}{0.00025} - f(0.001/2, \, 0.0001)(0.00025) + 1$$

$$= 16{,}466$$

Again we could never really design such a high-order filter, but, for theoretical purposes, we can compute its multiplication rate (again employing symmetry in both the decimator and interpolator) to give

$$R_1^* = \frac{N_1}{2(100)} (2) = 165 \quad \text{MPS}$$

resulting in a potential savings (of multiplications per second) of about 47.2 to 1 over the direct-form implementation.

For a two-stage implementation of the decimator and interpolator, a reasonable set of ratios for decimation (see Chapter 5) is $M_1 = 50$, $M_2 = 2$, resulting in filter orders of $N_1 = 423$ and $N_2 = 347$. The total multiplication rate for the two-stage structure is

$$R_2^* = 2\left[\frac{N_1}{2(50)} + \frac{N_2}{2(100)} \right] = 11.9 \quad \text{MPS}$$

resulting in a potential savings of about 655 to 1 over the direct-form structure.

Finally, if we use a three-stage implementation of the decimator and interpolator, a reasonable set of ratios is $M_1 = 10$, $M_2 = 5$, $M_3 = 2$, resulting in filter orders of $N_1 = 50$, $N_2 = 44$, and $N_3 = 356$. The total multiplication rate is then

$$R_3^* = 2\left[\frac{N_1}{2(10)} + \frac{N_2}{2(50)} + \frac{N_3}{2(100)} \right] = 9.4 \quad \text{MPS}$$

resulting in a savings of 829 to 1 over the direct-form implementation. (Even greater efficiencies might be possible by considering alternative design procedures, discussed in Chapter 5, for this particular example.)

By way of comparison, an elliptic filter meeting the given filter design specifications is of fourteenth order and, in a cascade realization, requires 22 MPS. This shows that for the given design example, the three-stage FIR design is about 2.3 times more efficient than a single-stage, fixed-rate IIR filter (and it has linear phase). However, it requires a significantly larger amount of storage for coefficients and data than the IIR design.

A summary of these results (for the FIR filters) is presented in Table 6.1. The key point to note is the spectacular gains in efficiency that are readily achievable by a multistage, multirate implementation over that of a direct FIR filter implementation. Furthermore, we see that, for the three-stage structure, the resulting FIR designs can be readily achieved and implemented without problem due to excessive length of the impulse response, or the excessive round-off noise in the implementation. Thus we have achieved both an efficient implementation *and* an efficient method of designing very long, very narrow bandwidth lowpass filters with tight ripple tolerances.

TABLE 6.1 COMPARISONS OF FILTER CHARACTERISTICS FOR SEVERAL MULTISTAGE IMPLEMENTATIONS OF A LOWPASS FILTER WITH SPECIFICATIONS $F = 1.0$, $f_p = 0.00475$, $f_s = 0.005$, $\delta_p = 0.001$, $\delta_s = 0.0001$

	Direct Form	One-Stage	Two-Stage	Three-Stage
Decimation Rates	—	100	50 2	10 5 2
Filter Lengths	15,590	16,466	423 347	50 44 356
MPS	7,795	165	11.9	9.4
Rate Reduction (MPS)	1	47.2	655	829
Total Storage for Filter Coefficients	7,795	8,233	385	225

6.2 MULTIRATE IMPLEMENTATION OF A BANDPASS FILTER

The techniques presented in Section 6.1 for multirate implementations of a lowpass filter can be extended directly to the bandpass case (with essentially trivial modifications) for one simple case — when the band edges of the bandpass filter are exact multiples of half the lowest sampling frequency, π/M (i.e., the case of integer-band decimation discussed in Section 2.4.2). In this case the model of Fig. 6.1 can be applied directly, and the filters $h_1(n)$ and $h_2(n)$ are appropriately designed bandpass rather than lowpass filters.

The operation of such an integer-band, multirate, bandpass filtering structure [using the model of Fig. 6.1(b)] is illustrated in Fig. 6.6. Figure 6.6(a) shows the output spectrum of the first bandpass filter. This output is nonzero over the integer-band from $\omega = l\pi/M$ to $\omega = (l + 1)\pi/M$ where l is an integer in the range $0 \leqslant l \leqslant M - 1$. This bandpass signal is modulated by the sampler whose spectrum is shown in Fig. 6.6(b). The modulated signal, $u(n)$, is of the form

$$u(n) = s(n) \cdot w(n) \qquad (6.27)$$

so its transform, $U(e^{j\omega})$, is of the form

$$U(e^{j\omega}) = S(e^{j\omega}) \circledast W(e^{j\omega}) \qquad (6.28)$$

where $\circledast$ is circular convolution. The spectrum of $U(e^{j\omega})$ is given in Fig. 6.6(c) and is seen to consist of a periodic repetition of the spectrum of the integer-band signal. Finally, filtering of $u(n)$ by the bandpass filter $h_2(n)$ [with passband from $\omega = l\pi/M$ to $\omega = (l+1)\pi/M$] restores the original signal as shown in Fig. 6.6(d).

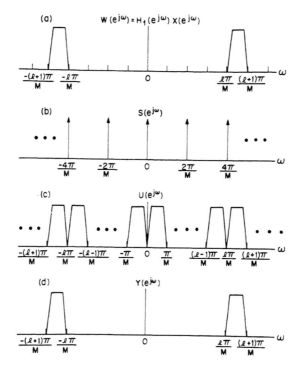

Figure 6.6 Filtering considerations for an integer-band structure for bandpass filtering.

Based on the discussion above we see that a multirate structure (using the efficient structures in Chapter 3 or 5) can perform bandpass filtering with efficiencies similar to that of the lowpass filtering structure whenever the bandpass filters satisfy a set of integer-band relations. If we assume that an FIR bandpass filter is desired with specifications of the type shown in Figure 6.7, then the integer-band specifications are:

$$M = \frac{\pi}{\omega_{s_u} - \omega_{s_l}} \qquad\qquad (6.29\text{a})$$

$$\omega_{s_l} = \frac{l\pi}{M}, \ l \text{ an integer, } 0 \leqslant l \leqslant M - 1 \qquad\qquad (6.29\text{b})$$

$$\omega_{s_u} = \frac{(l + 1)\pi}{M} \qquad\qquad (6.29\text{c})$$

Equation (6.29a) specifies the potential decimation rate of the structure, and Eq. (6.29b) specifies the band number l and location of the lower stopband edge. If the desired bandpass filter can meet the specifications of Eqs. (6.29), the techniques of Section 6.1 can be applied directly.

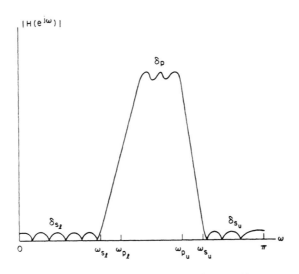

Figure 6.7   The frequency domain specifications of a bandpass filter.

6.2.1 Pseudo Integer-Band, Multirate Bandpass Filter Implementations

In the case where a desired bandpass filter does not exactly meet the specifications of Eqs. (6.29), a pseudo integer-band problem can be defined such that there is some set of M and l that satisfy Eqs. (6.29). The way in which the pseudoband is defined is to generalize (and weaken) the integer-band constraints to the form

$$M \leqslant \frac{\pi}{\omega_{s_u} - \omega_{s_l}} \tag{6.30a}$$

$$\omega_{s_l} \geqslant \frac{l\pi}{M}, \ l \text{ an integer} \tag{6.30b}$$

$$\omega_{s_u} \leqslant \frac{(l + 1)\pi}{M} \tag{6.30c}$$

In this manner, by accepting a decimation rate M less than the optimum rate M^* (where Eq. (6.30a) is solved with equality) a set of integer-bands can usually be found such that the constraints of Eqs. (6.30b) and (6.30c) are met. This situation is illustrated in Fig. 6.8. Figure 6.8(a) shows the desired passband of the bandpass filter, and shows where the band edges of the nearest integer-band fall with respect to the filter band edges. It can be seen that the filter band cannot be made simply to lie within a fixed integer-band at the maximum decimation rate (M^*). Hence the solution of Eq. (6.30a) is to lower the decimation rate until the desired band can be made to lie within an integer-band, as illustrated in Fig. 6.8(b). For most practical cases of interest, some value of decimation rate exists such that the pseudo integer-band can be defined. In this manner some of the expected efficiency of the realization is traded for ease of implementation.

The important question about the pseudo integer-band realizations is how to find suitable values of M such that Eqs. (6.30) are satisfied. Clearly, for small bandwidth filters, a large set of possible values of M must be checked, and for each possible M we must check if a pseudo integer-band can be found. Furthermore, for multistage implementations, the search for an acceptable value of M for each stage is even more tedious and time consuming. Mintzer and Liu [6.3] have carefully investigated this problem and have concluded that a computerized search is

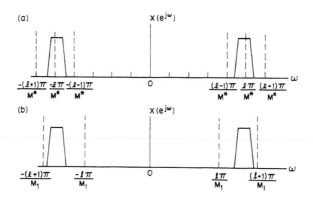

Figure 6.8 Creation of a pseudo integer-band by employing a decimation factor M_1, smaller than the optimal rate M^*.

required to find all possible (reasonable) sets of solutions to Eqs. (6.30) for the general case.

6.2.2 Alternative Multirate Implementations of Bandpass Filters

As discussed in Sections 2.4.3 and 2.4.4, alternative approaches to integer-band sampling for bandpass signals include quadrature and single-sideband modulation. Each of these procedures can be adapted to yield a multirate implementation of a bandpass filter, as long as the desired filter obeys some single design constraints (to be described later).

Figure 6.9 illustrates the quadrature modulation realization of a bandpass filter, and Fig. 6.10 shows the required filter specifications. The desired passband of the bandpass filter is the region

$$\omega_{p_l} \leqslant \omega \leqslant \omega_{p_u} \quad \text{passband} \tag{6.31a}$$

and the lower and upper stopbands are defined from

$$0 \leqslant \omega \leqslant \omega_{s_l} \quad \text{lower stopband} \tag{6.31b}$$

$$\omega_{s_u} \leqslant \omega \leqslant \pi \quad \text{upper stopband} \tag{6.31c}$$

The filter specifications for these bands are

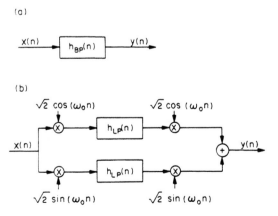

Figure 6.9 A quadrature modulation structure for bandpass filtering using lowpass filter prototypes.

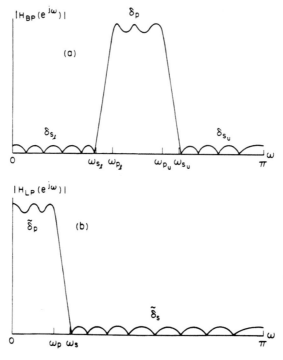

Figure 6.10 Frequency-response specifications for (a) the bandpass, and (b) the lowpass filters of a quadrature modulation scheme for bandpass filtering.

$$0 \leqslant \left| H_{\mathrm{BP}}(e^{j\omega}) \right| \leqslant \delta_{s_l} \qquad\qquad 0 \leqslant \omega \leqslant \omega_{s_l} \qquad (6.32\mathrm{a})$$

$$1 - \delta_p \leqslant \left| H_{\mathrm{BP}}(e^{j\omega}) \right| \leqslant 1 + \delta_p \qquad \omega_{p_l} \leqslant \omega \leqslant \omega_{p_u} \qquad (6.32\mathrm{b})$$

$$0 \leqslant \left| H_{\mathrm{BP}}(e^{j\omega}) \right| \leqslant \delta_{s_u} \qquad\qquad \omega_{s_u} \leqslant \omega \leqslant \pi \qquad (6.32\mathrm{c})$$

The structure for realizing the bandpass filter as a lowpass filter is shown in Fig. 6.9(b). The signal $x(n)$ is modulated so that the desired band is centered at $\omega = 0$. Examination of Fig. 6.10(a) shows that we require

$$\omega_0 = \frac{\omega_{s_u} + \omega_{s_l}}{2} \qquad (6.33)$$

for the desired passband to be centered at $\omega = 0$ following the quadrature modulation. The modulated signal is then lowpass filtered by a filter whose specifications are given in Fig. 6.10(b) For the lowpass filter to be "equivalent" to the desired bandpass filter, the following constraints must be made on the bandpass response:

1. Symmetry around $\omega = \omega_0 = (\omega_{s_u} + \omega_{s_l})/2$

$$\omega_0 - \omega_{p_l} = \omega_{p_u} - \omega_0 \qquad (6.34\text{a})$$

$$\omega_0 - \omega_{s_l} = \omega_{s_u} - \omega_0 \qquad (6.34\text{b})$$

or, equivalently,

$$\omega_{p_l} - \omega_{s_l} = \omega_{s_u} - \omega_{p_u} \qquad (6.35)$$

Thus the widths of the lower and upper transitions bands must be equal.

2. Ripple symmetry:

$$\delta_{s_l} = \delta_{s_u} \qquad (6.36)$$

If the constraints of Eqs. (6.34) to (6.36) are met, the lowpass filter specifications become

$$\widetilde{\delta}_p = \delta_p \qquad (6.37\text{a})$$

$$\widetilde{\delta}_s = \delta_{s_l} = \delta_{s_u} \qquad (6.37\text{b})$$

$$\omega_p = \omega_{p_u} - \omega_0 \qquad (6.37\text{c})$$

$$\omega_s = \omega_{s_u} - \omega_0 \qquad (6.37\text{d})$$

and the desired bandpass filtering is achieved by the structure of Figure 6.9(b).

To achieve efficiency, the lowpass filters of Fig. 6.9(b) are realized in a multirate, multistage structure, as discussed in Section 6.1.

In summary, standard narrow-band bandpass filters can be implemented in a multirate, multistage structure using any of the bandpass sampling techniques of Chapter 2, whereby the bandpass filtering is either restricted to a band of frequencies related to the frequency of the equivalent modulation structure (the integer-band case) or the bandpass filtering is converted to an equivalent lowpass filtering problem (via a suitable modulation technique) and the efficient lowpass filter structures are utilized.

6.2.3 Multirate Implementation of Narrow-Band Highpass and Bandstop Filters

In the preceding sections we have shown how multirate techniques can be applied to the implementation of standard lowpass and bandpass filters, yielding structures whose efficiency increases as the signal bandwidth (i.e., the width of the passband) decreases. It should be clear that we could equally well apply these techniques to narrow stopband highpass and bandstop filters, by realizing such filters as

$$H_{\mathrm{HP}}(e^{j\omega}) = 1 - H_{\mathrm{LP}}(e^{j\omega}) \tag{6.38a}$$

$$H_{\mathrm{BS}}(e^{j\omega}) = 1 - H_{\mathrm{BP}}(e^{j\omega}) \tag{6.38b}$$

where HP, LP, BS, and BP refer to highpass, lowpass, bandstop, and bandpass designs, respectively. The computation of Eqs. (6.38), as illustrated in Fig. 6.11, consists of lowpass (or bandpass) filtering the signal $x(n)$ and then subtracting the filtered signal from the unfiltered input. The lowpass or bandpass filtering shown in Fig. 6.11, when implemented via a multirate structure, leads to an efficient structure for highpass and bandstop filters. The net result is a structure whose efficiency increases as the width of the stopband decreases. Note that in the definition above we are assuming that the filter $h_{\mathrm{LP}}(n)$ or $h_{\mathrm{BP}}(n)$ is a zero-phase (i.e., zero group delay) FIR filter. In practice, the signal $x(n)$ in Fig. 6.11 must be delayed by the delay of the filter before the difference is taken. Thus it is important to design the filter $h_{\mathrm{LP}}(n)$ or $h_{\mathrm{BP}}(n)$ to have a flat delay of an integer number of samples for this method to work.

The equivalent filter characteristics, when using the structures of Fig. 6.11, of the highpass and bandstop filters have inverted roles for the ripple factors δ_p and δ_s;

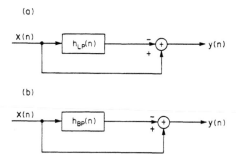

Figure 6.11 Implementations for performing (a) highpass filtering; and (b) bandstop filtering using lowpass and bandpass filters respectively.

for example, δ_p of the lowpass is the equivalent δ_s of the highpass, and vice versa. Thus for correct ripple specification the equivalent lowpass specifications must be properly chosen.

One important application of the bandstop filtering scheme of Fig 6.11(b) is the case of narrow-band notch filtering, where a narrow band of frequencies (e.g., 60-Hz hum) is removed from a signal. In such a case the equivalent bandpass bandwidth is quite small and a multirate structure can achieve very high efficiency compared to that of a standard notch filter.

6.3 DESIGN OF FRACTIONAL SAMPLE PHASE SHIFTERS BASED ON MULTIRATE CONCEPTS

In many signal processing applications there is a need for a network which essentially delays the input signal by a fixed number of samples. When the desired delay is an integer number of samples, at the current sampling rate, such a network is trivially realized as a cascade of unit delays. However, when delays of a fraction of a sample are required, the processing required to achieve such a delay is considerably more difficult. We show here how multirate signal processing concepts can be used to greatly simplify the process required to design noninteger delays as long as the desired delay is a rational fraction of a sample.

Consider the ideal delay network $h_{AP}(n)$ of Fig. 6.12(a). The desired all-pass network processes $x(n)$ to give the output $y(n)$ so that the relationship between their Fourier transforms is

$$Y(e^{j\omega}) = e^{-j\omega l/M} X(e^{j\omega}) \qquad (6.39)$$

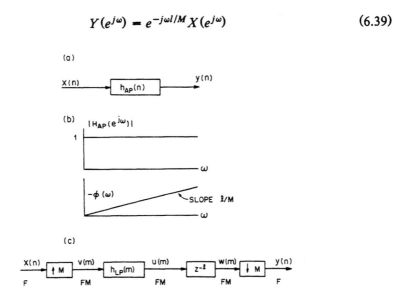

Figure 6.12 A multirate structure for realizing a fixed delay of l/M samples.

where l and M are any integers. In the time domain this amounts to a delay of the envelope of the signal $x(n)$ by a fraction of a sample l/M. The magnitude and phase responses of $h_{\text{AP}}(n)$, shown in Fig. 6.12(b), are of the form

$$H_{\text{AP}}(e^{j\omega}) = |H_{\text{AP}}(e^{j\omega})|e^{j\phi(\omega)} \tag{6.40a}$$

where

$$|H_{\text{AP}}(e^{j\omega})| = 1 \tag{6.40b}$$

$$\phi(\omega) = -\frac{l\omega}{M} \tag{6.40c}$$

and where $\phi(\omega)$ denotes the phase of $H_{\text{AP}}(e^{j\omega})$. (At this point the observant reader should notice the similarity between the desired response and that of a polyphase filter. We return to this equivalence later in this section.) It should be clear that for arbitrary values of l and M, the desired frequency response of Eq. (6.40) cannot be achieved exactly by either an FIR or an IIR filter. An FIR filter (all zeros) cannot achieve an exact all-pass magnitude response, and an IIR filter (poles and zeros) cannot achieve an exact linear phase response. Thus the desired noninteger delay network $h_{\text{AP}}(n)$ cannot be realized exactly but only approximated through some design procedure.

Using multirate principles, this design problem can be clearly defined as illustrated in Fig. 6.12(c). The key to this procedure is the realization that a delay of l/M samples at rate F is equivalent to a delay of l samples (i.e., an integer delay) at rate FM. Hence the structure in Fig. 6.12(c) first raises the sampling rate of the signal to FM, filters the signal with a lowpass filter $h_{\text{LP}}(m)$ to eliminate images of $x(n)$, delays the signal by l samples, and then decimates it back to the original sampling rate.

A simple analysis of the structure of Fig. 6.12(c) gives

$$V(e^{j\omega}) = X(e^{j\omega M}) \tag{6.41}$$

$$U(e^{j\omega}) = H_{\text{LP}}(e^{j\omega})X(e^{j\omega M}) \tag{6.42}$$

$$W(e^{j\omega}) = U(e^{j\omega})e^{-j\omega l}$$

$$= H_{\text{LP}}(e^{j\omega})X(e^{j\omega M})e^{-j\omega l} \tag{6.43}$$

$$Y(e^{j\omega}) = \frac{1}{M}\sum_{r=0}^{M-1} W(e^{-j2\pi r/M}e^{j\omega/M}) \tag{6.44}$$

We assume that $H_{\text{LP}}(e^{j\omega})$ sufficiently attenuates the images of $X(e^{j\omega})$ so that they are negligible in Eq. (6.44), thereby giving only the $r = 0$ term, that is

$$Y(e^{j\omega}) \simeq \frac{1}{M} W(e^{j\omega/M})$$

$$\simeq \frac{1}{M} H_{LP}(e^{j\omega/M}) e^{-j\omega l/M} X(e^{j\omega}) \qquad (6.45)$$

We further assume that $H_{LP}(e^{j\omega})$ is an FIR filter with exactly linear phase, whose delay (at the high rate) is $(N-1)/2$ samples, and this value is chosen to be an integer delay at the low rate, that is,

$$\frac{N-1}{2} = IM \qquad (6.46a)$$

or

$$N = 2IM + 1 \qquad (6.46b)$$

We also require $H_{LP}(e^{j\omega})$ to have a magnitude response essentially equal to M (to within a small tolerance) in the passband, thereby giving for $Y(e^{j\omega})$

$$Y(e^{j\omega}) \simeq e^{-j\omega l} e^{-j\omega l/M} X(e^{j\omega}) \qquad (6.47)$$

or, as an equivalent network,

$$\frac{Y(e^{j\omega})}{X(e^{j\omega})} \simeq e^{-j\omega l} e^{-j\omega l/M} \qquad (6.48)$$

Thus the structure of Fig. 6.12(c) is essentially an all-pass network with a fixed integer delay of I samples, and a variable, noninteger delay of l/M samples.

One efficient implementation of the multirate allpass filter of Fig. 6.12(c) is given in Fig. 6.13. A polyphase structure is used to realize the sampling rate increase and lowpass filtering based on the counterclockwise commutator structure of Fig. 3.25, where

$$p_\rho(n) = h_{LP}(nM + \rho), \quad 0 \leqslant \rho \leqslant M - 1 \qquad (6.49)$$

The delay of l samples is implemented as a new initial position of the commutator corresponding to the $m = 0$ sample. Finally, the decimation by M is implemented as a fixed arm position of the commutator, since each M samples it is back at the original position.

Thus for a single *fixed* delay of l/M samples only one branch (corresponding to the $\rho = l$th polyphase filter branch) is required, and for a network that requires all possible values of l (from 0 to $M-1$) the entire network of Fig. 6.13 is required, i.e., it represents a selectable choice of M different fractional delays.

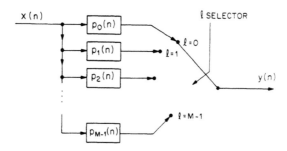

Figure 6.13 A polyphase network implementation of a fractional sample delay network.

If one wanted to consider a network that implemented phase advances of l/M samples (rather than delays), relative to some fixed overall delay, the clockwise commutator model of the polyphase network realization could be used where

$$\bar{p}_\rho(n) = h_{\text{LP}}(nM - \rho),\ 0 \leqslant \rho \leqslant M - 1 \tag{6.50}$$

An equivalent structure to that of Fig. 6.13 results from this approach.

A key point made in this section is that the design of an N-tap time-invariant filter with a noninteger delay (and flat magnitude response), such as $h_{\text{AP}}(n)$ in Fig. 6.12(a), can be readily transformed to that of an NM-tap lowpass filter design. This transformation is accomplished by means of an appropriate multirate interpretation of the problem and an application of the concept of polyphase filters. Any of the conventional lowpass filter designs discussed in Chapter 4 can be used. Also the same lowpass filter design gives a complete family of M phase shifter designs for fractional delays of l/M, $0 \leqslant l \leqslant M - 1$.

6.3.1 Design of Phase Shifter Networks with Fixed Phase Offsets

If we consider replacing the lowpass filter of the phase shifter network of Fig. 6.12(c) with an integer-band, bandpass filter, as shown in Fig. 6.14(a), then the phase shifter network is essentially converted to a phase offset network in addition to a phase shifter network. This idea is illustrated in Fig. 6.14(b), where we assume that that the passband of the filter is the integer-band $r\pi/M \leqslant \omega \leqslant (r + 1)\pi/M$. Therefore, one of the harmonic images of $X(e^{j\omega})$ is filtered and phase shifted rather than the baseband signal itself [i.e., one of the terms other than the $r = 0$ term in Eq. (6.44) is used]. We will assume, for the sake of discussion, that the bandpass filter $h_{\text{BP}}(n)$ has zero delay. Then the phase response of the output of the l-sample delay is as shown in Fig. 6.14(b):   that is, an initial phase offset of $lr\pi/M$, at the lower edge of the band and a phase slope of

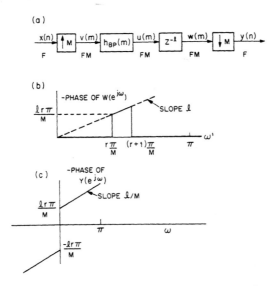

Figure 6.14 A multirate structure for realizing a fixed phase offset plus a fixed delay of l/M samples.

l. Following decimation the phase response of $y(n)$ is as shown in Fig. 6.14(c). The resulting lowpass signal has a phase offset of $lr\pi/M$ with a phase slope of l/M. Thus a phase offset, between positive and negative frequencies of magnitude $2lr\pi/M$, is achieved by the network of Fig. 6.14(a). The utility of such a network will be seen in the next section, where we discuss Hilbert transform networks where a phase shift of $\pi/2$ is required for positive frequencies, and $-\pi/2$ for negative frequencies. For the case $lr = M/2$, such a phase shift (together with the delay of l/M samples) is achieved. Another application of the phase offset network is in speech coding, as described by Crochiere [6.10].

6.4 MULTIRATE IMPLEMENTATION OF A HILBERT TRANSFORMER

The Hilbert transform is a set of mathematical relationships between the real and imaginary parts, or the magnitude and phase of the Fourier transform of certain signals (e.g., causal and minimum phase sequences) [6.6, 6.7]. Based on the properties of the discrete Hilbert transform, a network called a *Hilbert transformer* can be defined, with the ideal frequency response

$$H_{\text{HT}}(e^{j\omega}) = \begin{cases} -j, & 0 \leqslant \omega < \pi \\ +j, & -\pi \leqslant \omega < 0 \end{cases} \tag{6.51}$$

and with the ideal impulse response

$$h_{HT}(n) = \begin{cases} \dfrac{2}{\pi}\dfrac{\sin^2(\pi n/2)}{n}, & n \neq 0 \\ 0, & n = 0 \end{cases} \tag{6.52}$$

These responses are shown in Fig. 6.15. It can readily be seen that the ideal Hilbert transformer is noncausal and infinite in duration, and therefore cannot be realized exactly (just as an ideal lowpass or bandpass filter cannot be realized exactly). Thus to implement a Hilbert transformer network some approximation technique is required. A variety of such approximation techniques, based on FIR design techniques, have been proposed [6.6-6.8]. We will not discuss these methods directly, but instead we will show in this section one possible implementation using half-band filters [6.11]. If we denote the input to a Hilbert transformer network as $x(n)$, and the output of the network as $\hat{x}(n)$, then, for many communications problems, the signal of interest is the so-called (complex) analytic signal

$$x_A(n) = x(n) + j\hat{x}(n) \tag{6.53}$$

which has the property that

$$X_A(e^{j\omega}) = \begin{cases} 2X(e^{j\omega}), & 0 \leqslant \omega < \pi \\ 0, & -\pi \leqslant \omega \leqslant 0 \end{cases} \tag{6.54}$$

Figure 6.16 illustrates the spectral properties of typical signals $X(e^{j\omega})$, $\hat{X}(e^{j\omega})$, and $X_A(e^{j\omega})$. The signal bandwidth $(2\omega_0)$ can be as large as π for practical applications. The complex signal $x_A(n)$ can be reduced in sampling rate by at least a factor of 2 to 1, since $X_A(e^{j\omega}) = 0$ for $\omega < 0$.

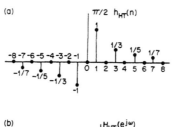

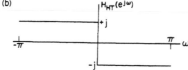

Figure 6.15   (a) Ideal impulse; and (b) frequency responses of a Hilbert network.

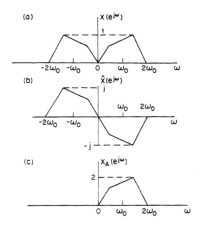

Figure 6.16 (a) Frequency responses of a signal; (b) its Hilbert transform; (c) the analytic signal.

The conventional network for obtaining the analytic signal $x_A(n)$ is shown in Fig. 6.17(a). The network $\hat{h}_{HT}(n)$ is an approximation to the ideal Hilbert transformer network. (We will assume, for the sake of discussion, that it has zero delay.) An alternative structure for obtaining the analytical signal $x_A(n)$ is shown in Fig. 6.17(b). This network is seen to consist of a modulator (whose frequency is the midpoint of the signal band) followed by a lowpass filter $h_{LP}(n)$ followed by a demodulator.

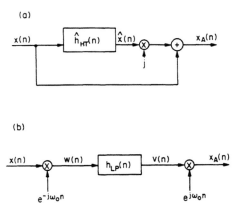

Figure 6.17 Two network structures for obtaining an analytic signal: (a) using a Hilbert network; (b) using a modulator and a half-band lowpass filter.

An analysis of the spectral properties of the signals $x(n)$, $w(n)$, $v(n)$, and

$x_A(n)$ is given in Fig. 6.18 for the case when $h_{LP}(n)$ is an ideal lowpass filter with cutoff frequency ω_0. It can be seen that the output of the network is indeed the analytic signal $x_A(n)$ associated with the real signal $x(n)$.

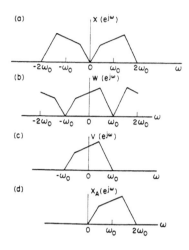

Figure 6.18 Frequency responses of typical signals in the modulator structure for obtaining the analytic signal.

A formal analysis of the network of Fig. 6.17(b) leads to the following set of equations. If we define the complex signal $v(n)$ as

$$v(n) = a(n) + jb(n) \tag{6.55}$$

where $a(n)$ and $b(n)$ are real sequences, we have

$$a(n) + jb(n) = \sum_{m=-\infty}^{\infty} x(m)e^{-j\omega_0 m}h_{LP}(n-m)$$

$$= e^{-j\omega_0 n} \sum_{m=-\infty}^{\infty} x(m)e^{j\omega_0(n-m)}h_{LP}(n-m)$$

$$= e^{-j\omega_0 n}\left\{ x(n) * [h_{LP}(n)e^{j\omega_0 n}] \right\} \tag{6.56}$$

Then the output of the network is given as

$$x_A(n) = x(n) + j\hat{x}(n)$$

$$= [a(n) + jb(n)]e^{j\omega_0 n}$$

$$= e^{-j\omega_0 n} \left\{ x(n) * [h_{LP}(n)e^{j\omega_0 n}] \right\} e^{j\omega_0 n}$$

$$= x(n) * [h_{LP}(n)e^{j\omega_0 n}] \tag{6.57}$$

or, equating real and imaginary parts, we get

$$x(n) = x(n) * [h_{LP}(n) \cos(\omega_0 n)] \tag{6.58}$$

$$\hat{x}(n) = x(n) * [h_{LP}(n) \sin(\omega_0 n)] \tag{6.59}$$

If we consider the case where $\omega_0 = \pi/2$ (i.e., a full band input signal), we have

$$\cos(\omega_0 n) = \begin{cases} (-1)^{n/2}, & n \text{ even} \\ 0, & n \text{ odd} \end{cases} \tag{6.60}$$

and

$$\sin(\omega_0 n) = \begin{cases} 0, & n \text{ even} \\ (-1)^{(n-1)/2}, & n \text{ odd} \end{cases} \tag{6.61}$$

By comparing Eqs. (6.58) and (6.60), we see that in order for the real part, $x(n)$, to remain unchanged, we require that

$$h_{LP}(n) = \begin{cases} 1, & n = 0 \\ 0, & n \text{ even}, n \neq 0 \\ \text{arbitrary}, & n \text{ odd} \end{cases} \tag{6.62}$$

Then Eq. (6.58) becomes

$$x(n) = x(n) * u_0(n) = x(n) \tag{6.63}$$

where $u_0(n)$ is the unit sample function. The condition in Eq. (6.62) is identical to that for the half-band filters discussed in Chapters 4 and 5 for sampling rate conversion by a factor of 2. Thus $h_{LP}(n)$ must be a half-band lowpass filter.

By comparing Eqs. (6.59) and (6.61) and the Hilbert transform relation [see Fig. (6.17)], we get

$$\hat{x}(n) = x(n) * h_{HT}(n) \tag{6.64}$$

We see that the coefficients of the Hilbert transform satisfy the condition

$$h_{HT}(n) = \begin{cases} (-1)^{(n-1)/2} h_{LP}(n), & n \text{ odd} \\ 0, & n \text{ even} \end{cases} \tag{6.65a}$$

$$= (-1)^{(n-1)/2} \hat{p}_1(n) \tag{6.65b}$$

where $\hat{p}_1(n)$ is the $\rho = 1$ polyphase filter of $h_{LP}(n)$ expanded by 2 to 1. Thus the coefficients of the Hilbert network are seen to be identical to half-band filter designs discussed in Chapter 4 with appropriate sign changes according to Eq. (6.65) and with $h_{HT}(0) = 0$.

The fact that the even-valued samples of $h_{HT}(n)$ are zero can be applied in an efficient way in the direct-form FIR filter realization (see Fig. 6.17) by removing multiplications and summations associated with even taps of the direct-form structure. An alternative way is by recognizing that even samples of $\hat{x}(n)$ are computed from odd samples of $x(n)$ and odd samples of $\hat{x}(n)$ are computed from even samples of $x(n)$. This leads to the multirate realization shown in Fig. 6.19, in which even and odd samples of $x(n)$ are filtered separately at the reduced rate by the filters $(-1)^n p_1(n)$, where $p_1(n)$ is the $\rho = 1$ polyphase filter (i.e., the odd samples) of the lowpass filter $h_{LP}(n)$. This structure can be compared with the Hilbert network of Fig. 6.17. Although both structures can potentially be realized with the same efficiency, the manner in which the data are accessed and stored in each structure is different.

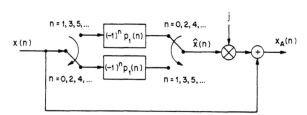

Figure 6.19 Multirate network for realizing the Hilbert transform.

6.5 NARROW-BAND, HIGH-RESOLUTION SPECTRAL ANALYSIS USING MULTIRATE TECHNIQUES

One of the most fundamental signal processing operations is spectral analysis: that is, evaluation of the z-transform of a sequence over a band of frequencies on the unit circle. Formally, if we define an N-point sequence, $x(n)$, $n = 0, 1, ..., N-1$, with z-transform

$$X(z) = \sum_{n=0}^{N-1} x(n) z^{-n} \tag{6.66}$$

then the spectrum of $x(n)$ at frequency ω is

$$X(e^{j\omega}) = X(z)\Big|_{z=e^{j\omega}} = \sum_{n=0}^{N-1} x(n) e^{-j\omega n} \tag{6.67}$$

Although a variety of techniques have been proposed for evaluating $X(e^{j\omega})$ of Eq. (6.67), perhaps the most commonly used one is the class of FFT algorithms. These algorithms evaluate, in an efficient manner, the z-transform of a finite (or periodic) sequence at a set of equispaced frequencies

$$\omega_k = k\frac{2\pi}{N}, \quad k = 0, 1, ..., N-1 \tag{6.68}$$

giving

$$X(e^{j\omega_k}) = X(k) = \sum_{n=0}^{N-1} x(n) e^{-j(2\pi/N)nk} \tag{6.69}$$

From the set of FFT coefficients $X(k)$, $k = 0, 1, ..., N-1$, the original sequence can be calculated via an inverse FFT relation of the form

$$x(n) = \frac{1}{N} \sum_{k=0}^{N-1} X(k) e^{j(2\pi/N)nk} \tag{6.70}$$

The set of FFT and inverse FFT relations, of Eqs. (6.69) and (6.70), provide a powerful framework for wide-band digital spectral analysis of signals, and have been used in this manner in a wide variety of applications.

One major disadvantage of the FFT algorithm is that it automatically provides spectral information over a uniformly spaced set of frequencies covering the entire band of the signal of interest. For some applications only the spectral information in a narrow band of frequencies is of interest, and for such cases it is highly inefficient to make all the calculations of Eq. (6.69) and then to use a small subset of them. By using the techniques of multirate bandpass filtering, the signal band of interest can be shifted to low frequencies, the sampling rate can be reduced to twice the bandwidth of the desired signal, and a low-order FFT analysis can be performed, giving the desired high-resolution, narrow-band spectral data.

Figure 6.20 illustrates the principles used to derive a multirate structure for high-resolution, narrow-band spectral analysis. In Fig. 6.20(a) we show the multirate structure for bandpass filtering for an integer-band signal. (We first assume that this is the case. Later we discuss noninteger band signals.) The decimation rate is M, defined as

$$M = \frac{\pi}{\omega_{s_u} - \omega_{s_l}} \qquad (6.71)$$

where ω_{s_l} and ω_{s_u} are the lower and upper frequencies of interest in the bandpass signal. The signal is first bandpass filtered to isolate the desired band of interest, and the resulting signal is decimated by a factor of M. Since we assumed integer-bands, we have

$$\omega_{s_l} = \frac{r\pi}{M} \qquad (6.72a)$$

$$\omega_{s_u} = \frac{(r+1)\pi}{M} \qquad (6.72b)$$

for some value of r, $1 \leqslant r \leqslant M - 1$. The resulting signal, $w(m)$, has a lowpass spectrum as shown in Fig. 6.20(b). For the lowpass signal (assuming negligible aliasing), the frequency ω' corresponds to the frequency ω in $X(e^{j\omega})$ such that

$$\omega' = \frac{(\omega - \omega_{s_l})\pi}{\omega_{s_u} - \omega_{s_l}} \qquad (6.73)$$

To obtain the desired high-resolution, narrow-band spectrum, the system of Fig. 6.20(c) is used, in which an N'-point FFT is performed on $w(m)$, yielding spectral values for frequencies

$$\omega' = \frac{2\pi k}{N'}, \quad k = 0, 1, ..., N' - 1 \qquad (6.74)$$

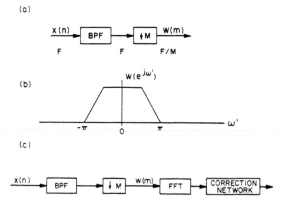

Figure 6.20 Multirate structure for performing high-resolution, narrow-band spectral analysis.

and these values are translated back to the ω range using Eq. (6.73). The value of N' is, approximately,

$$N' = \frac{N}{M} \tag{6.75}$$

since the decimated signal has only N/M samples. A correction network is shown which can correct for the imperfect passband response of the bandpass filter.

The improvement in efficiency derives from the use of a considerably smaller value of N' than N, whenever M is large, thereby reducing the FFT computation, which is proportional to $N \log N$. However, there is extra computation due to the bandpass filter. Liu and Mintzer [6.9] have shown that for most cases of interest (bands where $M \geqslant 10$), there is a real reduction in computation using the multirate structure of Fig. 6.20(c) over the direct FFT computation.

For the case in which the desired band of interest is not an integer-band, any of the techniques of Section 6.2 can be applied: that is, a pseudo-integer band can be created, or a modulation technique can be used to create a lowpass signal. The FFT processing can then be applied to the resulting lowpass signal to provide the desired spectral analysis.

6.6 SAMPLING RATE CONVERSION BETWEEN SYSTEMS WITH INCOMMENSURATE SAMPLING RATES

Throughout the book we have considered multirate systems in which the various sampling rates within the system are related by exact integer or rational fraction ratios. That is, the sampling rates within the process are all generated from some common (higher rate) clock. In practice, however, it is sometimes desired to interface digital systems that are controlled by *independent* (incommensurate) clocks. For example, it may be desired to exchange signals between two different systems, both of which are intended to be sampled at a rate F. However, due to practical limitations on the accuracy of the clocks the first system may be sampled at an actual rate of $F + \epsilon_1$ and the second system may be sampled at an actual rate of $F + \epsilon_2$ where ϵ_1 and ϵ_2 represent slowly varying components of drift as a function of time. If we simply exchange digital signals between these two systems (e.g. by means of a sample-and-hold process) samples may be either lost or repeated in the exchange due to the relative "slippage" of the two clocks. If the signals are highly uncorrelated from sample-to-sample (i.e., sampled at their Nyquist rates) this process can introduce large spikes or errors into the signals. The rate of occurrence of these errors is directly related to the amount of sampling rate slippage; that is, to the ratio $(\epsilon_1 - \epsilon_2)/F$.

One way to avoid the above problem, as discussed in Chapter 2, is to interface the two digital systems through an analog connection. That is, to convert the signals to analog signals and then resample them with the new clock. In principle this process provides an error-free interface. In practice, however, it is limited by the practical capabilities and expense of the A/D and D/A conversion process as well as the dynamic range of the analog connection.

A more attractive all-digital approach to this problem can be accomplished by applying the multirate techniques discussed in Chapters 2 through 5 to, in effect, duplicate the analog process in digital form [6.12]. Figure 6.21 shows an example of a system for transferring a digital signal $x(n)$ from system 1 (with sampling rate $F + \epsilon_1$) to system 2 (with sampling rate $F + \epsilon_2$). The signal $x(n)$ is first interpolated by a 1-to-L interpolator (where $L \gg 1$) to produce the highly oversampled signal $y(m)$ [at sampling rate $L(F + \epsilon_1)$]. This signal is then converted to a signal $\hat{y}(m)$ [at the sampling rate $L(F + \epsilon_2)$] through a digital sample-and-hold which interfaces the two incommensurate sampling rates. It is then decimated by an L-to-1 decimator to produce the signal $\hat{x}(n)$ for input to system 2. The interface between the oversampled digital signals $y(m)$ and $\hat{y}(m)$ plays the same role as that of an analog interface and as $L \rightarrow \infty$ it is equivalent. The advantage is that the system is all-digital and the accuracy of the conversion can be designed with any degree of desired precision.

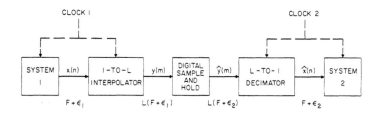

Figure 6.21 A multirate approach to sampling rate conversion between systems with incommensurate sampling rates.

Although samples may be repeated or dropped in the sample-and-hold conversion process between $y(m)$ and $\hat{y}(m)$ it can be shown that the effects of these errors become small as L becomes large. This can be seen by considering the sample-to-sample difference of the signal $y(m)$ [or $\hat{y}(m)$] relative to its actual value as a function of L. This difference determines the relative energy in the error spike when a sample is dropped or repeated. To see this we can define the ratio of the energy in the sample-to-sample difference signal to that in the signal itself as

$$E = \frac{\sum_{m=-\infty}^{\infty} [y(m) - y(m-1)]^2}{\sum_{m=-\infty}^{\infty} [y(m)]^2} \tag{6.76}$$

Furthermore we can note that the sample-to-sample correlation of the signal $y(m)$

can be defined as

$$C \triangleq \frac{\sum\limits_{m=-\infty}^{\infty} y(m)y(m-1)}{\sum\limits_{m=-\infty}^{\infty} [y(m)]^2} \qquad (6.77)$$

Applying this to Eq. (6.76) then leads to the relation

$$E = 2(1 - C) \qquad (6.78)$$

which shows that as the signal becomes highly correlated from sample-to-sample the relative energy in the difference signal becomes small, that is as $C \to 1$, it follows that $E \to 0$.

We can now show the relationship between the sample-to-sample correlation and the oversampling ratio L by applying Parsevals relation [6.6]

$$\sum_{m=-\infty}^{\infty} y(m)v(m) = \frac{1}{2\pi} \int_{-\pi}^{\pi} Y(e^{j\omega}) V^*(e^{j\omega}) d\omega \qquad (6.79)$$

where $Y(e^{j\omega})$ and $V(e^{j\omega})$ are the Fourier transforms of $y(m)$ and $v(m)$ respectively. By letting $v(m) = y(m-1)$ in the numerator and $v(m) = y(m)$ in the denominator of Eq. (6.77) and applying Eq. (6.79) twice, it can be shown that C can be expressed in the form

$$C = \frac{\dfrac{1}{2\pi} \int_{-\pi}^{\pi} |Y(e^{j\omega})|^2 e^{j\omega} d\omega}{\dfrac{1}{2\pi} \int_{-\pi}^{\pi} |Y(e^{j\omega})|^2 d\omega}$$

$$= \frac{\int_{0}^{\pi} |Y(e^{j\omega})|^2 \cos(\omega) d\omega}{\int_{0}^{\pi} |Y(e^{j\omega})|^2 d\omega} \qquad (6.80)$$

where it has been assumed that $y(m)$ is a real signal.

Consider now the case in Fig. 6.21 where the signal $x(n)$ is a broadband signal such that its spectral magnitude $|X(e^{j\omega'})|$ is flat; that is

$$|X(e^{j\omega'})| = 1 \quad \text{for } |\omega'| \leqslant \pi \qquad (6.81)$$

Then the spectral magnitude of the interpolated signal $y(m)$ is bandlimited to the range

$$|Y(e^{j\omega})| = \begin{cases} 1, & \text{for } |\omega| < \pi/L \\ 0, & \text{otherwise} \end{cases} \qquad (6.82)$$

The correlation coefficient C in Eq. (6.80) then becomes

$$C = \frac{\displaystyle\int_0^{\pi/L} \cos(\omega)\,d\omega}{\displaystyle\int_0^{\pi/L} d\omega} = \frac{\sin\left[\dfrac{\pi}{L}\right]}{\left[\dfrac{\pi}{L}\right]} \qquad (6.83)$$

This form clearly reveals the relationship between the oversampling ratio L and the sample-to-sample correlation of the signal $y(m)$. As L becomes large (i.e., as $L \to \infty$) it can be seen that $C \to 1$ and $E \to 0$. Thus the energy in the transients due to the loss or repeat of a sample in converting from $y(m)$ to $\hat{y}(m)$ in Fig. 6.21 becomes small. Alternatively, if the sampling rate is not increased in the connection between the two systems (i.e. if $L = 1$) then $C = 0$ and $E = 2$ and it is seen that the relative energy in the transients due to sample repeating or dropping with drifting of the clocks is significant.

The sampling rate conversion system in Fig. 6.21 can be efficiently realized using the structures and filter designs in Chapters 2 through 5. In particular, for large values of L, the multistage designs in Chapter 5 are appropriate. Also it should be noted that although the system in Fig. 6.21 is described in terms of a conversion process between two "equivalent" but incommensurate sampling rates $F + \epsilon_1$ and $F + \epsilon_2$, it can be readily extended to a process of conversion between any two incommensurate sampling rates by choosing different ratios for the interpolator and decimator. It is only necessary that the signals $y(m)$ and $\hat{y}(m)$ have "equivalent" rates to minimize the amount of sample repeating or dropping in the sample-and-hold.

6.7 SUMMARY

In this chapter we have shown, briefly, how some standard digital signal processing algorithms can be implemented as multirate processes. It was shown that in many cases a significant computational advantage is achieved using multirate techniques, over the standard, single-rate implementations. Alternatively, a multirate viewpoint to the problem sometimes leads to a new and novel way of realization or design. We have not dwelled on the details of the implementations because the reader should be able to provide these on his own based on the material in Chapters 2 to

5. Instead, we have chosen to show how the algorithms can be implemented, and to discuss the advantages and disadvantages of the multirate approach.

In the next chapter we study, in great detail, the topic of how to implement filter bank spectrum analyzers and synthesizers, and we endeavor to show how multirate techniques essentially provide the key to understanding some common structures used to implement filtering: for example, the overlap-add and overlap-save methods of FFT convolution.

REFERENCES

6.1   R. E. Crochiere and L. R. Rabiner, "Optimum FIR Digital Filter Implementations for Decimation, Interpolation, and Narrow Band Filtering," *IEEE Trans. Acoust. Speech Signal Process.*, Vol. ASSP-23, No. 5, pp. 444-456, October 1975.

6.2   L. R. Rabiner and R. E. Crochiere, "A Novel Implementation for Narrow Band FIR Digital Filters," *IEEE Trans. Acoust Speech Signal Process.*, Vol. ASSP-23, No. 5, pp. 457-464, October 1975.

6.3   F. Mintzer and B. Liu, "The Design of Optimal Multirate Bandpass and Bandstop Filters," *IEEE Trans. Acoust. Speech Signal Process.*, Vol. ASSP-26, No. 6, pp. 534-543, December 1978.

6.4   R. E. Crochiere, L. R. Rabiner, and R. R. Shively, "A Novel Implementation of Digital Phase Shifters," *Bell Syst. Tech. J.*, Vol. 54, No. 8, pp. 1497-1502, October 1975.

6.5   B. Gold, A. V. Oppenheim, and C. M. Rader, "Theory and Implementation of the Discrete Hilbert Transform," *Proc. Sym. Comput. Process. Comm.*, pp. 235-250, 1969.

6.6   A. V. Oppenheim and R. W. Schafer, *Digital Signal Processing.* Englewood Cliffs, N.J.: Prentice-Hall, 1975.

6.7   L. R. Rabiner and B. Gold, *Theory and Application of Digital Signal Processing.* Englewood Cliffs, N.J.: Prentice-Hall, 1975.

6.8   L. R. Rabiner and R. W. Schafer, "On the Behavior of Minimax FIR Digital Hilbert Transformers," *Bell Syst. Tech. J.*, Vol. 53, No. 2, pp. 363-390, February 1974.

6.9   B. Liu and F. Mintzer, "Calculation of Narrow Band Spectra by Direct Decimation," *IEEE Trans. Acoust. Speech Signal Process.*, Vol. ASSP-26, No. 6, pp. 529-534, December 1978.

6.10 R. E. Crochiere, "A Novel Approach for Implementing Pitch Prediction in Sub-Band Coding," *Proc. 1979 ICASSP*, Tulsa, Okla., April 1979.

6.11   L. B. Jackson, "On the Relationship Between Digital Hilbert Transformers and Certain Lowpass Filters," *IEEE Trans. Acoust. Speech Signal Process.*, Vol. ASSP-23, No. 4, pp. 381-383, August 1975.

6.12   D. J. Goodman and M. J. Carey, "Nine Digital Filters for Decimation and Interpolation," *IEEE Trans. Acoust. Speech Signal Process.*, Vol. ASSP-25, No. 2, pp. 121-126, April 1977.

7

Multirate Techniques in Filter Banks and Spectrum Analyzers and Synthesizers

7.0 INTRODUCTION

Digital filter bank and spectrum analysis and synthesis concepts arise in many areas of science and engineering [7.1-7.25]. They are extremely important, for example, in systems for speech analysis, bandwidth compression, radar and sonar processing, and spectral parameterization of signals. Nearly all of these types of systems have as their basis some form of filter bank decomposition or reconstruction of a signal in which the filter bank components occur in a decimated form. Thus these systems are inherently multirate and involve the operations of decimation and interpolation.

In many cases considerable efficiency can be obtained in the implementation of these filter banks or spectrum analyzers by carefully applying the multirate techniques discussed in Chapters 1 to 6 and by designing their structures so that computation can be shared among channels of the filter bank. In this chapter we consider several issues involved in such designs. In Section 7.1 we first briefly review a number of basic issues and definitions concerning filter banks and spectrum analyzers. Section 7.2 covers basic models and multirate structures for uniform DFT filter banks. In Section 7.3 we discuss filter design criteria for uniform DFT filter banks and in Section 7.4 we consider effects of modifications of the filter bank signals. In Section 7.5 we consider a more general model of the DFT filter bank based on a generalized definition of the DFT. This definition is used in Section 7.6 to develop multirate structures for single-sideband filter banks. Finally, Section 7.7 considers multistage structures and tree structures for both uniform and nonuniform filter bank designs.

7.1 GENERAL ISSUES AND DEFINITIONS

Figure 7.1(a) illustrates the basic framework for à K-channel filter bank *analyzer* and Fig. 7.1(b) illustrates a similar framework for a K-channel *synthesizer*. In the analyzer the input signal $x(n)$ is divided into a set of K-channels where $X_k(m)$, $k = 0, 1,..., K - 1$, denotes the channel signals. In the synthesizer the corresponding signals will be denoted by the "hat" notation. Thus $\hat{X}_k(m)$, $k = 0, 1,..., K - 1$, denotes the input channel signals to the synthesizer and $\hat{x}(n)$ denotes the output signal, where the sampling rates for the channel signals and the input and output signals will, in general, be different. The signals $X_k(m)$ and $\hat{X}_k(m)$ may also, in general, be different depending on the application. It should be noted that in this chapter we will refer to systems of the form of Fig. 7.1 *interchangeably* as filter banks or spectrum analyzers (or synthesizers) since the basic mathematical models and structures can be applied to both types of applications.

(a)

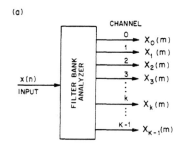

(b)

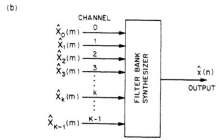

Figure 7.1 Basic framework for; (a) a K-channel filter bank analyzer; and (b) a K-channel filter bank synthesizer.

There are many ways to define the channel signals in the configuration above. For example, one important characteristic that distinguishes different classes of filter banks is the manner in which the channel signals are modulated. In this chapter we consider mainly two types of modulation as discussed in Chapters 2 and 6: *complex* (or *quadrature*) *modulation* and *single-sideband* (*SSB*) *modulation*.

Figure 7.2 briefly reviews these two forms of modulation, where ω_Δ denotes the width of the frequency band of interest for a particular filter bank channel centered at frequency ω_k. In the case of complex modulation [Fig. 7.2(a)] it is clear from the discussion in Chapter 2 that the channel signals $X_k(m)$ are complex-valued signals, whereas for single-sideband modulation [Fig. 7.2(b)] they are real valued. Complex modulation is generally preferred in systems such as spectrum analyzers [7.1, 7.2, 7.6, 7.10-7.12, 7.19, 7.20], whereas single-sideband modulation is generally preferred in systems such as communication systems [7.4, 7.5, 7.8, 7.9, 7.16-7.18] and coding systems [7.13-7.15, 7.24].

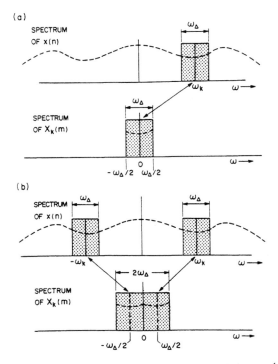

Figure 7.2 Comparison of filter bank modulation techniques; (a) complex (quadrature) modulation; (b) single-sideband modulation.

Another major difference in filter bank systems results from the manner in which the filters are designed for the channels. Figure 7.3 illustrates three examples of filter designs which result in nonoverlapping [part (a)], slight overlapping [part (b)], and substantial overlapping [part (c)] of the bands of the filter bank. Nonoverlapping designs are of interest, for example, in time division to frequency division (TDM-FDM) transmultiplexers for communication systems. In such applications the channel signals $X_k(m)$ and $\hat{X}_k(m)$ may be sets of telephone signals and $x(n)$ and $\hat{x}(n)$ may be frequency-multiplexed forms of these telephone signals [7.4, 7.5, 7.8, 7.9, 7.16-7.18]. In an application such as this, it is desired to keep the channel signals completely separated with guard bands between channels to avoid *crosstalk*. Crosstalk also occurs due to the fact that the stopband regions of the filter designs can never have infinite attenuation. Figure 7.4(a) illustrates this fact: namely, that crosstalk is directly affected by the degree of stopband attenuation in the filter design.

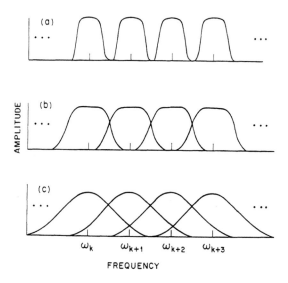

Figure 7.3 Comparison of filter bank designs; (a) nonoverlapping; (b) slight amount of overlapping; (c) substantial amount of overlapping.

Slightly overlapping filter bank designs are sometimes desired in systems, such as bandwidth compression systems, where an input signal $x(n)$ is analyzed in terms of its spectral components $X_k(m)$ for encoding into a compressed digital form [7.2, 7.3, 7.6, 7.7, 7.10, 7.13-7.15, 7.19, 7.24]. In reconstruction, the spectral components, $\hat{X}_k(m)$, are decoded and used to synthesize an output signal $\hat{x}(n)$ which is intended to be a replica of the input signal $x(n)$. Thus, in the absence of encoding, it is desired to have an analyzer/synthesizer system such that if the output of the analyzer is applied directly (i.e., without any channel distortion) to the input of the synthesizer, that is, if

$$\hat{X}_k(m) = X_k(m) \quad \text{for all } k \text{ and } m \tag{7.1a}$$

then

$$\hat{x}(n) = x(n) \tag{7.1b}$$

That is, the output, $\hat{x}(n)$, should be an exact replica of the input $x(n)$.[1] This can be accomplished if the bands overlap slightly as illustrated in Fig. 7.3(b), so that the reconstructed output $\hat{x}(n)$ contains no spectral "holes" between bands. For the sake of efficient bandwidth compression, however, it may be desired to keep this overlap minimized so that the channel signals are as independent as possible.

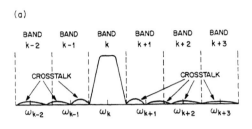

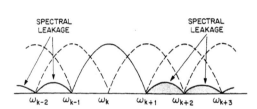

Figure 7.4 Illustration of crosstalk in nonoverlapping filter banks and spectral leakage in overlapping filter bank designs for spectral analysis.

Finally, heavily overlapped filter bank designs, as illustrated in Fig. 7.3(c), are of interest in systems such as spectrum analyzers, where a high-resolution (i.e., many channel) analysis and a smooth or well-interpolated spectral model may be desired [7.10-7.12]. In this case each filter bank channel conveys information about the amount of spectral energy in $x(n)$ located near the center frequency of the band. Filter designs in this case are often referred to as *windows*. The fact that these designs cannot have infinite attenuation outside the main band of interest again leads to the phenomenon of crosstalk as illustrated in Fig. 7.4(b). In spectral analysis this is often referred to as *spectral leakage*. The smaller the sidelobes of the analysis window, the lower this spectral leakage and the more accurately the spectral analyzer can model the true spectrum. Also, as the main lobe of the analysis window is made narrower in frequency, the frequency resolution of the spectrum analyzer becomes sharper. However, the time resolution (the length of the window) must be made wider.

---

[1] In practice this is usually accomplished with some flat delay in the system due to the filters, i.e. $\hat{x}(n) \cong x(n - n_o)$ where n_o is the delay. For the sake of discussion in this chapter we will often assume that $n_o = 0$.

A third major consideration in the design of filter banks and spectrum analyzers is the rate at which the channel signals $X_k(m)$ and $\hat{X}_k(m)$ are sampled. If the bandwidths of the channels are ω_Δ, as illustrated in Fig. 7.2, then, theoretically, the sampling rate of each of the channel signals can be reduced by a factor M, where

$$M \leqslant \frac{2\pi}{\omega_\Delta} \quad \text{complex modulation} \tag{7.2a}$$

or

$$M \leqslant \frac{\pi}{\omega_\Delta} \quad \text{SSB modulation} \tag{7.2b}$$

If the equality applies in Eqs. (7.2), the bands are said to be *critically sampled*, and if it does not apply, they are *oversampled*. For example, if the bands in a K-channel complex filter bank are uniformly and contiguously spaced such that

$$\omega_\Delta = \frac{2\pi}{K} \tag{7.3}$$

and they are critically sampled, then

$$M = K \tag{7.4}$$

and the filter bank is referred to as a *critically sampled filter bank*. In this case the total number of samples in the channel signals $X_k(m)$, $k = 0, 1,..., K - 1$, is equal to the total number of samples in the signal $x(n)$. Later we will see examples of such systems.

As seen from the discussion in Chapter 2, when the sampling rate of a channel is reduced by a factor M, signal components outside the bandwidth π/M are *aliased* back into the band in the decimated signal. Thus each of the channels contains some aliasing components from other channels due to overlap or spectral leakage of the filters. Similarly, in the filter bank synthesizer each band imposes some *imaging* in other bands due to the overlap or spectral leakage from the interpolation filters. In an appropriately designed analysis/synthesis framework, where the output of an analyzer is applied directly to the input of a synthesizer, it is possible to choose the design such that the aliasing that occurs in the analyzer is exactly canceled by the imaging that occurs in the synthesizer. The net result is a reconstructed signal $\hat{x}(n)$ which contains no aliasing or imaging due to the analysis/synthesis. Designs of this type are particularly important for bandwidth compression and coding systems. Details of this design will be discussed in the course of this chapter.

Another important characteristic that distinguishes the design and behavior of
different classes of filter banks is the manner in which the spacing and widths of
the frequency bands are chosen. Of particular interest is the class of *uniform filter
banks* in which all the channels have the same bandwidths and sampling rates.
From the point of view of practical implementation, uniform filter banks are often
very appealing because they can be realized with the aid of fast transform
algorithms such as the FFT (fast Fourier transform). In combination with single-
stage decimation or interpolation techniques, these implementations often take the
form of *block processing* algorithms.

Two types of *channel stacking* arrangements are often considered in uniform
filter banks, *even* types and *odd* types [7.16]. For even types the $k = 0$ channel is
centered at $\omega = 0$ and the centers of the bands are at frequencies

$$\omega_k = \frac{2\pi k}{K}, \quad k = 0, 1, ..., K - 1 \qquad (7.5)$$

as illustrated in Fig. 7.5a. For odd types the $k = 0$ channel is centered at $\omega = \pi/K$
and the bands are at frequencies

$$\omega_k = \frac{2\pi k}{K} + \frac{\pi}{K}, \quad k = 0, 1, ..., K - 1 \qquad (7.6)$$

as illustrated in Fig. 7.5(b). In addition to even and odd types there are skewed

(a)

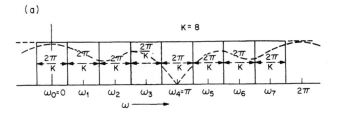

(b)

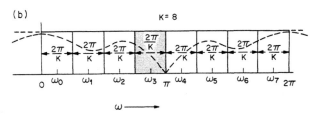

Figure 7.5 Filter stacking arrangements for uniform filter banks; (a) even types;
(b) odd types for the case of $K = 8$ band, uniform, contiguous, complex filter bank.

types, which are discussed in more detail in Sections 7.5 and 7.6. In the case of uniform SSB filter banks, even-type channel stacking is not possible since channels must occur in conjugate pairs.

In addition to uniform filter banks there are, of course, innumerable ways to define *nonuniform* channel stacking arrangements. For most of these choices efficient transform techniques for implementation do not exist. However, in many cases, efficient methods of realization may be possible by using multistage techniques for interpolation or decimation as discussed in Chapter 5. One important class of nonuniform filter banks that can be implemented in this way are *octave-spaced* designs [7.23, 7.24]. Figure 7.6 illustrates an example of a $K = 6$-band octave band stacking arrangement in which each band k is twice the width of the preceding band $k - 1$ (except for the $k = 0, 1$ bands). Section 7.7 covers multistage designs for this type of arrangement.

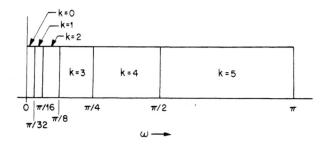

Figure 7.6   Octave-band stacking arrangement for filter bank channels.

7.2 UNIFORM DFT FILTER BANKS AND SHORT-TIME FOURIER ANALYZERS AND SYNTHESIZERS

One of the most important classes of filter banks is the DFT (discrete Fourier transform) filter bank; thus a major portion of this chapter is devoted to this topic. The concepts of DFT filter banks evolved primarily through the need to do spectral analysis and synthesis of quasi-stationary signals such as speech [7.1-7.3, 7.6, 7.7, 7.11, 7.12, 7.19, 7.20, 7.25]. The use of the Fourier transform, in particular, is a result of its providing a unique representation for signals in terms of the eigenfunctions of linear time-invariant systems, namely complex exponentials. Historically, this development evolved from two points of view: a filter bank interpretation and a block transform interpretation. These interpretations were later shown to be two different methods of realization of the same mathematical framework [7.19, 7.20]. Another realization, based on a polyphase decomposition of the filter bank, evolved through interest in channel bank filters for communication purposes [7.8, 7.9] and it also has the same basic mathematical framework.

This body of theory generally evolved around the use of an "even-type" uniform channel stacking arrangement and complex modulation because of the influence of the FFT (fast Fourier transform) algorithm for implementing the DFT.

We begin in Section 7.2.1 by discussing the filter bank interpretation of the Fourier analyzer and synthesizer and the mathematical framework behind it. In Section 7.2.2 we consider an alternative complex bandpass filter interpretation of the filter bank. We then discuss an efficient polyphase realization of this filter bank in Sections 7.2.3 and 7.2.4 and finally an efficient block transform or weighted overlap-add realization in Sections 7.2.5 and 7.2.6.

7.2.1 Filter Bank Interpretation Based on the Complex Modulator

Perhaps the simplest interpretation or model of the DFT filter bank is that in which each channel is separately bandpass modulated by a complex modulator as discussed in Chapter 2. This model is illustrated in Fig. 7.7 for the kth channel of a K-channel filter bank. Figure 7.7 will be used to identify the basic parameters and elements of the system. Signal paths with double lines in this figure refer to paths with complex signals.

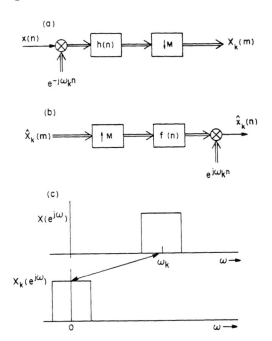

Figure 7.7 (a) Single channel of a DFT filter bank analyzer; (b) single channel of a DFT filter bank synthesizer; (c) a spectral interpretation.

In the analyzer [Fig. 7.7(a)] the input signal $x(n)$ is modulated by the function $e^{-j\omega_k n}$ and lowpass filtered by the filter $h(n)$. It is then reduced in sampling rate by a factor M to produce the channel signal $X_k(m)$. The filter $h(n)$ in this system is often referred to as the *analysis filter* since it determines the width and frequency response of each of the channels. We will assume that the same filter, $h(n)$, is used in each channel, and in Section 7.3 we deal with issues of the design of this filter.

Since the DFT filter bank has uniformly spaced filters with an even-type stacking arrangement, the center frequencies of the channels are

$$\omega_k = \frac{2\pi k}{K}, \quad k = 0, 1, ..., K - 1 \tag{7.7}$$

It is convenient to define

$$W_K = e^{j(2\pi/K)} \tag{7.8a}$$

and the complex modulation function as

$$e^{-j\omega_k n} = e^{-j(2\pi kn/K)} = W_K^{-kn} \tag{7.8b}$$

The channel signals can then be expressed as (see Chapter 2)

$$X_k(m) = \sum_{n=-\infty}^{\infty} h(mM - n)x(n)W_K^{-kn}, \quad k = 0, 1, ..., K - 1 \tag{7.9}$$

The DFT filter bank synthesizer interpolates all the channel signals back to their high sampling rate and modulates them back to their original spectral locations. It then sums all the channel signals to produce a single output. Figure 7.7(b) illustrates the basic complex modulator model for a single channel, k, of this synthesizer and Fig. 7.8 shows the overall analyzer/synthesizer configuration. The input signal $\hat{X}_k(m)$ is interpolated by a factor M with interpolation filter $f(n)$, which is often referred to as the *synthesis filter*. It is then modulated by the function $W_K^{nk} = e^{j\omega_k n}$ to shift the channel signal back to its original location ω_k [see Fig. 7.7(c)]. The resulting output of the channel is denoted as $\hat{x}_k(n)$. Finally, the output of the synthesizer is the sum all the channel signals, that is,

$$\hat{x}(n) = \frac{1}{K} \sum_{k=0}^{K-1} \hat{x}_k(n) \tag{7.10}$$

where the scale factor of $1/K$ is inserted for convenience in later discussion. From Fig. 7.7(b) and the discussion in Chapter 2 it can be shown that each channel signal can be expressed as

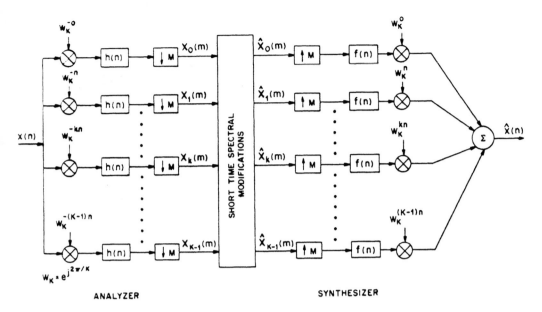

Figure 7.8 Complex modulator model of the DFT filter bank.

$$\hat{x}_k(n) = W_K^{kn} \sum_{m=-\infty}^{\infty} \hat{X}_k(m) f(n - mM), \quad k = 0, 1, ..., K - 1 \qquad (7.11)$$

Applying Eq. (7.11) to Eq. (7.10) and interchanging sums then gives the resulting overall expression for the DFT filter bank synthesizer in the form

$$\hat{x}(n) = \sum_{m=-\infty}^{\infty} f(n - mM) \frac{1}{K} \sum_{k=0}^{K-1} \hat{X}_k(m) W_K^{kn} \qquad (7.12)$$

Equations (7.9) and (7.12) form the mathematical basis for all the realizations of DFT filter bank analyzer and synthesizers discussed in this section. The properties of the analyzer and synthesizer are determined by the choice of the number of bands K, the decimation ratio M, and the designs of the analysis filter $h(n)$ and the synthesis filter $f(n)$. These design parameters, however, are not entirely independent. For example, if we wish to connect the outputs $X_k(m)$ of the analyzer, respectively, to the inputs $\hat{X}_k(m)$ of the synthesizer to obtain the result $\hat{x}(n) = x(n)$, the design of $h(n)$ and $f(n)$ are interdependent. This can be seen by applying Eq. (7.9) to Eq. (7.12) to get

$$\hat{x}(n) = \sum_{m=-\infty}^{\infty} f(n - mM) \frac{1}{K} \sum_{k=0}^{K-1} W_K^{kn} \sum_{r=-\infty}^{\infty} h(mM - r) x(r) W_K^{-kr}$$

$$= \sum_{m=-\infty}^{\infty} \sum_{r=-\infty}^{\infty} x(r)f(n-mM)h(mM-r)\frac{1}{K}\sum_{k=0}^{K-1} W_K^{k(n-r)} \tag{7.13}$$

Noting that

$$\frac{1}{K}\sum_{k=0}^{K-1} W_K^{k(n-r)} = \frac{1}{K}\sum_{k=0}^{K-1} e^{j(2\pi k(n-r))/K} = \begin{cases} 1, & \text{if } r=n-sK \\ 0, & \text{otherwise} \end{cases} \tag{7.14}$$

where s is any arbitrary integer, we can express Eq. (7.13) as

$$\hat{x}(n) = \sum_{s=-\infty}^{\infty} x(n-sK)\sum_{m=-\infty}^{\infty} f(n-mM)h(mM-n+sK) \tag{7.15}$$

Since we wish to have $\hat{x}(n) = x(n)$ it is clear that the filter designs $h(n)$ and $f(n)$ must be related according to [7.19]

$$\sum_{m=-\infty}^{\infty} f(n-mM)h(mM-n+sK) = u_0(s) = \begin{cases} 1, & s=0 \\ 0, & \text{otherwise} \end{cases} \tag{7.16}$$

for all n. Alternatively, if we wish to have $\hat{x}(n)$ be equal to a delayed version of $x(n)$ such as $\hat{x}(n) = x(n-s'K)$, then Eq. (7.16) must be zero for all values of s except s' (where it should be equal to 1). The interpretation of this condition will become clearer as we discuss the various realizations and interpretations of Eqs. (7.9) and (7.12).

7.2.2 Complex Bandpass Filter Interpretation

An alternative interpretation to the DFT filter bank can be developed by a slight modification of Eqs. (7.9) and (7.12). For the analyzer this modification is obtained by multiplying the right side of Eq. (7.9) by $W_K^{kmM} \cdot W_K^{-kmM} = 1$ and recombining terms to get the form

$$X_k(m) = W_K^{-kmM}\sum_{n=-\infty}^{\infty} [h(mM-n)W_K^{k(mM-n)}]x(n)$$

$$= W_K^{-kmM}\underline{X}_k(m), \quad k=0,1,...,K-1 \tag{7.17a}$$

where

$$\underline{X}_k(m) = \sum_{n=-\infty}^{\infty} h(mM-n)W_K^{k(mM-n)}x(n) \tag{7.17b}$$

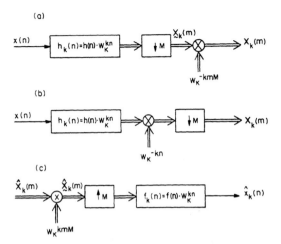

Figure 7.9 (a) Complex bandpass filter and modulator interpretations of the kth channel of the DFT filter bank analyzer; (b) analyzer with commuted modulator; (c) synthesizer.

This form suggests the bandpass filter bank interpretation shown in Fig. 7.9(a) for a single channel k (the interpretation of the signal $\underline{X}_k(m)$ will become clearer in the discussion in Section 7.2.5). Each filter in the channel consists of a complex bandpass filter with center frequency $\omega_k = 2\pi k/K$ and an impulse response that can be expressed as the product

$$h_k(n) = h(n)W_K^{kn} = h(n)e^{j(2\pi kn/K)} \tag{7.18}$$

The resulting bandpass-filtered signal is then decimated and modulated by the function W_K^{-kmM}. The function of this modulation is clearer if we apply one of the commutative identities in Chapter 3 to get the structure in Fig. 7.9(b). Here it is more clearly seen that the function of this modulation is the same as that for the lowpass filter bank of the preceding section: namely, to shift the center of the band from $\omega = \omega_k = 2\pi k/K$ to $\omega = 0$. The only difference in the structure in Fig. 7.9(a) is that the modulation is being performed at the low sampling rate after the decimation process.

A similar bandpass filter interpretation can be developed for the synthesizer by multiplying the right side of Eq. (7.11) by $W_K^{kmM}W_K^{-kmM} = 1$ and recombining terms to get

$$\hat{x}_k(n) = \sum_{m=-\infty}^{\infty} [\hat{X}_k(m)W_K^{kmM}][f(n-mM)\cdot W_K^{k(n-mM)}], \quad k = 0, ..., K-1 \tag{7.19}$$

This form of the equation suggests the bandpass synthesis structure in Fig. 7.9(c). The modulator, operating at the low sampling rate, translates the band back to its original frequency location (actually one of the periodic images of the baseband is translated to the original band location). The sampling rate expander then increases the sampling rate of the modulated channel signal, $\hat{X}_k(m)$, by the factor M (by filling in $M - 1$ zeros between each pair of samples) and the complex filter $f_k(n) = f(n)W_K^{kn}$ acts as a bandpass interpolating filter to filter out the appropriate harmonic images of the signal after sampling rate expansion.

A particular case of interest is the *critically sampled* case, where the decimation and interpolation ratio M is equal to the number of bands, K. In this case it is seen that the frequency bands are located at center frequencies

$$\omega_k = \frac{2\pi k}{K} = \frac{2\pi k}{M}, \quad k = 0, 1, ..., K - 1, K = M \tag{7.20}$$

and the modulation operation reduces to

$$W_K^{kmM}\big|_{M=K} = W_K^{-kmM}\big|_{M=K} = 1 \tag{7.21}$$

The structures in Fig. 7.9(a) and (c) then simplify to the structures illustrated in Fig. 7.10, and they can be recognized as being complex versions of the integer-band structures discussed in Chapter 2. Similarly, Eqs. (7.17) and (7.19) simplify to the respective equations

$$X_k(m) = \sum_{n=-\infty}^{\infty} h_k(mM - n)x(n)\big|_{M=K} \tag{7.22a}$$

and

$$\hat{x}_k(n) = \sum_{m=-\infty}^{\infty} \hat{X}_k(m)f_k(n - mM)\big|_{M=K} \tag{7.22b}$$

which imply the structures of Fig. 7.10(a) and (b), respectively.

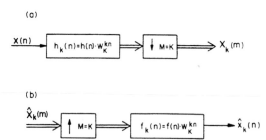

Figure 7.10 Integer-band complex filter model of the kth channel of the DFT filter bank: (a) analyzer; (b) synthesizer.

7.2.3 Polyphase Structures for Efficient Realization of Critically Sampled DFT Filter Banks

Thus far we have considered two interpretations or block diagram models of the DFT filter bank. The structures suggested by these block diagrams (Figs. 7.7 to 7.10) are highly inefficient, however, if implemented directly. In the next four sections we discuss more efficient structures which can be used to realize these filter banks. This section and Section 7.2.4 consider polyphase structures, and Sections 7.2.5 and 7.2.6 consider weighted overlap-add structures.

Several techniques are utilized in these filter bank structures for improving the efficiency of practical realizations. The first is the use of the more efficient decimation and interpolation structures discussed in Chapter 3. A second is that of sharing computation among channels. Finally, a third technique involves implementing the channel modulation with the aid of a fast algorithm such as the FFT (fast Fourier transform). The details of these mechanisms will be discussed as we consider the various structures.

The polyphase structure for implementing the DFT filter bank is based on applying the polyphase decimator and interpolator structures, discussed in Chapter 3, to the filter bank models in Fig. 7.10 [7.8, 7.9]. This realization is most easily seen for the case of critically sampled filter banks where $M = K$. Designs for other choices of M (or K) are not as straightforward as the critically sampled case [7.21, 7.22]. Such designs for the case $K = IM$, where I is an integer, are considered in the next section. Implementations for arbitrary choices of M and K will be considered in Section 7.2.5 using the weighted overlap-add structure. It is convenient in the derivation of the polyphase filter bank to use the clockwise commutator polyphase structure (see Chapter 3) for the filter bank analyzer and the counterclockwise commutator structure for the synthesizer. The reasons for this will become apparent in the following discussion.

From the discussion in Chapter 3 we can define $\bar{p}_{\rho,k}(m)$, the unit sample responses for the polyphase branches of the bandpass filters $h_k(n)$, as

$$\bar{p}_{\rho,k}(m) = h_k(mM - \rho), \quad \rho = 0, 1, ..., M - 1 \tag{7.23}$$

and the branch input signals as

$$x_\rho(m) = x(mM + \rho), \quad \rho = 0, 1, ..., M - 1 \tag{7.24}$$

where ρ refers to the ρth branch of the polyphase structure and k refers to the kth channel of the filter bank. Recall that the overbar notation refers to the clockwise commutator model for the polyphase structure. Noting that

$$h_k(n) = h(n) W_M^{kn} \tag{7.25}$$

and applying Eq. (7.23) leads to

$$\bar{p}_{\rho,k}(m) = h(mM - \rho)W_M^{k(mM-\rho)}$$

$$= h(mM-\rho)W_M^{-k\rho} \tag{7.26}$$

Now we can define the ρth polyphase branch for the lowpass analysis filter $h(n)$ as

$$\bar{p}_\rho(m) = h(mM - \rho) \tag{7.27}$$

and see that the form of the bandpass polyphase filters $\bar{p}_{\rho,k}(m)$ are separable as the product

$$\bar{p}_{\rho,k}(m) = \bar{p}_\rho(m)W_M^{-k\rho} \quad \text{for } \rho,k = 0, 1, ..., M-1 \tag{7.28}$$

The first term, $\bar{p}_\rho(m)$, is only a function of the sample time m. The second term, $W_M^{-\rho k}$, is only a function of the polyphase branch ρ and the channel k. This separation property is extremely important in the development of the polyphase filter bank structure.

Applying the definitions above to the filter bank structure in Fig. 7.10(a) or equivalently to Eq. (7.22a) leads to the structure of Fig. 7.11(a) and to the equation

$$X_k(m) = \sum_{n=-\infty}^{\infty} h_k(mM - n)x(n)$$

$$= \sum_{n=-\infty}^{\infty} h(mM-n)W_M^{k(mM-n)}x(n) \tag{7.29}$$

By a change of variables, Eq. (7.29) can be expressed in the form

$$X_k(m) = \sum_{n=-\infty}^{\infty} h(n)W_M^{kn}x(mM - n) \tag{7.30}$$

Furthermore, letting n be expressed as

$$n = rM - \rho \tag{7.31}$$

and summing on both r and ρ leads to

$$X_k(m) = \sum_{\rho=0}^{M-1} \sum_{r=-\infty}^{\infty} h(rM - \rho)W_M^{k(rM-\rho)}x((m - r)M + \rho) \tag{7.32}$$

Finally applying the polyphase definition in Eqs. (7.24), (7.27) and (7.28) leads to

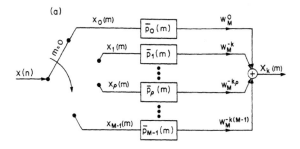

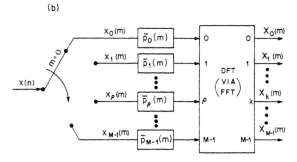

Figure 7.11 (a) Polyphase structure for the kth channel of a DFT filter bank analyzer; (b) the total polyphase DFT filter bank structure with an FFT modulator.

the form

$$X_k(m) = \sum_{p=0}^{M-1} \sum_{r=-\infty}^{\infty} \bar{p}_p(r) W_M^{-kp} x_p(m-r)$$

$$= \sum_{p=0}^{M-1} W_M^{-kp} [\bar{p}_p(m) * x_p(m)] \qquad (7.33)$$

where $*$ denotes discrete convolution. This form suggests the structure of Fig. 7.11(a) for a single filter bank channel, k.

Equation (7.33) suggests that $X_k(m)$ is of the form of a discrete Fourier transform (DFT) of the convolved outputs $\bar{p}_p(m) * x_p(m)$ of the polyphase branches. In fact, these output signals are *independent* of the filter channel number k due to the separable form of Eq. (7.28). Therefore, it is clear from Eq. (7.33) or Fig. 7.11(a) that computations involving the polyphase filters $\bar{p}_p(m)$ can be shared among all the filter bank channels, saving a factor of M in the total computation. An additional factor of M is gained from the polyphase form of the structure as discussed in Chapter 3. Finally, it can be recognized that the DFT in Eq. (7.33) can be performed with $M \log_2 M$ efficiency (as opposed to M^2 efficiency) by using the fast Fourier transform (FFT) algorithm [7.26, 7.27] or with an algorithm of similar efficiency [7.28]. When the input signals to the M-point DFT are real, it is also well known [7.29] that the DFT computation can be performed more efficiently using an $N/2$-point complex DFT. This then leads to the overall polyphase filter bank structure of Fig. 7.11(b), which is considerably more efficient than the structure implied by Fig. 7.10(a) for large M. Note that the input indices to the DFT are with respect to the polyphase branch number ρ and the output indices are with respect to filter bank channel number k.

The polyphase filter bank structure for the synthesizer can be developed in a similar manner by using the counterclockwise commutator model of the polyphase structure. This polyphase structure can be applied directly to the system of Fig. 7.10(b) [or equivalently to Eq. (7.22b)]. The polyphase branches for the bandpass filters $f_k(n)$ are defined as

$$q_{\rho,k}(m) = f_k(mM + \rho) \quad \text{for } \rho, k = 0, 1, ..., M - 1 \tag{7.34a}$$

where

$$f_k(n) = f(n)W_M^{kn}, \quad k = 0, 1, ..., M - 1 \tag{7.34b}$$

and $f(n)$ is the lowpass synthesis filter. Therefore, as in the case of the analyzer, $q_{\rho,k}(m)$ is seen to have the separable form

$$q_{\rho,k}(m) = f(mM + \rho)W_M^{k(mM+\rho)}$$

$$= q_\rho(m)W_M^{k\rho} \tag{7.35}$$

where $q_\rho(m)$ is the ρth polyphase filter of the lowpass synthesis filter

$$q_\rho(m) = f(mM + \rho) \tag{7.36}$$

Applying these definitions to Eqs. (7.10) and (7.22b) then leads to the form

$$\hat{x}(n) = \frac{1}{M} \sum_{k=0}^{M-1} \sum_{m=-\infty}^{\infty} \hat{X}_k(m)f_k(n - mM)$$

$$= \frac{1}{M} \sum_{k=0}^{M-1} \sum_{m=-\infty}^{\infty} \hat{X}_k(m)f(n-mM)W_M^{k(n-mM)} \tag{7.37}$$

Next, we can define M sets of polyphase output signals of the form

$$\hat{x}_\rho(r) = \hat{x}(rM + \rho) \tag{7.38}$$

which consist of subsets of the overall output signal $\hat{x}(n)$. Then from Eqs. (7.37) and (7.38) we get

$$\hat{x}_p(r) = \frac{1}{M} \sum_{k=0}^{M-1} \sum_{m=-\infty}^{\infty} \hat{X}_k(m) f((r-m)M + \rho) W_M^{k((r-m)M+\rho)} \qquad (7.39)$$

Finally, by applying the definition of the polyphase filter in Eq. (7.36) and interchanging sums, we get

$$\hat{x}_p(r) = \frac{1}{M} \sum_{k=0}^{M-1} \sum_{m=-\infty}^{\infty} \hat{X}_k(m) q_\rho(r-m) W_M^{k\rho}$$

$$= \sum_{m=-\infty}^{\infty} q_\rho(r-m) \left\{ \frac{1}{M} \sum_{k=0}^{M-1} \hat{X}_k(m) W_M^{k\rho} \right\} \qquad (7.40)$$

This form of the equation suggests the form of the polyphase synthesis structure shown in Fig. 7.12. First the inverse DFT is taken of the channel signals $\hat{X}_k(m)$, with respect to the index k at each sample time m. The output signals from the DFT are designated with the index ρ and are filtered respectively with the polyphase filters $q_\rho(m)$, with m as the running time index. Thus in Eq. (7.40) the sum on k corresponds to the DFT and the sum on m is associated with the convolution of $q_\rho(m)$ with the ρth channel signal of the DFT output. The resulting signal $\hat{x}_p(r)$ is the output signal for the ρth polyphase branch and the final output signal $\hat{x}(n)$ is the net result of interleaving these polyphase branch outputs through the use of the counterclockwise commutator of the polyphase structure.

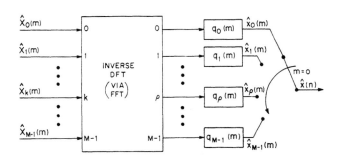

Figure 7.12 Synthesis structure for the polyphase DFT filter bank.

An alternative way of deriving the structure of Fig. 7.12 is through structural manipulations as discussed in Chapter 3. The structure of Fig. 7.10(b) is first replaced by the polyphase interpolator structure. Each polyphase filter is then recognized as being separable into the product of a filter and a fixed term $W_M^{k\rho}$. Finally, the set of coefficients $W_M^{k\rho}$ is seen to form the terms for a DFT structure which then leads to the overall structure of Fig. 7.12. A third way to derive this structure is by developing the counterclockwise polyphase structure for the filter

bank analyzer and then taking its Hermitian transpose as discussed in Chapter 3 to
get the synthesizer structure. As in the case of the analyzer structure, the
polyphase synthesis structure is considerably more efficient than a direct
implementation of the block diagram discussed in Section 7.2.1 or 7.2.2.

As discussed in Section 7.2.1, it is sometimes desired to perform an analysis
followed by a synthesis such that the output of the synthesizer $\hat{x}(n)$ is identical to
the analyzer input $x(n)$ in this back-to-back configuration. It was seen that this
requirement imposes the condition given by Eq. (7.16) on the relationship of the
design of the analysis and synthesis filters $h(n)$ and $f(n)$, respectively. This
condition has a particularly revealing form when viewed in the framework of the
critically sampled polyphase structure, as will be seen in the following discussion.

Figure 7.13(a) shows a block diagram of the back-to-back connection of the
polyphase analyzer/synthesizer structure. Clearly, the functions of the DFT and
inverse DFT are inverse operations and can be removed from the circuit, as shown
by Fig. 7.13(b), without affecting the input-to-output performance of the filter
bank. The structure is now divided into M independent structures consisting of
cascaded connections of the respective polyphase branches of the analyzer and
synthesizer. Alternatively, the same property can be derived by combining Eqs.
(7.33) and (7.40) to obtain

$$\hat{x}_\rho(r) = \sum_{m=-\infty}^{\infty} q_\rho(r-m) \frac{1}{M} \sum_{k=0}^{M-1} W_M^{k\rho} \left\{ \sum_{\nu=0}^{M-1} W_M^{-k\nu} [\bar{p}_\nu(m) * x_\nu(m)] \right\} \qquad (7.41)$$

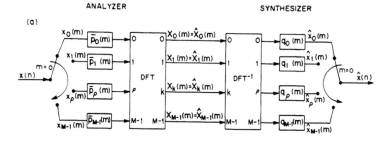

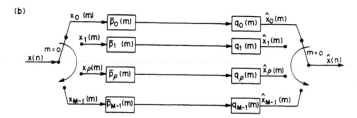

Figure 7.13 (a) Back-to-back polyphase analyzer/synthesizer structure; (b) the
structure with the DFT and inverse DFT removed.

Noting that

$$\frac{1}{M} \sum_{k=0}^{M-1} W_M^{k(\rho-\nu)} = \begin{cases} 1, & \text{if } \rho = \nu \\ 0, & \text{otherwise} \end{cases} \tag{7.42}$$

then leads to the independent form

$$\hat{x}_\rho(r) = \sum_{m=-\infty}^{\infty} q_\rho(r - m)[\bar{p}_\rho(m) * x_\rho(m)]$$

$$= q_\rho(r) * \bar{p}_\rho(r) * x_\rho(r) \quad \text{for } \rho = 0, 1, ..., M - 1 \tag{7.43}$$

where * denotes discrete convolution. It is easily seen that Eq. (7.43) describes the individual branches in Fig. 7.13(b) for each value ρ.

The condition for exact resynthesis [Eq. (7.16)] can now be specified in terms of the polyphase filters by requiring that for each polyphase branch ρ, the output $\hat{x}_\rho(r)$ be equal to the input $x_\rho(r)$ for all inputs. Thus from Eq. (7.43), or equivalently from Fig. 7.13(b), it is seen that this condition is of the form

$$q_\rho(m) * \bar{p}_\rho(m) = u_0(m) \quad \text{for all } \rho \tag{7.44}$$

where $u_0(m)$ is the unit sample function.

Equation (7.16) can then be derived directly from Eq. (7.44) by expressing it in the form

$$\sum_{m=-\infty}^{\infty} q_\rho(s - m)\bar{p}_\rho(m) = u_0(s) \quad \text{for all } \rho \tag{7.45}$$

Applying Eqs. (7.27) and (7.36) leads to

$$\sum_{m=-\infty}^{\infty} f((s - m)M + \rho))h(mM - \rho) = u_0(s) \quad \text{for all } \rho \tag{7.46}$$

Then applying the change of variables

$$n = sM + \rho \tag{7.47}$$

gives the form

$$\sum_{m=-\infty}^{\infty} f(n - mM)h(mM - n + sM) = u_0(s) \quad \text{for all } n \tag{7.48}$$

which is identical to Eq. (7.16) for $M = K$.

By Fourier transforming the condition for exact resynthesis, given by Eq. (7.44), we get the form

$$Q_\rho(e^{j\omega'})\overline{P}_\rho(e^{j\omega'}) = 1 \quad \text{for each } \rho \tag{7.49}$$

where ω' denotes the frequency response with respect to the decimated sampling rate. From this form it is clear that this condition for exact resynthesis requires that the polyphase filters $Q_\rho(e^{j\omega'})$ and $\overline{P}_\rho(e^{j\omega'})$ in each branch ρ must be inverse filters. In the case of ideal polyphase filters for decimation and interpolation, as discussed in Chapter 4, this is indeed the case. From the discussion in Chapter 4 it is seen that ideal polyphase filters based on the counterclockwise commutator model are all-pass linear phase filters with a phase advance ρ/M for the ρth branch, that is,

$$Q_\rho(e^{j\omega'}) = e^{j\omega'\rho/M} \tag{7.50}$$

This is based on the assumption that the synthesis filter has the ideal form

$$F(e^{j\omega}) = \begin{cases} 1, & 0 \leqslant |\omega| \leqslant \dfrac{\pi}{M} \\ 0, & \text{otherwise} \end{cases} \tag{7.51}$$

Similarly, it can be shown that if the analysis filter has the ideal form

$$H(e^{j\omega}) = \begin{cases} 1, & 0 \leqslant |\omega| \leqslant \dfrac{\pi}{M} \\ 0, & \text{otherwise} \end{cases} \tag{7.52}$$

then the polyphase filter $\overline{P}_\rho(e^{j\omega'})$ (based on the clockwise commutator model) has the form of a ρ/M phase delay for the ρth branch, that is,

$$\overline{P}_\rho(e^{j\omega'}) = e^{-j\omega'\rho/M} \tag{7.53}$$

Under these (ideal) conditions $\overline{P}_\rho(e^{j\omega'})$ and $Q_\rho(e^{j\omega'})$ clearly satisfy the inverse relation in Eq. (7.49).

In practical filter designs, these ideal conditions cannot be met exactly and filter design techniques that approximate these conditions must be considered. These design issues are discussed in more detail in Section 7.3.

7.2.4 Polyphase Filter Bank Structures for $K = MI$

In the previous section we saw how a DFT filter bank can be efficiently realized by combining the polyphase structure with the DFT. The principle assumption in this derivation is that the filter bank is critically sampled, that is, $K = M$, so that the separation property in Eq. (7.28) can be applied to separate the computation of the polyphase structure from that of the DFT. In many applications, however, it is desired to have an oversampled filter bank where $M < K$ so that the filter bank signals $X_k(m)$ are sampled at a higher rate [7.21, 7.22]. In this section we show how this can be achieved for the more general case

$$K = MI \tag{7.54}$$

where I is a positive integer, for example, $I = 1, 2, 3, \ldots$. We refer to I in this case as the *oversampling ratio* since it determines the amount by which the filter bank signals are oversampled from their theoretical minimum rate (that is, if $I = 1$ it is critically sampled and if $I = 2$ it is oversampled by a factor of two).

To develop this class of structures, consider again the basic mathematical form of the DFT filter bank analyzer given in Eq. (7.9). By making the change of variables

$$n = rK + \rho, \quad \rho = 0, 1, \ldots, K - 1 \tag{7.55}$$

we get

$$X_k(m) = \sum_{r=-\infty}^{\infty} \sum_{\rho=0}^{K-1} h(mM - rK - \rho)x(rK + \rho)W_K^{-k\rho} \tag{7.56}$$

As in the critically sampled case, the input signal $x(n)$ can be considered as an interleaved set of K polyphase signals of the form

$$x_\rho(r) = x(rK + \rho), \quad \rho = 0, 1, \ldots, K - 1, \quad r = 0, \pm 1, \ldots \tag{7.57}$$

By applying this definition along with Eq. (7.54) to Eq. (7.56) we then get

$$X_k(m) = \sum_{r=-\infty}^{\infty} \sum_{\rho=0}^{K-1} h((m - rI)M - \rho)x_\rho(r)W_K^{-k\rho} \tag{7.58}$$

The form of Eq. (7.58) can now be made clearer by defining a modified or *extended* set of polyphase filters of the form

$$\bar{p}_\rho(m) = h(mM - \rho), \quad \rho = 0, 1, \ldots K - 1 \tag{7.59}$$

Note that this polyphase definition is inherently different from that in the previous section (except when $M = K$) in that it is extended over a group of K interleaved samples of the impulse response $h(n)$ instead of M samples (that is, ρ takes on K distinct values rather than M distinct values). However the samples are spaced M apart. Applying this definition to Eq. (7.58) then leads to the form

$$X_k(m) = \sum_{\rho=0}^{K-1} W_K^{-k\rho} \left[\sum_{r=-\infty}^{\infty} \bar{p}_\rho(m - rI)x_\rho(r) \right] \qquad (7.60)$$

By recalling the basic principles in Chapter 2 it can now be recognized that the term in brackets in Eq. (7.60) simply defines an I sample interpolator. Letting $y_\rho(m)$ be the output, the basic mathematical model of this interpolator is given in Fig. 7.14 where the polyphase filter $\bar{p}_\rho(m)$ acts as the interpolating filter. Applying this model to that of the basic polyphase structure in Eq. (7.60) then leads to the structure in Fig. 7.15. Clearly the interpolators in this structure (shown in dashed boxes) can be realized with any of the structures discussed in Chapter 3. For example, if we use polyphase structures then we must consider the I sample polyphase decompositions of the filters $\bar{p}_\rho(m)$, that is, the polyphase filters of the polyphase filters.

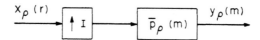

Figure 7.14 Integer interpolator using the extended polyphase filter $\bar{p}_\rho(m)$ as the interpolation filter.

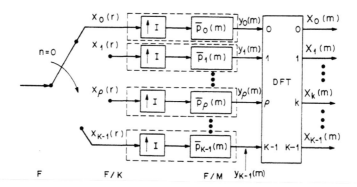

Figure 7.15 Polyphase structure for the filter bank analyzer for $K = IM$.

It can also be noted that since the polyphase filters $\bar{p}_\rho(m)$ in this definition are an *extended* set of filters they are related within the set. That is if we define ρ_o as the *principle subset* of decompositions such that $\rho_o = 0, 1, ..., M - 1$ then it can be shown from Eq. (7.59) that for $\rho = \rho_o + iM$, $i = 0, 1, ..., I - 1$ that

$$\bar{p}_{\rho_o+iM}(m) = \bar{p}_{\rho_o}(m - i), \quad \rho_o = 0, 1, ... M - 1, \quad i = 0, 1, ... I - 1 \quad (7.61)$$

Thus there are I subsets of polyphase filters $\bar{p}_\rho(m)$ (with M unique filters in each subset) such that they differ only by a delay of i samples from subset to subset.

The polyphase synthesis structure for the case $K = MI$ can be derived in a similar manner. Starting with the basic mathematical form in Eq. (7.12), and applying the change of variables in Eq. (7.55) and the definition in Eq. (7.54), leads to the form

$$\hat{x}(rK + \rho) = \sum_{m=-\infty}^{\infty} f((rI - m)M + \rho) \left[\frac{1}{K} \sum_{k=0}^{K-1} \hat{X}_k(m) W_K^{k\rho} \right] \quad (7.62)$$

Then defining the *extended* set of polyphase filters (using the counterclockwise model)

$$q_\rho(m) = f(mM + \rho), \quad \rho = 0, 1, ..., K - 1 \quad (7.63)$$

and applying it to Eq. (7.62) leads to the form

$$\hat{x}(rK + \rho) = \sum_{m=-\infty}^{\infty} q_\rho(rI - m) \left[\frac{1}{K} \sum_{k=0}^{K-1} \hat{X}_k(m) W_K^{k\rho} \right] \quad (7.64)$$

By careful consideration of the principles discussed in Chapter 2, it can be shown that this form leads to the structure in Fig. 7.16. In this case the structures in dashed boxes represent integer decimators which can again be realized using any of the efficient methods discussed in Chapter 3. The filters used in these decimators are the extended polyphase filters defined by Eq. (7.63).

7.2.5 Weighted Overlap-Add Structures for Efficient Realization of DFT Filter Banks

A second basic structure for realizing DFT filter banks is the weighted overlap-add structure [7.20]. This structure is based on an interpretation of the DFT filter bank in terms of a block-by-block transform analysis (or synthesis) of the signal and it represents a somewhat different point of view from that of the polyphase structure. This block-by-block transform interpretation is often used in the context of spectral analysis [7.2, 7.6, 7.10-7.12, 7.19]. The weighted overlap-add structure is somewhat more general than that of the polyphase structure in that it can be more easily applied to cases where the channel decimation/interpolation ratio, M, is unrelated to the number of frequency channels, K. That is, there are no restrictions to the relationship between M and K.

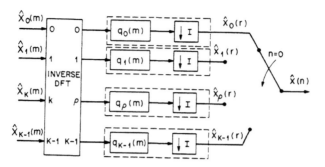

Figure 7.16 Polyphase structure for the filter bank synthesizer for $K = IM$.

To develop the weighted overlap-add analyzer structure, we begin with the basic filter bank model derived in Section 7.2.1. Recall that the output signal $X_k(m)$ for the kth channel of the filter bank analyzer can be expressed in the form

$$X_k(m) = \sum_{n=-\infty}^{\infty} h(mM - n)x(n)W_K^{-kn}, \quad k = 0, 1, ..., K - 1 \qquad (7.65)$$

In the block-by-block transform interpretation, the decimated sample time m denotes the block number (or the block time) and $X_k(m)$ is referred to as the *short-time spectrum* of the signal at time $n = mM$. It is interpreted, according to Eq. (7.65), as the discrete Fourier transform of the modified sequence

$$y_m(n) = h(mM - n)x(n) \qquad (7.66a)$$

such that

$$X_k(m) = \sum_{n=-\infty}^{\infty} y_m(n)W_K^{-kn} \qquad (7.66b)$$

In the interpretation above the filter $h(n)$ acts as a sliding *analysis window* which selects out and weights the short-time segment of the signal $x(n)$ to be analyzed. This interpretation is illustrated in Fig. 7.17 for the case of a symmetric FIR analysis window (filter), $h(n)$. Figure 7.17(a) illustrates an example of the signal $x(n)$ and the locations of the sliding window at block times m and $m + 1$. Figure 7.17(b) and (c) illustrate the resulting short-time signal sequences $y_m(n)$ and $y_{m+1}(n)$, respectively. The discrete Fourier transforms of these sequences then produce the short-time spectra $X_k(m)$ [sampled in frequency at frequencies $\omega_k = 2\pi k/K$, $k = 0, 1, ..., K - 1$, as illustrated in Fig. 7.17(d)]. The size of the analysis window, $h(n)$, and the number of samples, K, in the discrete Fourier transform determine the respective time and frequency resolution of the resulting short-time spectrum. Issues concerning these parameters are discussed in more detail in Section 7.3.

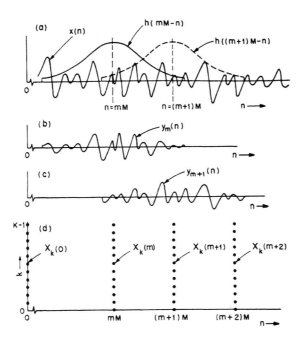

Figure 7.17 Block-by-block transform interpretation of the DFT filter bank.

Note that the point of origin of the discrete Fourier transform $X_k(m)$ in the interpretation above is referred to the origin, $n = 0$, of the fixed time frame, n, and the window is sliding linearly in this time frame. This is just the opposite of what we generally observe in a practical implementation, such as a filter, where the data are sliding by in time (as in a shift register) and the filter is invariant. Thus, for a practical implementation of this process it is convenient to convert our point of reference from that of a fixed time frame to that of a sliding time frame which is centered at the origin of the analysis window. This is accomplished by the change of variables

$$r = n - mM \qquad (7.67)$$

where n corresponds to the fixed time frame and r corresponds to the sliding time frame of the observer. Then the short-time transform $X_k(m)$ can be expressed as

$$X_k(m) = \sum_{r=-\infty}^{\infty} h(-r)x(r + mM)W_K^{-k(r+mM)}$$

$$= W_K^{-kmM}\underline{X}_k(m) \qquad (7.68)$$

where $\underline{X}_k(m)$ is defined as the short-time Fourier transform referenced to the origin of the sliding (linearly increasing) time frame r of the observer, that is,

$$\underset{\sim}{X}_k(m) = \sum_{r=-\infty}^{\infty} h(-r)x(r + mM)W_K^{-kr} \qquad (7.69)$$

Note that in the following discussion we will use the tilde $(\sim)$ notation under the variable to denote signals referenced to the sliding time frame. This change of time frame is illustrated in Fig. 7.18, where Fig. 7.18(a) shows the fixed time frame, Fig. 7.18(b) shows the sliding time frame for block m, and Fig. 7.18(c) shows the sliding time frame for block $m + 1$. Note that in the sliding time frame the analysis window $h(-r)$ (inverted in time) now appears to be stationary and the data appear to be sliding to the left. As seen by Eq. (7.68), the linearly increasing phase term W_K^{-kmM} converts the transform $\underset{\sim}{X}_k(m)$, computed in the sliding time frame, to the transform $X_k(m)$ for the fixed time frame.

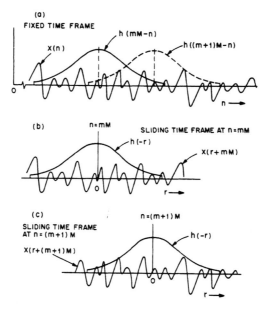

Figure 7.18 Illustration of the conversion of the fixed time frame to the sliding time frame.

An alternative interpretation to this sliding time frame point of view is obtained by considering the complex bandpass filter interpretation of the DFT filter bank in Section 7.7.2 and Fig. 7.9. It can be seen from this model that the sliding to fixed time frame conversion in Eq. (7.68) is equivalent to the modulation of the decimated complex bandpass filter output by W_K^{-kmM} [see Fig. 7.9(a)]. That is, the short-time Fourier transform referenced to the sliding time frame, $\underset{\sim}{X}_k(m)$, is identical to the decimated output of the complex bandpass filter $h_k(n)$ prior to modulation by W_K^{-kmM}. The sliding time frame interpretation therefore only provides another way of looking at this signal and it provides additional insight into the block transform point of view of short-time spectral analysis.

In terms of a practical implementation, the sliding transform in Eq. (7.69) can be efficiently realized with the aid of a fast algorithm such as the FFT (fast Fourier transform). The form of Eq. (7.69), however, is not quite in the form necessary for a K-point FFT since the summation on r may be over more than K samples of data. This problem can be converted to one that is appropriate for FFT computation by the well-known method of time aliasing the signal $h(-r)x(r+mM)$ [7.25-7.27]. Define $y_m(r)$ as the short-time sequence

$$\underline{y}_m(r) = h(-r)x(r+mM) \tag{7.70}$$

so that Eq. (7.69) becomes

$$\underline{X}_k(m) = \sum_{r=-\infty}^{\infty} \underline{y}_m(r)W_K^{-kr} \tag{7.71}$$

The sequence $y_m(r)$ can then be time aliased into the form of a K-point sequence $x_m(r)$ by subdividing it into blocks of K samples and stacking and adding (i.e., time aliasing) these blocks to produce the form

$$\underline{x}_m(r) = \sum_{l=-\infty}^{\infty} \underline{y}_m(r+lK)$$

$$= \sum_{l=-\infty}^{\infty} h(-r-lK)x(r+lK+mM), \quad r = 0, 1, ..., K-1 \tag{7.72}$$

Then the transform in Eq. (7.69) can be expressed in the form

$$\underline{X}_k(m) = \sum_{r=0}^{K-1} \underline{x}_m(r)W_K^{-kr} \tag{7.73}$$

which is an appropriate form for a K-point FFT algorithm (or other fast algorithms for implementing a K-point DFT). In this approach it is clear from Eq. (7.72) that the window $h(n)$ must be finite in duration; that is, it must be an FIR filter, so that the sum on l is finite in a practical realization.

Figure 7.19 summarizes the foregoing sequence of operations at block time m, necessary to implement the weighted overlap-add structure for the DFT filter bank analyzer. Input data, $x(n)$, are shifted into an N_h-sample shift register in blocks of M samples, where N_h is the number of taps in the analysis window $h(n)$. Here

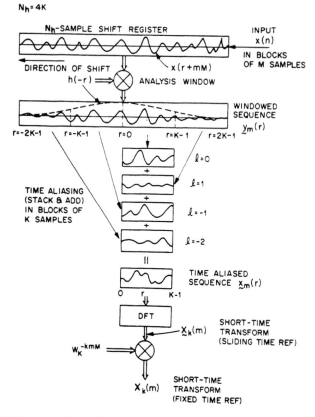

Figure 7.19 Sequence of operations for the weighted overlap-add spectral analyzer for block index m.

it is assumed that $N_h = 4K$; that is, the size of the window is four times the size of the transform. The data in the shift register are then weighted with the time-reversed window $h(-r)$, according to Eq. (7.70), to produce the N_h-sample short-time sequence $\underline{y}_m(r)$. This sequence is then divided into blocks of K samples, starting at $r = 0$, which are time aliased according to Eq. (7.72) to produce the K-sample, aliased sequence $\underline{x}_m(r)$. The K-point DFT of $\underline{x}_m(r)$ is then computed to obtain the short-time Fourier transform $\underline{X}_k(m)$, $k = 0, 1, \ldots K - 1$, [Eq. (7.73)], which is referenced to the sliding time frame of the observer. Finally, this transform is multiplied by the factor W_K^{-kmM}, according to Eq. (7.68), to convert it to the desired fixed time frame transform $X_k(m)$. For the next block time $m + 1$ the shift register is updated by M new samples of the input, $x(n)$, and the process is repeated to compute the next short-time transform $X_k(m + 1)$, $k = 0,1, \ldots K - 1$.

The structure described above is considerably more efficient than a direct implementation of the DFT filter bank as implied by the models in Sections 7.2.1 and 7.2.2. It obtains a factor of M in efficiency due to the use of the direct-form decimation structure (see Chapter 3) in the implementation of the window. Another factor of K is obtained by sharing the same windowing process with the K filter bank channels. Finally, the modulation is achieved with the $K \log_2 K$ efficiency of the FFT.

The weighted overlap-add synthesis structure for implementing the DFT filter bank is in effect a reversal (i.e., the transpose) of the operations discussed above. Recall from Section 7.2.1 that the basic equation for the synthesizer can be expressed in the form

$$\hat{x}(n) = \sum_{m=-\infty}^{\infty} f(n - mM) \frac{1}{K} \sum_{k=0}^{K-1} \hat{X}_k(m) W_K^{kn} \qquad (7.74)$$

This equation can be translated to the point of view of the sliding time frame for block time $m = m_0$ by considering the change of variables

$$r = n - m_0 M \qquad (7.75)$$

Applying Eq. (7.75) to Eq. (7.74) then leads to the form

$$\hat{x}(r + mM) \bigg|_{m=m_0} = f(r) \frac{1}{K} \sum_{k=0}^{K-1} \hat{X}_k(m) W_K^{kmM} W_K^{kr} \bigg|_{m=m_0} + \text{(terms for } m \neq m_0 \text{)} \qquad (7.76)$$

The first term on the right of Eq. (7.76) represents the contribution to the output signal $\hat{x}(r + m_0 M)$ due to the short-time spectrum $\hat{X}_k(m_0)$ at time m_0 and therefore it defines the synthesis structure. The second term denotes the contributions obtained from processing at all other block times $m \neq m_0$. Equation (7.76) can be further simplified to the form

$$\hat{x}(r + mM) \bigg|_{m=m_0} = f(r) \, \hat{\underline{x}}_m(r) \bigg|_{m=m_0} + \text{(terms for } m \neq m_0 \text{)} \qquad (7.77a)$$

by defining the translation of $X_k(m)$ from the fixed to the sliding time frame as

$$\hat{\underline{X}}_k(m) = \hat{X}_k(m) W_K^{kmM} \qquad (7.77b)$$

and defining $\hat{\underline{x}}_m(r)$ as the inverse DFT of $\hat{\underline{X}}_k(m)$ so that

$$\hat{\underline{x}}_m(r) = \frac{1}{K} \sum_{k=0}^{K-1} \hat{\underline{X}}_k(m) W_K^{kr} \qquad (7.77c)$$

From the form of Eqs. (7.76) and (7.77) it can be seen that the filter $f(r)$ appears to be stationary in the sliding time frame and again the synthesized data are sliding by to the left in this time frame. It is also seen that the output, $\hat{x}(n)$, is simply the sum of (time-shifted) inverse short-time transforms $\hat{\underline{x}}_m(r)$ weighted by $f(r)$. Consequently, $f(r)$ is referred to as the *synthesis window* in this framework. For each new block of time, m, a new term is added to this output. Since the sum on m in Eq. (7.76) is infinite, it is clear that the window $f(r)$ must be a finite-duration window [i.e., $f(r)$ must be an FIR filter] in a practical realization. If the length of $f(r)$ is N_f samples, there are only N_f/M terms that can contribute to any given output sample $\hat{x}(n)$.

It must also be noted that for the case where the length of the synthesis window, $f(r)$, is greater than the number of samples in the transform (i.e., where $N_f > K$), the sequence $\hat{\underline{x}}_m(r)$ must be interpreted as a periodically extended sequence (with a period of K samples). Its duration is then limited by the synthesis window $f(r)$.

Figure 7.20 depicts a block diagram of the sequence of operations described above for implementing the weighted overlap-add structure for the DFT filter bank synthesizer at block time m. It is assumed that the filter bank signals $\hat{X}_k(m)$ (i.e., the short-time transforms) are available. These short-time transforms are first converted to the sliding time frame by multiplication by the linearly increasing phase factor W_K^{kmM}. The inverse DFT (via the FFT algorithm or a similar fast algorithm) is then taken to produce the K-sample sequence $\hat{\underline{x}}_m(r)$ according to Eq. (7.77). This sequence is then periodically extended and windowed by the synthesis window $f(r)$. This process of periodic extension is in effect the transpose of the process of windowing and aliasing in the analysis structure of Fig. 7.19. The resulting windowed sequence $f(r) \, \hat{\underline{x}}_m(r)$ represents one term (for each block m) in Eq. (7.76). This term is added to the running sum of terms illustrated at the bottom of Fig. 7.20. Since the processing is performed sequentially in a block-by-block fashion the bottom box in Fig. 7.20 can be interpreted as a shift-register accumulator. Zeros are shifted into the shift register from the right and the direction of shift is to the left. At each block time, m, a new term $f(r) \, \hat{\underline{x}}_m(r)$ is added to the contents of this shift register in a parallel fashion. It is then shifted M samples to the left.

The term "weighted overlap-add" associated with this structure refers to this process of weighting the sequence $\hat{\underline{x}}_m(r)$ by $f(r)$ and overlapping and adding it to the contents of the shift register accumulator. The length of the shift register must be N_f, the length of $f(r)$. Once the data in the shift register have passed beyond the range of the synthesis window, the finite sum in Eq. (7.76) is complete. Thus the output signal $\hat{x}(n)$ are taken out of the left side of the shift register in blocks of M samples as its data are shifted left between frames.

As in the case of the weighted overlap-add analysis structure, the synthesis structure of Fig. 7.20 efficiently realizes the DFT filter bank synthesis by taking

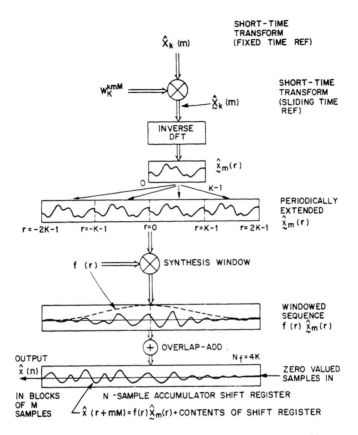

Figure 7.20 Sequence of operations for the weighted overlap-add spectral synthesizer for block index m.

advantage of an efficient method of interpolation, sharing filter computation among channels, and using a fast inverse transform algorithm to do the modulation.

There are a number of ways in which additional reductions in computation in the weighted overlap-add structures of Figs. 7.19 and 7.20 can be achieved in specific applications. For example, the modulation by the term W_K^{-kmM} to convert the short-time spectrum from the sliding to the fixed time references can be achieved without multiplication by recognizing that a modulo K shift of the (time domain) sequence $\underset{\sim}{x}_m(r)$ at the input of the DFT introduces a linear phase shift of the DFT output (frequency domain) [7.10]. This can be seen by expressing $X_k(m)$ as

$$X_k(m) = W_K^{-kmM} \underset{\sim}{X}_k(m)$$

$$= W_K^{-kmM} \sum_{r=0}^{K-1} \underline{x}_m(r) W_K^{-kr}$$

$$= \sum_{r=0}^{K-1} \underline{x}_m(r) W_K^{-k(r+mM)}$$

$$= \sum_{r=0}^{K-1} \underline{x}_m(r) W_K^{-k((r+mM))_K} \qquad (7.78)$$

where $((\cdot))_K$ denotes the quantity $((\cdot))$ modulo K. If we denote the change of variables

$$n' = ((r + mM))_K \qquad (7.79a)$$

then

$$r = ((n' - mM))_K \qquad (7.79b)$$

and Eq. (7.78) becomes

$$X_k(m) = \sum_{n'=0}^{K-1} x_m(n') W_K^{-kn'} \qquad (7.80)$$

where $x_m(n')$ is the circularly rotated sequence

$$x_m(n') = \underline{x}_m((n' - mM))_K \qquad (7.81)$$

This process is illustrated by the equivalence in Fig. 7.21. The change of variable

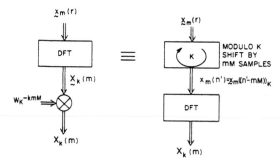

Figure 7.21 Equivalence of phase modification of $\hat{\underline{X}}_k(m)$ and circular (modulo K) rotation of $\hat{\underline{x}}_m(r)$.

in Eq. (7.79) is in effect a modulo K translation of the time frame back to the fixed time frame. That is, it is a modulo K version of Eq. (7.67), where the modulo K is due to the fact that in the DFT framework it is implied that the DFT input is either one period of a periodic sequence or a K-point sequence appended with zero-valued samples.

Similarly, in the synthesizer the phase shift W_K^{mM} can be realized as a modulo K rotation of the DFT output by computing

$$\hat{x}_m(n') = \frac{1}{K} \sum_{k=0}^{M-1} \hat{X}_k(m) W_K^{kn'} \tag{7.82}$$

and then obtaining $\hat{\underline{x}}_m(r)$ (see Fig. 7.20) by circular rotation of $\hat{x}_m(n')$, that is,

$$\hat{\underline{x}}_m(r) = \hat{x}_m((r + mM))_K \tag{7.83}$$

In many applications the change of time reference from $\underline{X}_k(m)$ to $X_k(m)$ may not be necessary and the phase modifications W_K^{-mK} or W_K^{mK} in the structures of Fig. 7.19 or 7.20 [or equivalently the circular rotation of $\underline{x}_m(r)$ as discussed above] can be eliminated. This is the case, for example, in a spectrum analyzer where only the magnitude of the transform is desired. Since

$$|X_k(m)| = |\underline{X}_k(m)| \tag{7.84}$$

the phase modification is unnecessary.

Another class of applications where the phase modification may be unnecessary occurs in analysis/synthesis, where the output of the analyzer is modified and then resynthesized. Consider the modification

$$\hat{X}_k(m) = G(X_k(m)) \tag{7.85}$$

where $G(\cdot)$ is some specified function. If the function $G(\cdot)$ is commutable (see Chapter 3) with the phase shift W_K^{mk}, then the phase shifts W_K^{mk} and W_K^{-mk} in the analyzer and synthesizer cancel. The modification can be applied directly to the transform in the sliding time frame to get

$$\hat{\underline{X}}_k(m) = G(\underline{X}_k(m)) \tag{7.86}$$

This occurs, for example, in filtering, where the function $G(\cdot)$ is the product of a filter spectrum times $\underline{X}_k(m)$.

7.2.6 A Simplified Weighted Overlap-Add Structure for Windows Shorter Than the Transform Size

A particularly important simplification occurs in the weighted overlap-add analysis and synthesis structures when the duration of the analysis and synthesis windows are less than the size of the transform [7.20], that is, when

$$N_h \leqslant K \tag{7.87a}$$

and

$$N_f \leqslant K \tag{7.87b}$$

This is generally the case for applications in spectral analysis and synthesis, often where $f(n)$ is reduced to a rectangular window [7.3, 7.6, 7.7, 7.10-7.12, 7.19, 7.25].

For convenience, assume that the windows $h(n)$ and $f(n)$ are symmetrical FIR windows centered at $n = 0$. Then from Fig. 7.19 it is seen that only the $l = 0$ and $l = -1$ terms are used in the stacking and adding process and that they contribute nonoverlapping signal components due to the condition in Eq. (7.87a). Furthermore, it can be recognized that this stacking and adding process can be eliminated entirely by circularly rotating the center of the windowed data to the center (the $K/2$ point) of the transform. According to the identity illustrated by Fig. 7.21, this time shift can be compensated for in the frequency domain by multiplying the resulting transform by a phase factor. Thus a shift of $K/2$ samples in time results in a compensating phase shift of

$$W_K^{-kK/2} = e^{-j(2\pi/K)(kK/2)} = e^{-j\pi k} = (-1)^k \tag{7.88}$$

that is, by a sign reversal of odd-numbered spectral components. A similar process can be performed in the synthesizer structure to avoid periodically extending the data before windowing them with $f(n)$ (see Fig. 7.19).

By applying these principles we get the simplified weighted overlap-add structure shown in Fig. 7.22, where the upper part of the figure shows the analyzer and the lower part of the figure shows the synthesizer. For convenience the input and output shift registers may be chosen to be equal to the size of the transform. Again, if the spectral modifications commute with the phase factors $(-1)^k W_K^{kmM}$, they may be eliminated in the structure. Also, if the synthesis window $f(n)$ is a rectangular window equal to the size of the transform, that is, if

$$f(n) = \begin{cases} 1, & -\dfrac{K}{2} \leqslant n < \dfrac{K}{2} \\ 0, & \text{otherwise} \end{cases} \tag{7.89}$$

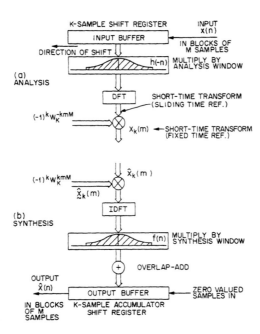

Figure 7.22   A simplified weighted overlap-add structure when the lengths of the analysis and synthesis windows are less than or equal to the size of the transform.

then the multiplication by $f(n)$ can be eliminated from the structure. This leads to the overlap-add method for spectral analysis/synthesis [7.11, 7.12]. Further details on this approach are discussed in Section 7.3.4.

7.2.7 Comparison of the Polyphase Structure and the Weighted Overlap-Add Structure

In this section we have discussed two practical structures for efficient implementation of the uniform DFT filter bank: the polyphase structure and the weighted overlap-add structure. Both structures follow from the mathematical filter bank models discussed in Sections 7.2.1 and 7.2.2 and are simply more efficient ways of implementing these models.

The polyphase structure is limited primarily to the critically sampled case where $M = K$ or to cases when $MI = K$, where I is an integer [7.21, 7.22]. Either FIR or IIR filter designs can be used in the polyphase structure although special considerations must be made in the case of IIR designs [7.8, 7.9].

In contrast, the weighted overlap-add structure is generally restricted to the use of FIR filter designs; however, there are no restrictions on the relationship between M and K. For the critically sampled case, when $M = K$, the weighted overlap-add structure has a form similar to that of the polyphase structure. This can be seen by comparing Fig. 7.19 to Fig. 7.11(b) and Fig. 7.20 to Fig. 7.12. The phase shift terms W_K^{-kmM} and W_K^{kmM} become unity and the M-sample block structure in the shift registers aligns with the K-sample stacking or periodically extending operations. Thus in the analyzer it can be seen that the weighting and stacking (time aliasing) operations are equivalent to the decimation and polyphase filtering operations of the polyphase structure. In the weighted overlap-add structure the filter weighting is applied first and then the stacking and adding of blocks is performed. In the polyphase structure these operations are performed in reverse order: the partitioning of the signal by the commutator is performed first and then these polyphase branch signals are filtered. Mathematically these operations are identical for $M = K$ (only the order in which they are performed is different) and therefore both structures have the same efficiency.

It is also interesting to note that in the derivation of the polyphase structure it is convenient to use the clockwise commutator polyphase structure for the analyzer and the counterclockwise polyphase structure for the synthesizer. This compares with the property in the weighted overlap-add structure that the window $h(n)$ is time reversed in the analyzer, whereas the window $f(n)$ in the synthesizer is not reversed in time.

A final consideration in comparing these structures is that of control and data flow in the structures. From our experience in speech processing we have often found that the polyphase structure is more amenable to a stream processing computing environment whereas the weighted overlap-add structure is more amenable to a block processing (array processor) environment. However it is difficult to make more specific statements than this since these issues are usually highly dependent on the nature of the application and on the hardware and software architecture in which they are used.

7.3 FILTER DESIGN CRITERIA FOR DFT FILTER BANKS

As pointed out earlier, the properties of the DFT filter bank or spectral analyzer/synthesizer are strongly dependent on the choice of the number of bands K, the decimation ratio M, and the designs of the analysis and synthesis filters (windows) $h(n)$ and $f(n)$. These choices are in turn strongly affected by the intended purpose of the filter bank. In this section we consider a number of issues involved in these designs and in particular we consider the design requirements on the filters $h(n)$ and $f(n)$ for control of the aliasing and imaging terms in the filter bank. Two types of aliasing and imaging effects can be identified, those in the time domain and those in the frequency domain. Also, it is necessary to distinguish between effects of aliasing within the filter bank, at the output of each interpolated channel, and at the output of a back-to-back analyzer/synthesizer configuration.

In Sections 7.3.1 and 7.3.2 we consider the effects of frequency domain aliasing and discuss filter bank designs that are based solely on a set of frequency domain specifications. In Sections 7.3.3 and 7.3.4 we consider the effects of time domain aliasing and discuss an alternative set of filter bank designs based solely on an analogous set of time domain constraints. Finally, in Section 7.3.5 we discuss the interaction between these time and frequency domain constraints.

7.3.1 Aliasing and Imaging in the Frequency Domain

Recall that a single channel, k, of the back-to-back DFT filter bank can be modeled as shown in Fig. 7.23(a), where the signal variables are expressed in terms of their Fourier transforms. That is, $X(e^{j\omega})$ is the transform of $x(n)$, $X_k(e^{j\omega'}) = \hat{X}_k(e^{j\omega'})$ are the transforms of the decimated channel signals $X_k(m) = \hat{X}_k(m)$, and $\hat{X}_{BPk}(e^{j\omega})$ is the transform of the kth interpolated channel signal $\hat{x}_k(n)$. Also define $H(e^{j\omega})$ and $F(e^{j\omega})$ as the transforms of $h(n)$ and $f(n)$ respectively. Then from Chapter 2 we can show that

$$X_k(e^{j\omega'}) = \frac{1}{M} \sum_{l=0}^{M-1} H(e^{j(\omega'-2\pi l)/M}) X(e^{j(\omega_k+(\omega'-2\pi l)/M)}) \tag{7.90}$$

where $\omega_k = 2\pi k/K$ is the center frequency of the band and ω' denotes the frequency with respect to the decimated sampling rate. Since the $l = 0$ term in Eq. (7.90) represents the desired term and terms for $l = 1, ..., M - 1$ represent undesired terms, it is clear that to avoid aliasing in the frequency domain of the channel signals the analysis filter $H(e^{j\omega})$ must satisfy the usual requirement for decimation by M, that is,

$$|H(e^{j\omega})| = \begin{cases} 1, & 0 \leqslant |\omega| \leqslant \dfrac{\pi}{M} \\ 0, & \text{otherwise} \end{cases} \tag{7.91a}$$

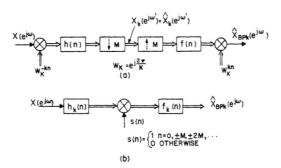

(a)

(b)

Figure 7.23 (a) Back-to-back complex modulation model for a single channel of the filter bank; and (b) an equivalent complex bandpass filter model.

or equivalently its stopband cutoff frequency, ω_{ch}, must satisfy the constraint

$$\omega_{ch} \leqslant \frac{\pi}{M} \qquad (7.91b)$$

In a similar manner we can express the interpolated output of the kth channel of the synthesizer as

$$\hat{X}_{\text{BP}k}(e^{j\omega}) = MF(e^{j(\omega-\omega_k)})\hat{X}_k(e^{j(\omega-\omega_k)M}) \qquad (7.92)$$

where the scale factor M is used to normalize the input and output signal levels in the 1-to-M interpolation process (see Chapter 2). Since $\hat{X}_k(e^{j\omega})$ is periodic with period 2π and the term $(\omega - \omega_k)M$ in Eq. (7.92) traverses through M periods as ω goes from 0 to 2π, it is clear that there are M images of $\hat{X}_k(e^{j\omega})$ in the channel output $\hat{X}_{\text{BP}k}(e^{j\omega})$, only one of which is desired. These images can be removed by applying the usual condition on the interpolation filter

$$|F(e^{j\omega})| = \begin{cases} 1, & 0 \leqslant |\omega| < \dfrac{\pi}{M} \\ 0, & \text{otherwise} \end{cases} \qquad (7.93a)$$

or equivalently its stopband cutoff frequency ω_{cf} must satisfy the constraint

$$\omega_{cf} \leqslant \frac{\pi}{M} \qquad (7.93b)$$

In a nonoverlapping filter bank such as that shown in Fig. 7.3(a), the conditions in Eqs. (7.91) and (7.93) can be applied and the filters $h(n)$ and $f(n)$ can be designed using the techniques discussed in Chapter 4.

In an overlapped filter bank design such as in Figs. 7.3(b) or (c), where a back-to-back analysis/synthesis framework is desired, a number of additional constraints are imposed on the filter design. In such cases it is generally desired that the analysis/synthesis system be an identity system when no modifications are made to the channel signals. Thus the regions of overlap must be carefully designed to satisfy this constraint.

If we consider a back-to-back analysis/synthesis system of this type for a K-channel filter bank, then we may express the output of the filter bank as the Fourier transform of Eq. (7.10),

$$\hat{X}(e^{j\omega}) = \frac{1}{K}\sum_{k=0}^{K-1}\hat{X}_{\text{BP}k}(e^{j\omega}) \qquad (7.94)$$

Then applying Eqs. (7.90) and (7.92) gives the input-to-output frequency domain relationship for the system in the form

$$\hat{X}(e^{j\omega}) = \frac{1}{K} \sum_{k=0}^{K-1} \sum_{l=0}^{M-1} F(e^{j(\omega-\omega_k)}) H(e^{j(\omega-\omega_k-(2\pi l/M))}) X(e^{j(\omega-(2\pi l/M))}) \quad (7.95)$$

Equation (7.95) can be further simplified by applying the complex bandpass filter interpretation of Section 7.2.2. According to this interpretation each channel k in the filter bank analyzer can be modeled as a complex bandpass filter

$$h_k(n) = h(n) W_K^{kn} \quad (7.96)$$

followed by a sampling rate compressor and a modulation of W_K^{-kmM} as shown in Fig. 7.9(a). Similarly, each channel k of the synthesizer can be modeled as a modulation by W_K^{kmM} followed by a sampling rate expander and a complex bandpass filter

$$f_k(n) = f(n) W_K^{kn} \quad (7.97)$$

as shown in Fig. 7.9(c). This leads to the equivalent structure shown in Fig. 7.23(b) for a single channel of the back-to-back filter bank where the modulator terms W_K^{-kmM} and W_K^{kmM} are canceled in the back-to-back structures of Fig. 7.9 and the remaining back-to-back compressor/expander is replaced by a modulator (sampler)

$$s(n) = \begin{cases} 1, & n = 0, \pm M, \pm 2M,... \\ 0, & \text{otherwise} \end{cases} \quad (7.98)$$

according to the identities in Chapter 3. By Fourier transforming Eqs. (7.96) and (7.97), we get

$$H_k(e^{j\omega}) = H(e^{j(\omega-\omega_k)}) \quad (7.99)$$

and

$$F_k(e^{j\omega}) = F(e^{j(\omega-\omega_k)}) \quad (7.100)$$

Applying these relations to Eq. (7.95) then leads to the simplified form

$$\hat{X}(e^{j\omega}) = X(e^{j\omega}) \frac{1}{K} \sum_{k=0}^{K-1} F_k(e^{j\omega}) H_k(e^{j\omega})$$

$$+ \sum_{l=1}^{M-1} X(e^{j(\omega-(2\pi l/M))}) \frac{1}{K} \sum_{k=0}^{K-1} F_k(e^{j\omega}) H_k(e^{j(\omega-(2\pi l/M))}) \quad (7.101)$$

The first term of Eq. (7.101) expresses the desired input-to-output relation of the back-to-back filter bank. From this term it is seen that in order for the system to be an identity system such that $\hat{X}(e^{j\omega}) = X(e^{j\omega})$, it is necessary that the bandpass filter products $(1/K)F_k(e^{j\omega})H_k(e^{j\omega})$ sum to 1, that is,

$$\frac{1}{K}\sum_{k=0}^{K-1} F_k(e^{j\omega})H_k(e^{j\omega}) = 1 \quad \text{for all } \omega \qquad (7.102a)$$

Alternatively, in a practical system, if there is some delay or phase distortion associated with the filters (e.g., as with IIR filters) it may be sufficient that

$$\frac{1}{K}\left|\sum_{k=0}^{K-1} F_k(e^{j\omega})H_k(e^{j\omega})\right| = 1 \quad \text{for all } \omega \qquad (7.102b)$$

The requirement above is illustrated in Fig. 7.24 and it implies a constraint not only on the passband and stopband regions of the filters but also a constraint on how their transition bands behave.

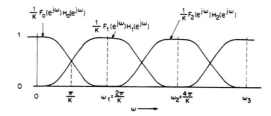

Figure 7.24 Illustration of the frequency response overlap condition in a DFT filter bank for a back-to-back identity system.

If only two adjacent filters overlap, as shown in Fig. 7.24, then only two terms in Eqs. (7.102) interact at any frequency and the condition can be simplified to

$$\frac{1}{K}\sum_{k=0}^{1} F_k(e^{j\omega})H_k(e^{j\omega}) = 1 \quad \text{for } 0 \leqslant \omega \leqslant \frac{2\pi}{K} \qquad (7.103a)$$

or, equivalently,

$$\frac{1}{K}\left[F(e^{j\omega})H(e^{j\omega}) + F(e^{j(\omega-(2\pi/K))})H(e^{j(\omega-(2\pi/K))})\right] = 1 \qquad (7.103b)$$

for $0 \leqslant \omega \leqslant (2\pi/K)$. This suggests that the net transition band of the product filter $F(e^{j\omega})H(e^{j\omega})$ must be antisymmetric about the crossover frequency $\omega = \pi/K$. At this frequency

$$\frac{1}{K}F(e^{j\pi/K})H(e^{j\pi/K}) = \frac{1}{2} \qquad (7.103c)$$

as illustrated in Fig. 7.24. In some cases, if the tolerance (ripple) requirements on the identity system are not too stringent, it may be sufficient to design the filters only to meet the crossover condition in Eq. (7.103c). Furthermore, if the filters $F(e^{j\omega})$ and $H(e^{j\omega})$ are identical, this becomes

$$\frac{1}{\sqrt{K}}F(e^{j\pi/k}) = \frac{1}{\sqrt{K}}H(e^{j\pi/k}) = \frac{1}{\sqrt{2}} \qquad (7.103d)$$

where

$$F(e^{jo}) = H(e^{jo}) = \sqrt{K} \qquad (7.103e)$$

If more than two channels of the filter bank overlap spectrally, as for example in Fig. 7.3(c), the design requirement becomes considerably more difficult. For example, if three channels overlap at any frequency, Eqs. (7.102) can be expressed as

$$\frac{1}{K}\sum_{k=0}^{2} F_k(e^{j\omega})H_k(e^{j\omega}) = 1 \text{ for } 0 \leqslant \omega \leqslant \frac{6\pi}{K} \qquad (7.104a)$$

and if R channels overlap, it becomes

$$\sum_{k=0}^{R-1} F_k(e^{j\omega})H_k(e^{j\omega}) = 1 \text{ for } 0 \leqslant \omega \leqslant \frac{2\pi R}{K} \text{ where } R \leqslant K \qquad (7.104b)$$

In Section 7.3.5, we show how this constraint can be satisfied, in general, by applying appropriate constraints in the time domain.

The second term in Eq. (7.101) expresses the fact that there are $M - 1$ possible undesired aliasing/imaging components in the back-to-back filter bank output corresponding to the form of weighted versions of $X(e^{j(\omega - (2\pi l/M))})$, $l = 1, ..., M - 1$. They are due to the decimation and interpolation process in the structure, and each channel contributes one term to each of the components. Thus, to eliminate effects of aliasing and imaging, it is required that

$$\sum_{k=0}^{K-1} F_k(e^{j\omega})H_k(e^{j(\omega - (2\pi l/M))}) = 0 \text{ for } l = 1, ..., M - 1, \text{ all } \omega \qquad (7.105)$$

The condition above is automatically satisfied (to within the stopband attenuation error of the filters) by applying the usual lowpass cutoff frequency specifications for decimation and interpolation in Eqs. (7.91) and (7.93) on the filters $h(n)$ and $f(n)$, respectively. This will be seen more clearly in the next section. However, these conditions are not necessarily essential to an identity system. In the next section we show an example of how the conditions in Eqs. (7.91) and (7.93) can be relaxed to obtain a more flexible set of frequency domain constraints that satisfy Eq. (7.105). In Section 7.3.5 we show how this condition can be achieved by an equivalent set of time domain constraints that effectively allow the aliasing terms from the K separate bands to cancel appropriately.

7.3.2 Filter-Bank-Design by Frequency Domain Specification — The Filter-Bank-Sum Method

Given the frequency domain conditions above, it is possible to design DFT filter banks that contain no frequency domain aliasing terms either in the channel signals or in the output of a back-to-back analysis/synthesis filter bank. In this section we consider a number of design techniques that can be used for this class of filter bank designs. We also show that in the limit as M goes to 1, (i.e., no decimation or interpolation), this approach leads to a technique referred to as the *filter-bank-sum method* [7.7, 7.11, 7.12, 7.25].

Consider first the example illustrated Fig. 7.25, where Fig. 7.25(a) shows the desired and undesired spectral terms associated with channel $k = 0$, Fig. 7.25(b) shows similar terms for $k = 1$, and Fig. 7.25(c) shows the terms for $k = 2$. The shaded regions show the spectral band of interest for each channel and the dashed lines indicate the spectral locations of the undesired aliasing components $H_k(e^{j(\omega - (2\pi l/M))})$, $l = 1, ..., M - 1$, in each channel. Also shown are the frequency responses of the filters $F_k(e^{j\omega})$ that can be used to remove these aliasing/imaging components in each band. Although the example in Fig. 7.25 illustrates the case for $M = K/2$, it applies to any situation where

$$M < K \qquad (7.106)$$

A number of issues and design techniques can be illustrated by this example and we will point them out in the following discussion.

The first observation that can be made is that if frequency domain aliasing is to be avoided in the channel signals, the cutoff frequency ω_{ch} of $H(e^{j\omega})$ must be less than π/M [Eq. (7.91)]. Also, if the overlap condition in Eqs. (7.102) is to be satisfied, the cutoff frequency must be larger than π/K; thus

$$\frac{\pi}{K} < \omega_{ch} \leqslant \frac{\pi}{M} \qquad (7.107)$$

This can be easily visualized by recognizing that $H(e^{j\omega}) = H_0(e^{j\omega})$ and $F(e^{j\omega}) = F_0(e^{j\omega})$ and focusing on channel $k = 0$ in Fig. 7.25(a).

When ω_{ch} is significantly less than π/M, gaps occur between the aliased components. This property can be taken advantage of in the synthesis process in

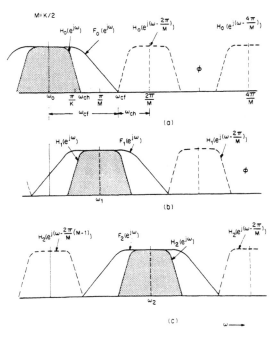

Figure 7.25 Example of a filter bank design based on frequency domain constraints.

two ways. First it allows a relaxation of the cutoff frequency requirements on the synthesis filter. That is, instead of the condition in Eq. (7.93), it is only necessary to satisfy the constraint

$$\omega_{ch} + \omega_{cf} \leqslant \frac{2\pi}{M} \qquad (7.108)$$

to avoid imaging within each of the interpolated channels. Furthermore, when M is significantly smaller than K and ω_{ch} is significantly smaller than π/M, the gaps between the images become large and they represent don't care regions in the spectrum where the design of $F(e^{j\omega})$ may be left unspecified. In this case the multiband filter designs discussed in Chapter 4 can be applied to the design of $F(e^{j\omega})$ for achieving greater efficiency. In the limit as M becomes small relative to K, the spacing between the aliasing components increases and the design of $F(e^{j\omega})$ approaches that of an M-sample comb filter (see Section 5.5). This is shown by the sequence of illustrations in Fig. 7.26. When M becomes unity (i.e., no decimation) no undesired images occur in the filter bank and the filter $f(n)$ reduces to a unit sample function, that is,

$$f(n) = u_o(n) \qquad (7.109a)$$

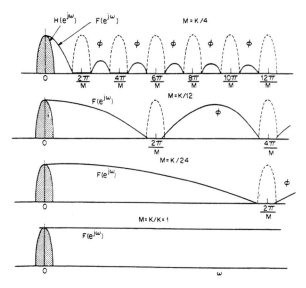

Figure 7.26 Illustration of the multiband constraints on $F(e^{j\omega})$ as M becomes small relative to K.

or

$$F(e^{j\omega}) = 1 \tag{7.109b}$$

This is exactly the case of the filter bank design technique referred to as the *filter-bank-sum method* [7.12].

Another important observation that can be made from the example in Fig. 7.25 is that if the design of the synthesis filter $F(e^{j\omega})$ is flat across the passband and transition band regions of $H(e^{j\omega})$, that is, if

$$F(e^{j\omega}) = 1 \quad \text{for } |\omega| < \omega_{ch} \tag{7.110}$$

then the overlap condition in Eqs. (7.102) simplifies to

$$\frac{1}{K} \sum_{k=0}^{K-1} H_k(e^{j\omega}) = 1 \quad \text{for all } \omega \tag{7.111}$$

Furthermore, as seen in the preceding section, if only two filters overlap in terms of their passband and transition bands at any given frequency, this condition further simplifies to the requirement that the frequency response of $H(e^{j\omega})$ be antisymmetric about the crossover frequency $\omega = \pi/K$.

It has been shown [7.7] that the antisymmetric condition above is very closely satisfied by the class of filter designs based on commonly used window designs (see

Chapter 4 for window designs). As an example, Fig. 7.27(a) shows a filter bank design based on an N_h = 175-tap Kaiser window, where the input sampling rate of the filter bank is 9600 Hz and channels below 200 Hz and above 3200 Hz are eliminated. The reconstruction error of this filter bank, based on Eq. (7.111), is shown in Fig. 7.27(b), and it is seen that this error is less than δ = 0.001 of the ideal value 1.0 across the 200 to 3200-Hz band of interest. That is, it is the error for which

$$\delta \geqslant \left| 1 - \frac{1}{K} \sum_{k=0}^{K-1} H_k(e^{j\omega}) \right| \qquad (7.112)$$

over the band of interest.

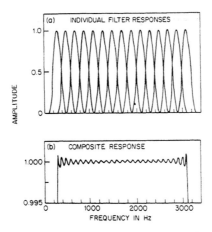

Figure 7.27 Example of a filter bank design based on a Kaiser window design and the filter-bank-sum method.

7.3.3 Aliasing and Imaging in the Time Domain

In the past two sections we have considered effects of aliasing and imaging in the frequency domain and methods of eliminating these effects based on frequency domain specifications on the analysis and synthesis filters of the filter bank. In this section we consider a similar set of effects in the time domain and an analogous set of time domain constraints that allows a back-to-back DFT filter bank to be designed as an identity system.

Recall from Eq. (7.15) that the input-to-output time domain relationship of a back-to-back DFT filter bank is

$$\hat{x}(n) = \sum_{m=-\infty}^{\infty} \sum_{s=-\infty}^{\infty} x(n - sK) f(n - mM) h(mM - n + sK) \qquad (7.113)$$

From this form we can observe that the $s = 0$ term corresponds to the desired output and all other values of s give rise to undesired aliasing or imaging components spaced K samples apart in time. Thus Eq. (7.113) is the time domain counterpart to Eq. (7.95). Note, however, that the spacing of the aliasing components in frequency are a function of M, whereas the spacing of the aliasing terms in time are a function of K. This is a consequence of the fact that frequency domain aliasing is a result of undersampling in time, whereas time domain aliasing is a result of undersampling in frequency [i.e., the fact that $X_k(e^{j\omega})$, $k = 0, 1, ..., K - 1$ corresponds to K samples of the ideal short-time spectrum] [7.11]. This interpretation will become clearer later.

The phenomenon of time domain aliasing can be visualized most easily from the weighted overlap-add structure illustrated in Figs. 7.19 and 7.20. In the analysis structure, time domain aliasing occurs in the stacking and adding process of generating the K-sample sequence $\underset{\sim}{x}_m(r)$ from the N_h-sample sequence $y_m(r)$. In the synthesis structure time domain imaging occurs in the process of periodically extending the K-sample sequence $\hat{x}_m(r)$ to produce the N_f-sample sequence that is weighted by the synthesis window.

Based on this interpretation, Eq. (7.113) can be simplified by defining the set of time-shifted windows

$$h_m(n) = h(n - mM) \tag{7.114}$$

and

$$f_m(n) = f(n - mM) \tag{7.115}$$

in a manner similar to that of the frequency-shifted filters in Eqs. (7.99) and (7.100). The reader should be cautioned that in this notation the subscript m refers to a time shift of mM samples, or a time slot m, whereas the subscript k refers to a frequency shift of $2\pi k/M$, or a frequency slot (channel) k. Applying Eqs. (7.114) and (7.115) to Eq. (7.113) leads to the form

$$\hat{x}(n) = x(n) \sum_{m=-\infty}^{\infty} f_m(n) h_m(-n)$$

$$+ \sum_{\substack{s=-\infty \\ s \neq 0}}^{\infty} x(n - sK) \sum_{m=-\infty}^{\infty} f_m(n) h_m(sK - n) \tag{7.116}$$

which is the time domain counterpart to Eq. (7.101).

From the first term in Eq. (7.116) it can be seen that for the system to be an identity system it is required that

$$\sum_{m=-\infty}^{\infty} f_m(n)h_m(-n) = 1 \quad \text{for all } n \qquad (7.117a)$$

or, equivalently,

$$\sum_{m=-\infty}^{\infty} f(n - mM)h(mM - n) = 1 \quad \text{for all } n \qquad (7.117b)$$

Equation (7.117) states that the sum of the product of time-shifted windows $f_m(n)h_m(-n)$ must add to 1 for all values of time n. This property is illustrated in Fig. 7.28 and it is seen to be the time domain analogy to Eq. (7.102) and Fig. 7.24. It can also be recognized that if only two adjacent window products overlap at any point in time, the product $f(n)h(-n)$ must be antisymmetric in time about the crossover time $n = M/2$ and that

$$f\left[\frac{M}{2}\right]h\left[\frac{-M}{2}\right] = \frac{1}{2} \qquad (7.118a)$$

and

$$f(0)h(0) = 1 \qquad (7.118b)$$

Furthermore, if only R adjacent window products overlap, the condition in Eqs. (7.117) becomes

$$\sum_{m=0}^{R-1} f_m(n)h_m(-n) = 1 \quad \text{for } 0 \leqslant n < RM \qquad (7.119)$$

The second term in Eq. (7.116) expresses the fact that there can be an infinite number of potential time-aliased terms in the back-to-back filter bank output spaced at intervals of sK samples. This is illustrated in Fig. 7.29 for time slot $m = 0$, where the shaded area depicts the signal of interest and the dashed lines

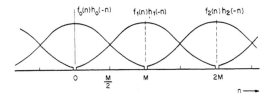

Figure 7.28 Illustration of the time response overlap condition in the DFT filter bank for a back-to-back identity system.

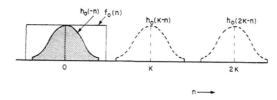

Figure 7.29 Illustration of time aliasing in the DFT filter bank for block time $m = 0$.

depict the undesired time-aliased terms. To eliminate these terms it is required that

$$\sum_{m=-\infty}^{\infty} f_m(n) h_m(sK - n) = 0 \quad \text{for all } s,\ s \neq 0 \text{ and for all } n \qquad (7.120)$$

This relation is the time domain analogy to Eq. (7.105). If N_h and N_f represent the lengths of the analysis and synthesis windows $f(n)$ and $h(n)$, respectively, this relation can be satisfied by the conditions

$$N_h \leqslant K \qquad (7.121a)$$

and

$$N_f \leqslant K \qquad (7.121b)$$

In this case the simplified weighted overlap-add structure in Fig. 7.22 can be used for the filter bank realization. Equations (7.121) represent the time domain analogy of Eqs. (7.91) and (7.93).

By careful examination of Eq. (7.120) and Fig. 7.29 it can be observed that with the appropriate choice of windows $h(n)$ and $f(n)$, an alternative and more general condition for eliminating time domain aliasing is that

$$N_h + N_f \leqslant 2K \qquad (7.122)$$

This is analogous to the frequency domain condition in Eq. (7.108). It implies that if the length of one of the windows is greater than K, time domain aliasing can still be eliminated at the output by appropriately making the length of the other window less than K and by carefully choosing the time locations of the windows with respect to each other and to the origin. In this case the more general structures of Figs. 7.19 and 7.20 are required. Also, it should be noted in the above discussion that windows do not necessarily have to be centered at the origin for the time domain designs to work.

7.3.4 Filter Bank Design by Time Domain Specification — The Overlap-Add Method

As seen in the preceding section, it is possible to design a DFT filter bank that contains no time domain aliasing and becomes a back-to-back identity system by designing the analysis and synthesis windows so that they meet the time domain specification in Eqs. (7.117) and (7.122). In practice, (in short-time spectral analysis/synthesis), a K-sample rectangular window

$$f(n) = \begin{cases} 1, & -\dfrac{K}{2} \leqslant n < \dfrac{K}{2} \\ 0, & \text{otherwise} \end{cases} \tag{7.123}$$

is often used for the synthesis window, and a symmetric window, $h(n)$, of length $N_h \leqslant K$, centered at $n = 0$, is often used for the analysis window. This is the condition illustrated in the example of Fig. 7.29. This choice leads to the analysis/synthesis method known as the *overlap-add method* [7.11, 7.12].

Several advantages are obtained from the above choice. First, it allows the use of the simplified analysis/synthesis structure of Fig. 7.22. A second advantage is that the synthesis window requires no multiplications; that is, the output of the DFT is simply overlap-added into the accumulator shift register. A third advantage is that it simplifies the design of the analysis filter. This is seen by the fact that in Eq. (7.117), $f(n) = 1$ over the duration of $h(-n)$ (see Fig. 7.29). Thus this requirement simplifies to the form

$$\sum_{m=-\infty}^{\infty} h_m(-n) = \sum_{m=-\infty}^{\infty} h(mM - n) = 1 \quad \text{for all } n \tag{7.124}$$

which is the time domain analog to Eq. (7.111). It has been shown that for $M \leqslant K/2$ the condition in Eq. (7.124) is well satisfied for several common classes of analysis windows such as Hamming, Hanning, and Kaiser windows [7.11, 7.12]. To demonstrate this, we can define the approximation error in Eq. (7.124) as the rms value over time:

$$\delta = \left[\sum_n [1 - \sum_m h_m(-n)]^2 \right]^{1/2} \tag{7.125}$$

and evaluate δ as a function of M, K, and N_h. Figure 7.30 shows this error (expressed in decibels) as a function of M, based on computer simulations [7.30], for the example $N_h = K = 128$ and for choices of a Hamming window, a Kaiser window, and a rectangular window for $h(n)$. As seen from this figure, Hamming

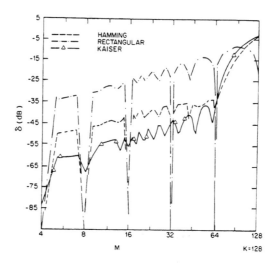

Figure 7.30 Example of the reconstruction error in the overlap-add method as a function of M for Hamming, Kaiser, and rectangular analysis windows.

and Kaiser windows can provide a good reconstruction for $M \leqslant K/2$. Also, it is seen that for cases where K/M is a power of 2, extremely good reconstruction is obtained with the Hamming window and perfect reconstruction is obtained with the rectangular window. For the trivial case of $M = K$, the rectangular window also gives perfect reconstruction, that is, it is simply a concatenated block transform technique.

In addition to the time domain constraints, noted above, it is often desired in the overlap-add method to constrain the cutoff frequency of the analysis window so that no frequency domain aliasing occurs in the channel signals $X_k(m)$. That is, the constraint in Eq. (7.91) is applied. This automatically rules out windows such as the rectangular window, which have extremely poor lowpass filtering capability. Common windows such as the Hamming, Hanning, and Kaiser windows, however, are well suited for this purpose. As an example, it can be observed that the cutoff frequency of an N_h-tap Hamming window (for a 42-dB stopband attenuation), where $N_h = K$, is

$$\omega_{ch} = \frac{4\pi}{N_h} = \frac{4\pi}{K} \bigg|_{\text{Hamming window}} \qquad (7.126)$$

By combining Eq. (7.126) with Eq. (7.91b), it is seen that

$$M \leqslant \frac{K}{4} \qquad (7.127)$$

Thus, to achieve a good back-to-back identity system with this approach, the channel signals must be oversampled by a factor of 4 or more.

The oversampling ratio above is typical of most designs in which both time and frequency domain aliasing terms are removed in the filter bank signals. To get a better idea of the trade-off between the analysis/synthesis reconstruction error and the oversampling ratio for other designs we can apply the design relation discussed in Chapter 4 for equiripple filters. Recall that this expression relates the filter length N, the passband and stopband cutoff frequencies ω_p and ω_s, and the passband and stopband tolerances δ_p and δ_s in the form

$$N \simeq \frac{D_\infty(\delta_p, \delta_s)}{(\omega_s - \omega_p)/2\pi} \tag{7.128}$$

where the function $D_\infty(\delta_p, \delta_s)$ is plotted in Fig. 4.11. If we let $\omega_{ch} = \omega_s \gg \omega_p$, and $N = K$ and apply Eqs. (7.91), we get the relation

$$M \leqslant \frac{K}{2D_\infty(\delta_p, \delta_s)} \tag{7.129}$$

which suggests that the oversampling ratio is determined by the factor $2D_\infty(\delta_p, \delta_s)$. Since δ_p and δ_s directly determine the degree of aliasing in the filter bank, this expression shows how the integrity of the analysis/synthesis can be traded for the oversampling ratio. If we choose $\delta_p = \delta_s = 0.01$, we get $2D_\infty(0.01, 0.01) \simeq 4$, which is essentially the same result as that of the Hamming window design.

7.3.5 Relationship of Time and Frequency Domain Specifications

In the past four sections we have discussed two alternative and analogous approaches to the design of DFT filter banks. The first approach is based on a frequency domain interpretation and it leads to designs in which no frequency domain aliasing occurs within the filter bank channels. However, since the length of the filters $h(n)$ and $f(n)$ are unspecified, such designs may allow time domain aliasing to occur within the channel signals. The second approach is based on a time domain interpretation and it leads to designs in which no time domain aliasing occurs in the channels. However, since no restriction need be placed on the bandwidths of $h(n)$ and $f(n)$, such designs may permit frequency domain aliasing to occur in the filter bank signals.

In both approaches, however, back-to-back filter banks can be designed so that the resulting system becomes an identity system (to within the tolerance limits to which the time or frequency domain constraints can be met). As such, it is clear that *both* the time and frequency domain aliasing terms at the final output of the back-to-back filter bank are eliminated with *either* approach. In the frequency domain approach, the frequency domain aliasing components are removed by filtering (frequency domain windowing) and the time domain aliasing terms which may occur in the channels are *canceled* in the process of summing the channel signals together. This time domain cancellation property is a consequence of the frequency domain specifications of this approach. Similarly in the time domain approach, the time domain aliasing terms are removed by time domain windowing, and the frequency domain aliasing terms which may occur in the channels are *canceled* in the process of summing overlapped time sequences together (e.g., in the accumulator shift register of the weighted overlap-add structure). This frequency domain cancellation property is a consequence of the time domain constraints imposed by this approach. In this section we examine these relationships in more detail and show how these cancellations take place.

Consider first the frequency domain approach. Recall that the constraints for this approach can be expressed as

$$\omega_{ch} \leqslant \frac{\pi}{M} \tag{7.130a}$$

$$\omega_{ch} + \omega_{cf} \leqslant \frac{2\pi}{M} \tag{7.130b}$$

and

$$\frac{1}{K} \sum_{k=0}^{K-1} F(e^{j(\omega-(2\pi k/K))}) H(e^{j(\omega-(2\pi k/K))}) = 1 \quad \text{for all } \omega \tag{7.131}$$

where ω_{ch} and ω_{cf} are lowpass cutoff frequencies of $H(e^{j\omega})$ and $F(e^{j\omega})$, respectively. For convenience, let

$$V(e^{j\omega}) = F(e^{j\omega}) H(e^{j\omega}) \tag{7.132}$$

and the inverse transform $v(n)$ be

$$v(n) = f(n) * h(n) \tag{7.133}$$

where * denotes discrete convolution. Then we can express the left side of Eq. (7.131) as

$$\frac{1}{K} \sum_{k=0}^{K-1} V(e^{j(\omega-(2\pi k/K))}) = \frac{1}{K} \sum_{k=0}^{K-1} \sum_{n=-\infty}^{\infty} v(n) e^{-j(\omega-(2\pi k/K))n}$$

$$= \sum_{n=-\infty}^{\infty} v(n) e^{-j\omega n} \frac{1}{K} \sum_{k=0}^{K-1} e^{j(2\pi k/K)n} \tag{7.134}$$

Recognizing that

$$\frac{1}{K} \sum_{k=0}^{K-1} e^{j(2\pi kn/K)} = \begin{cases} 1, & n = sK \\ 0, & \text{otherwise} \end{cases} \tag{7.135}$$

then leads to the form

$$\frac{1}{K} \sum_{k=0}^{K-1} V(e^{j(\omega - (2\pi k/K))}) = \sum_{s=-\infty}^{\infty} v(sK) e^{-j\omega sK} \tag{7.136}$$

Equation (7.136) is a frequency domain form of the Poisson summation formula [7.30]. It states that the sum of K frequency shifted terms $V(e^{j\omega})$, spaced $2\pi/K$ apart, produces a frequency domain function that is periodic in frequency with period $2\pi/K$. This function can be expressed in terms of a series expansion of terms $e^{-j\omega sK}$, where the coefficients of the expansion are time domain harmonic values of $v(n)$ at $n = sK$, $s = 0, \pm 1, \ldots$. Since, from Eq. (7.131), we wish to eliminate this periodicity and force the function to be a constant, it is desired that only the $s = 0$ term in this expansion have a nonzero coefficient, that is,

$$v(n) = f(n) * h(n) = \begin{cases} 1, & n = 0 \\ 0, & n = \pm K, \pm 2K, \ldots \\ \text{unspecified}, & \text{otherwise} \end{cases} \tag{7.137}$$

Thus the equivalent time domain constraint to the condition in Eq. (7.131) is that the convolution $f(n) * h(n)$ produce a sequence $v(n)$ that is zero for all values $n = sK$, $s = \pm 1, \pm 2, \ldots$, except $s = 0$.

If the length of the windows $h(n)$ and $f(n)$ are N_h and N_f, respectively, it is clear that the sequence $v(n)$ is $N_h + N_f - 1$ samples long. If, this sequence is centered at $n = 0$ then Eq. (7.137) is satisfied by the constraint $v(n) = 0$ for $|n| > K$, or, equivalently,

$$N_h + N_f \leqslant 2K \tag{7.138}$$

which is exactly the condition for the elimination of time domain aliasing in the time domain approach [see Section 7.3.3, Eq. (7.122)]. Thus if the designs of $h(n)$ and $f(n)$ satisfy both the time and frequency domain aliasing constraints in Eqs. (7.130b) and (7.138), the frequency domain overlap condition in Eq. (7.131) is satisfied automatically.

It was also seen in Section 7.3.2 that if the synthesis window $F(e^{j\omega})$ is designed to have a unity response over the passband and transition band frequency range of $H(e^{j\omega})$, then

$$F(e^{j\omega}) H(e^{j\omega}) \simeq H(e^{j\omega}) \tag{7.139a}$$

or

$$f(n) * h(n) = h(n) \tag{7.139b}$$

Under this constraint the overlap condition simplifies to the condition in Eq. (7.111) and Eq. (7.137) simplifies to

$$h(n) = \begin{cases} 1, & n = 0 \\ 0, & n = \pm K, \pm 2K, \dots \\ \text{unspecified}, & \text{otherwise.} \end{cases} \tag{7.140}$$

A similar set of analogous relations hold for the time domain approach to the DFT filter bank design. Recall that for this approach the design conditions are

$$N_h \leqslant K \tag{7.141a}$$

$$N_f + N_h \leqslant 2K \tag{7.141b}$$

and

$$\sum_{m=-\infty}^{\infty} f(n - mM) h(mM - n) = 1 \quad \text{for all } n \tag{7.142}$$

For convenience, let the product $w(n)$ be

$$w(n) = f(n)h(-n) \tag{7.143}$$

Then the Fourier transform of $w(n)$ can be expressed as

$$W(e^{j\omega}) = F(e^{j\omega}) \circledast H(e^{-j\omega}) \tag{7.144}$$

where $\circledast$ denotes circular convolution in the frequency domain. We can now express the left side of Eq. (7.142) as

$$w_M(n) = \sum_{m=-\infty}^{\infty} w(n-mM) = \sum_{m=-\infty}^{\infty} \frac{1}{2\pi} \int_{-\pi}^{\pi} W(e^{j\omega}) e^{j\omega(n-mM)} d\omega$$

$$= \frac{1}{2\pi} \int_{-\pi}^{\pi} W(e^{j\omega}) e^{j\omega n} d\omega \sum_{m=-\infty}^{\infty} e^{-j\omega mM} \tag{7.145}$$

Note that

$$\sum_{m=-\infty}^{\infty} e^{-j\omega mM} = \frac{2\pi}{M} \sum_{l=-\infty}^{\infty} \delta\left[\omega - \frac{2\pi l}{M}\right] \qquad (7.146)$$

where $\delta(\cdot)$ is the impulse function. Applying this relation to Eq. (7.145) leads to the form

$$w_M(n) = \sum_{m=-\infty}^{\infty} w(n - mM) = \frac{1}{M} \sum_{k=0}^{M-1} e^{j(2\pi kn/M)} W(e^{j(2\pi k/M)}) \qquad (7.147)$$

Equation (7.147) is the discrete time domain form of the Poisson summation formula [7.30]. It states that the discrete function $w_M(n)$ is periodic in n with period M and that it can be expressed in terms of an M-term series expansion of terms $e^{j(2\pi kn/M)}$, where the coefficients of these terms are harmonic values of the transform of $w(n)$ at frequencies $2\pi k/M$, respectively. Since from the condition in Eq. (7.142) we wish to set this function equal to a constant, independent of n, all terms in this expansion must be zero except the $k = 0$ term. Thus this leads to the condition

$$W(e^{j\omega}) = F(e^{j\omega}) \circledast H(e^{-j\omega}) = \begin{cases} M, & \omega = 0 \\ 0, & \omega = \frac{2\pi}{M}, \dots, \frac{(M-1)2\pi}{M} \end{cases} \qquad (7.148)$$

which specifies a set of M constraints on the circular convolution of $F(e^{j\omega})$ and $H(e^{j\omega})$ in frequency.

If the cutoff frequencies of the lowpass functions $F(e^{j\omega})$ and $H(e^{j\omega})$ are ω_{cf} and ω_{ch}, respectively, this convolution will produce a lowpass function which has a cutoff frequency of $\omega_{cf} + \omega_{ch}$. Thus one way of satisfying Eq. (7.148) is to require that

$$\omega_{cf} + \omega_{ch} \leqslant \frac{2\pi}{M} \qquad (7.149)$$

which is exactly the same condition as that for eliminating frequency domain aliasing in the frequency domain approach [Eq. (7.108) or (7.130)]. Thus if both the time and frequency domain aliasing constraints, Eqs. (7.141b) and (7.149), are observed, then the time domain overlap condition in Eq. (7.142) is automatically satisfied. Note also from earlier discussion that this same pair of conditions satisfies the frequency domain overlap condition in Eq. (7.131) automatically.

Several conclusions can be drawn from the discussion in Sections 7.2 and 7.3. First, if we wish to design a DFT filter bank which contains no internal aliasing either in time or frequency and is a back-to-back identity system, it is sufficient to constrain the lengths of the windows $h(n)$ and $f(n)$ and their cutoff frequencies according to Eqs. (7.130) and (7.141). The resulting back-to-back system will be an identity system to within a scale factor. As seen from the discussion in Section 7.3.3, this leads to a system that must be oversampled by approximately four or more. Alternatively, for the range $K \geqslant M \geqslant K/4$, systems that are identity systems can be designed which contain internal aliasing either in time or frequency or possibly both. For the special case $M = K$, the filter bank is critically sampled and can be conveniently defined in terms of the polyphase structure discussed in Section 7.2.3. Conditions for an identity system then simplify to a set of inverse filter relations for the polyphase filters as discussed in that section. Another closely related body of theory to these techniques is that of generalized sampling [7.31, 7.32]. The details of these methods are beyond the scope of this book.

7.4 EFFECTS OF MULTIPLICATIVE MODIFICATIONS IN THE DFT FILTER BANK AND METHODS OF FAST CONVOLUTION

An important application of the DFT filter bank is in the implementation of adaptive systems that are based on frequency domain criteria [7.2, 7.6, 7.10-7.13, 7.15, 7.19]. In such systems the input signal $x(n)$ is first analyzed with a DFT filter bank analyzer to produce the short-time sampled spectrum $X_k(m)$. Based on this spectrum, a modification is made to produce the signal $\hat{X}_k(m)$, which is used to synthesize an output $\hat{x}(n)$. A class of modifications that is of particular interest is that of multiplicative modifications of the form

$$\hat{X}_k(m) = G_k(m)X_k(m), \quad k = 0, 1, ..., K - 1 \qquad (7.150)$$

as illustrated in Fig. 7.31. Intuitively, one might expect, from classical Fourier theory, that modifications of this form appear as convolutional modifications in the time domain. This is indeed the case under a limited set of constraints that are governed by a choice of the filter bank parameters M, K, $h(n)$ and $f(n)$. These constraints are closely related to those discussed in the preceding section for control of aliasing at the output of the filter bank. In this section we discuss these issues in more detail and show additionally that, for the special case of rectangular windows, this approach leads to the conventional methods of overlap-add and overlap-save for fast convolution.

7.4.1 The General Model for Multiplicative Modifications

Recall from Section 7.2 that the equation for the filter bank analyzer is

$$X_k(m) = \sum_{n=-\infty}^{\infty} h(mM-n)x(n)W_K^{-kn}, \quad k = 0, 1, ..., K - 1 \qquad (7.151)$$

and the equation for the filter bank synthesizer is

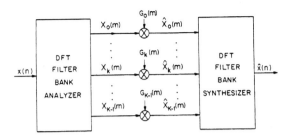

Figure 7.31 DFT filter bank analyzer/synthesizer with multiplicative modifications.

$$\hat{x}(n) = \sum_{m=-\infty}^{\infty} f(n - mM) \frac{1}{K} \sum_{k=0}^{K-1} \hat{X}_k(m) W_K^{kn} \qquad (7.152)$$

where K is the number of frequency channels (the size of the transform), M is the decimation ratio, $h(n)$ is the analysis window (filter), and $f(n)$ is the synthesis window (filter). Then applying Eq. (7.150), we get

$$\hat{x}(n) = \sum_{m=-\infty}^{\infty} \sum_{r=-\infty}^{\infty} f(n - mM) h(mM - r) x(r) \frac{1}{K} \sum_{k=0}^{K-1} G_k(m) W_K^{k(n-r)} \qquad (7.153)$$

The inverse DFT transform of the modification $G_k(m)$ can now be defined as the time domain impulse response

$$g_m(p) = \frac{1}{K} \sum_{k=0}^{K-1} G_k(m) W_K^{kp} \qquad (7.154)$$

where m refers to the time slot or block time in which the transform occurs and p refers to the time dimension of the inverse transform. Since $G_k(m)$ is described by K samples in frequency, only K unique samples of $g_m(p)$ can be described in time p; that is, the function $g_m(p)$ in Eq. (7.154) is periodic in p with period K. Alternatively, we can confine the range of p to be one period of this periodic function and define $\hat{g}_m(p)$ to be a finite-duration impulse response of length $N_g \leqslant K$. Then $g_m(p)$ can be expressed as

$$g_m(p) = \sum_{s=-\infty}^{\infty} \hat{g}_m(p + sK) \qquad (7.155a)$$

where

$$\hat{g}_m(p) = \begin{cases} \dfrac{1}{K} \displaystyle\sum_{k=0}^{K-1} G_k(m) W_K^{kp}, & n_{g_1} \leqslant p \leqslant n_{g_2} \\[2mm] 0, & \text{otherwise} \end{cases} \qquad (7.155b)$$

The variables n_{g_1} and n_{g_2} in Eq. (7.155b) define the beginning and ending sample times of $\hat{g}_m(p)$, respectively, where

$$n_{g_2} - n_{g_1} + 1 = N_g \leqslant K \qquad (7.156)$$

The choice of n_{g_1} and n_{g_2} are determined by the design of the windows $h(n)$ and $f(n)$ as will be seen in the subsequent discussion. To establish this more clearly, it is convenient to introduce the change of variables

$$p = n - r + sK \qquad (7.157)$$

where s is an integer that is chosen to confine p to the principal range $n_{g_1} \leqslant p \leqslant n_{g_2}$. Then Eq. (7.153) can be expressed as

$$\hat{x}(n) = \sum_{p=n_{g_1}}^{n_{g_2}} \sum_{s=-\infty}^{\infty} x(n - p + sK) \cdot$$

$$\sum_{m=-\infty}^{\infty} g_m(p)f(n - mM)h(mM - n + p - sK) \qquad (7.158)$$

Terms for $s \neq 0$ in Eq. (7.158) are equivalent to circular convolution terms in the structures of Figs. 7.19 and 7.20.

If it is desired that the modification above have the form of a linear convolution of $x(n)$ with an equivalent time-varying impulse response $\tilde{g}_n(p)$, then it must have the form (see Chapter 3)

$$\hat{x}(n) = \sum_{p=n_{g_1}}^{n_{g_2}} x(n - p)\tilde{g}_n(p) \qquad (7.159)$$

Comparing Eqs. (7.158) and (7.159) suggests that the equivalent impulse response, $\tilde{g}_n(p)$, of the system in Fig. 7.31 has the form

$$\tilde{g}_n(p) = \sum_{m=-\infty}^{\infty} g_m(p)f(n - mM)h(mM - n + p), \quad n_{g_1} \leqslant p \leqslant n_{g_2} \qquad (7.160)$$

It also suggests that to avoid time domain aliasing, the filter bank design must satisfy the condition

$$\sum_{m=-\infty}^{\infty} g_m(p)f(n - mM)h(mM - n + p - sK) = 0, \qquad (7.161)$$

$$\text{for } s \neq 0, \, n_{g_1} \leqslant p \leqslant n_{g_2}, \text{ and all } n$$

The constraints of the system are now clearly defined. The class of modifications that the filter bank system in Fig. 7.31 can realize is limited to that of finite-duration impulse responses of length $N_g \leqslant K$ and the filter bank must be carefully designed to avoid undesired aliasing terms.

The aliasing requirements are similar to those discussed in Section 7.3.3 except that they must apply for all values of p in the principal range $n_{g_1} \leqslant p \leqslant n_{g_2}$. These conditions can be defined by observing Eq. (7.161) at block time $m = 0$ and for the two extremes of p (i.e., $p = n_{g_1}$ and $p = n_{g_2}$). Under these conditions Eq. (7.161) becomes

$$f(n)h(n_{g_1} - n - sK) = 0 \quad \text{for } s \neq 0 \text{ and all } n \qquad (7.162a)$$

and

$$f(n)h(n_{g_2} - n - sK) = 0 \quad \text{for } s \neq 0 \text{ and all } n \qquad (7.162b)$$

Figure 7.32 illustrates this situation for terms $s = -1, 0, 1$ in Eqs. (7.162a) and (7.162b), respectively. The shaded regions refer to the desired terms $s = 0$, and the dashed lines refer to the undesired aliasing terms for $s = \pm 1$. By careful examination of this figure or equivalently of Eqs. (7.162), and applying Eq. (7.156), it can be shown that the general condition for elimination of the time domain aliasing is

$$N_h + N_f + N_g - 1 \leqslant 2K \qquad (7.163)$$

where N_h, N_f, and N_g are the durations of $h(n)$, $f(n)$, and $\hat{g}_m(p)$, respectively. The condition in Eq. (7.163) is achieved by appropriately designing the time alignment of $h(n)$ and $f(n)$ relative to the principal range of $g_m(p)$ so that the undesired aliasing terms are windowed out. In the limit where $g_m(p)$ becomes an identity [i.e., $g_m(p) = u_0(p)$], Eq. (7.163) reduces to that of Eq. (7.122), which is

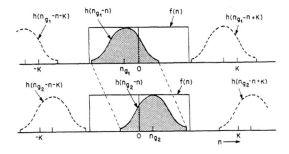

Figure 7.32 Time domain aliasing constraints in the filter bank in the presence of modifications.

the condition for elimination of time domain aliasing in a filter bank without modifications.

The interpretation of the equivalent time-varying impulse response in Eq. (7.160) can be simplified by defining the window

$$w_p(n) = f(n)h(p - n) \tag{7.164}$$

so that

$$\tilde{g}_n(p) = \sum_{m=-\infty}^{\infty} g_m(p)w_p(n - mM) \tag{7.165}$$

From the discussion in Chapter 2 this is seen to be the equation of a 1-to-M interpolator as illustrated in Fig. 7.33(a), where the window $w_p(n)$ acts as an interpolating filter to interpolate tap values of $g_m(p)$ to $\tilde{g}_n(p)$. By applying this model to Eq. (7.159), the equivalent tapped delay line model of the filter bank in Fig. 7.33(b) results, where it has been assumed that $h(n)$ and $f(n)$ are designed so that $n_{g_1} = 0$ and $n_{g_2} = N_g - 1$.

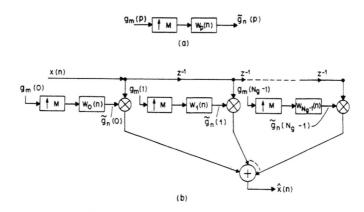

Figure 7.33 (a) Interpolator model of time-varying tap weights; and (b) tapped delay line model of the filter bank with time-varying modifications.

The model in Fig. 7.33 reveals that the filter bank design for adaptive modifications must not only satisfy the constraints for elimination of time domain aliasing but that the composite windows $w_p(n)$ for all principal values of p must satisfy the conditions of being good interpolator designs. If the variations of $g_m(p)$ are arbitrary functions of time m, then the cutoff frequency for the composite windows must be π/M as discussed in Chapter 2 on interpolator design. Alternatively, if $g_m(p)$ is slowly varying in time so that its transform (in time m) is limited to the bandwidth ω_{cg}, then the cutoff frequency ω_{cp} of the window $w_p(n)$

must be

$$\omega_{cp} + \frac{\omega_{cg}}{M} \leqslant \frac{2\pi}{M} \tag{7.166}$$

where ω_{cp} is defined relative to the high sampling rate and ω_{cg} is defined relative to the low sampling rate. More generally, this may be specified in terms of a multiband design according to the methods discussed in Chapter 4. In the limit as $\omega_{cg} = 0$ [i.e., when $G_k(m)$ or $g_m(p)$ are time invariant], this condition leads to that of an M-sample comb filter in which the transform of the window $w_p(n)$ must be zero at frequencies π/M, $2\pi/M$, $4\pi/M$, ... This will become clearer when we discuss methods of fast convolution for time-invariant modifications in Section 7.4.4.

7.4.2 Modifications in the Filter-Bank-Sum Method

In Section 7.3 we saw that two common methods of filter bank design are the filter-bank-sum method and the overlap-add method. In the filter-bank-sum method [assuming that the time aliasing requirement of Eq. (7.163) is satisfied] the bandwidth of $f(n)$ is much wider than the bandwidth of $h(n)$, as illustrated in Fig. 7.26. Therefore, in the time domain, the time width of $h(n)$ is much wider than the time width of $f(n)$. This is illustrated in Fig. 7.34. Therefore, $h(p-n) \approx h(p)$ over the duration of $f(n)$ and the window in Eq. (7.164) simplifies to the form

$$w_p(n) \approx f(n)h(p) \tag{7.167}$$

The modification for the filter-bank-sum method then takes the form

$$\widetilde{g}_n(p) \simeq h(p) \sum_{m=-\infty}^{\infty} g_m(p)f(n-mM) \quad \text{filter-bank-sum} \tag{7.168}$$

From this form it is seen that the filter-bank-sum modifications $g_m(p)$ are interpolated by the filter $f(n)$ in time m and time windowed by the window $h(p)$ in the lag time p.

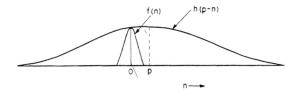

Figure 7.34 Relative time widths of $h(n)$ and $f(n)$ for the filter-bank-sum method.

7.4.3 Modifications in the Overlap-Add Method

In the overlap-add method of filter bank design, the window $f(n)$ is chosen to be unity over the duration of $h(p-n)$ for all principal values of p as illustrated in Fig. 7.32. The window $w_p(n)$ in Eq. (7.164) then takes the form

$$w_p(n) = h(p - n) \tag{7.169}$$

and the modification becomes

$$\widetilde{g}_n(p) = \sum_{m=-\infty}^{\infty} g_m(p)h(p - n + mM) \quad \text{overlap-add} \tag{7.170}$$

The resulting interpolation model then takes the form shown in Fig. 7.35. From this model it is seen that the filter $h(n)$ serves the dual purpose of acting as a decimation filter for decimating the channel signals and as an interpolation filter for interpolating the modifications $g_m(p)$.

Figure 7.35 Interpolator model for time-varying tap weights for the overlap-add design method.

7.4.4 Time-Invariant Modifications and Methods of Fast Convolution

If the modification $G_k(m)$ or $g_m(p)$ is a time-invariant modification, then it is independent of m and can be expressed simply as G_k or $g(p)$, respectively. Then Eq. (7.160) simplifies to the form

$$\widetilde{g}(p) = g(p) \sum_{m=-\infty}^{\infty} f(n - mM)h(mM - n + p), \quad n_{g_1} \leqslant p \leqslant n_{g_2} \tag{7.171}$$

and for $\widetilde{g}(p) = g(p)$ it is required that

$$\sum_{m=-\infty}^{\infty} f(n - mM)h(mM - n + p) = 1, \quad n_{g_1} \leqslant p \leqslant n_{g_2} \tag{7.172}$$

Recall also that in addition to the conditions above it is required that

$$N_h + N_f + N_g - 1 \leqslant 2K \qquad (7.173)$$

to avoid time aliasing at the output of the system.

As we have already seen in the preceding two sections, these conditions can be met in a number of ways and that in the limit as $g_m(p)$ becomes a time-invariant modification the equivalent interpolating filter that is used to interpolate $g_m(p)$ to $g_n(p)$ can be reduced to that of an M-sample rectangular window (i.e., an M-sample comb filter). If, in fact, both $h(n)$ and $f(n)$ are chosen to be rectangular windows, the foregoing method for time invariant modifications reduces to a generalization of the conventional methods of overlap-add and overlap-save for fast convolution.

In the overlap-add method [7.26, 7.27] $h(n)$ and $f(n)$ are chosen so that

$$N_h = M = K - N_g + 1 \qquad (7.174)$$

$$N_f = K \qquad (7.175)$$

and

$$h(n) = \begin{cases} 1, & -M + 1 \leqslant n \leqslant 0 \\ 0, & \text{otherwise} \end{cases} \qquad (7.176)$$

$$f(n) = \begin{cases} 1, & 0 \geqslant n \geqslant K - 1 \\ 0, & \text{otherwise} \end{cases} \qquad (7.177)$$

where it is assumed that

$$g(n) = \begin{cases} 0, & n < 0 \\ 0, & n \geqslant N_g \\ \text{unspecified}, & \text{otherwise} \end{cases} \qquad (7.178)$$

Under these conditions, Eq. (7.172) and the equality in Eq. (7.173) are exactly satisfied. The resulting sequence of operations are illustrated in Fig. 7.36. The filter $h(n)$ windows out M samples of $x(n)$ for each block. The product of $X_k(m)$ and G_k can be viewed in the time domain as the circular convolution of this block of data with $g(n)$ (see the weighted overlap-add structures in Figs. 7.19 and 7.20). However, because of the condition in Eq. (7.174) and the choice of the window $f(n)$, this appears as a K-sample linear convolution of the M-sample block of data with $g(n)$. By appropriately overlapping and adding these pieces of linear convolutions (e.g., in the accumulator shift register in Fig. 7.20), the resulting signal is the convolution of $x(n)$ and $g(n)$, as illustrated in Fig. 7.36.

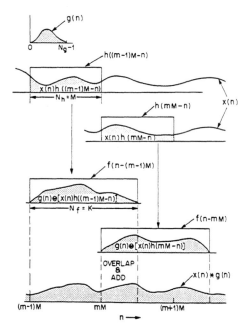

Figure 7.36 Overlap-add method of fast convolution.

In the overlap-save method of fast convolution the filters are chosen as

$$N_h = K \tag{7.179}$$

$$N_f = M = K - N_g - 1 \tag{7.180}$$

where

$$h(n) = \begin{cases} 1, & 0 \geqslant m \geqslant K - 1 \\ 0, & \text{otherwise} \end{cases} \tag{7.181}$$

and

$$f(n) = \begin{cases} 1, & -M + 1 \leqslant n \leqslant 0 \\ 0, & \text{otherwise} \end{cases} \tag{7.182}$$

Figure 7.37 illustrates this sequence of operations. In this case the convolution the K-sample windowed segment of $x(n)$ [due to the window $h(n)$] with $g(n)$ produces a circular convolution in which M samples of this result have the same

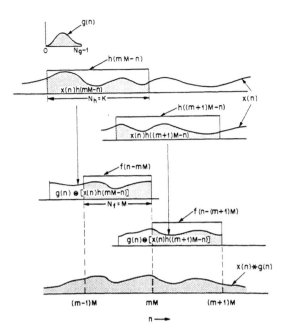

Figure 7.37   Overlap-save method of fast convolution.

form as a linear convolution. The window $f(n)$ selects only these portions of the blocks and concatenates them to form the final output, which has the form of a linear convolution of $x(n)$ with $g(n)$.

An interesting aspect of this method of convolution is that although the structure used to implement the system is a multirate structure, the final result is exactly that of a fixed-rate linear convolution. It is also well known that this method is more efficient than a direct-form convolution for large values of N_g, due to the efficiency of the fast Fourier transform algorithm [7.26, 7.27].

A final interesting observation that can be made from the comparison of $h(n)$ and $f(n)$ in the overlap-add and overlap-save methods is that they change roles in the two forms. From the discussion in Chapter 3 on network transposition it can be readily verified that this is a consequence of the fact that they are transpose methods. In fact, these two methods represent the two extremes of choices of rectangular windows for $h(n)$ and $f(n)$ that satisfy Eq. (7.172) and the equality in Eq. (7.173). A complete set of designs and their transposes for which $M = K - N_g - 1$ exist that fill the range between these two extremes.

7.4.5 Other Forms of Filter Bank Modifications and Systems

In this section we have focused almost exclusively on multiplicative modifications and the conditions under which they result in linear convolutional modifications in the time domain. Another important class of modifications is that of additive modifications, in which

$$\hat{X}_k(m) = X_k(m) + A_k(m) \tag{7.183}$$

where $A_k(m)$ is some additive component to the filter bank signals. In this case it is readily seen that if the filter bank is designed as a back-to-back identity system, then the modification in the time domain has the additive form

$$\hat{x}(n) = x(n) + a(n) \tag{7.184}$$

where $a(n)$ is the filter bank synthesis of $A_k(m)$ according to the form of Eq. (7.152).

Another form of modification that has been extensively studied is one in which the synthesis of a short-time spectrum $X_k(m)$ is performed with a different interpolation rate (i.e., where the value of M used in the synthesizer is different from that used in the analyzer). Such systems are of interest for rate modification of signals [7.2, 7.10, 7.19, 7.20, 7.33, 7.34] and for sampling rate conversion by ratios that are noninteger [7.35]. The details of these techniques are beyond the scope of this book.

7.5 GENERALIZED FORMS OF THE DFT FILTER BANK

7.5.1 The Generalized DFT (GDFT)

The DFT filter bank, as it has been defined and applied in the past three sections, is framed around the basic definition of the discrete Fourier transform (DFT). Recall that the DFT transform pair has the basic form

$$Y_k = \sum_{n=0}^{K-1} y(n) W_K^{-kn}, \quad k = 0, 1, ..., K - 1 \tag{7.185a}$$

and

$$y(n) = \frac{1}{K} \sum_{k=0}^{K-1} Y_k W_K^{kn}, \quad n = 0, 1, ..., K - 1 \tag{7.185b}$$

where $y(n)$ is a K-point sequence, Y_k, $k = 0, 1, ..., K$, corresponds to K samples of its Fourier transform, equally spaced in frequency, and W_K is defined as

$$W_K = e^{j(2\pi/K)} \tag{7.185c}$$

As a consequence of this definition it can be seen that the DFT filter bank is referenced to the time origin $n = 0$ and the frequency origin $\omega = 0$. This leads to a filter bank structure with an even-frequency channel stacking arrangement, as discussed in Section 7.2, as well as an even-time stacking arrangement in its structure. Although this form of the DFT filter bank is perhaps the most widely used, due to the availability of the FFT algorithm and due to its mathematical convenience, it is not always the most appropriate form. In some cases it is desirable to use other time and/or frequency origins as the basis for the filter bank design [7.4-7.6, 7.8, 7.9, 7.16-7.18, 7.36, 7.37]. For example, modification of the time origin is sometimes useful in the design of IIR filter banks for improving the frequency-response characteristics of the filter bank [7.6]. Alternatively, modification of the frequency origin leads to designs with different channel stacking arrangements. For example, by choosing the frequency origin $\omega = \pi/K$ for a K-channel filter bank, an odd-channel stacking arrangement results such that the center frequencies of the channels are located at frequencies

$$\omega_k = \frac{2\pi k}{K} + \frac{\pi}{K} = \frac{2\pi}{K}\left[k + \frac{1}{2}\right] \quad k = 0, 1, ..., K - 1 \qquad (7.186)$$

In this section we consider a somewhat more general definition of the DFT based on arbitrary time and frequency origins and show how this definition leads to a broader class of filter bank designs. Accordingly, we define this *generalized DFT* (GDFT) transform pair as

$$Y_k^{GDFT} = \sum_{n=0}^{K-1} y(n) W_K^{-(k+k_0)(n+n_0)}, \quad k = 0, 1, ..., K - 1 \qquad (7.187a)$$

and

$$y(n) = \frac{1}{K} \sum_{k=0}^{K-1} Y_k^{GDFT} W_K^{(k+k_0)(n+n_0)} \qquad (7.187b)$$

where n_0 corresponds to the new reference for the time origin, k_0 corresponds to the new reference for the discrete frequency origin, and W_K is defined as in Eq. (7.185c). It can be easily verified that Eqs. (7.187) form a valid transform and inverse transform pair. Typically, n_0 and k_0 are rational fractions that are less than 1 although they need not be. For example, when $n_0 = 0$ and $k_0 = 1/2$, this form of the GDFT is sometimes referred to as the odd-DFT [7.36] and when $n_0 = 1/2$ and $k_0 = 1/2$ it is sometimes referred to as the odd-squared-DFT [7.37].

By factoring the terms relating to k_0 and n_0 in Eqs. (7.187a) and (7.187b), the GDFT can be readily expressed in terms of the conventional DFT of an appropriately defined sequence, that is,

$$Y_k^{GDFT} = W_K^{-(k+k_0)n_0} \sum_{n=0}^{K-1} [y(n) \cdot W_K^{-k_0 n}] W_K^{-kn}, \quad k = 0, 1, ..., K - 1 \qquad (7.188a)$$

and

$$y(n) = W_K^{k_0 n} \frac{1}{K} \sum_{k=0}^{K-1} [Y_k^{GDFT} \cdot W_K^{(k+k_0)n_0}] W_K^{kn}, \quad n = 0, 1, ..., K-1 \qquad (7.188b)$$

Figure 7.38 shows the resulting relationships for computing the GDFT from the DFT. Alternatively, for special cases of the odd forms of the GDFT, more efficient methods exist for computing these transforms from the DFT [7.36, 7.37].

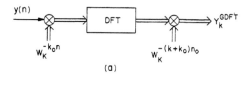

(a)

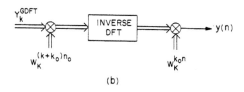

(b)

Figure 7.38 Relationship and method for computing the GDFT and inverse GDFT from the DFT and inverse DFT respectively.

7.5.2 The GDFT Filter Bank

By following a procedure similar to that in Section 7.2, it is possible to define a filter bank analysis/synthesis framework based on the definition of the GDFT in Eqs. (7.187) or (7.188). This leads to the filter bank analysis equation

$$X_k^{GDFT}(m) = \sum_{n=-\infty}^{\infty} h(mM - n)x(n) W_K^{-(k+k_0)(n+n_0)}, \quad k = 0, ..., K-1 \qquad (7.189)$$

and the filter bank synthesis equation

$$\hat{x}(n) = \sum_{m=-\infty}^{\infty} f(n - mM) \frac{1}{K} \sum_{k=0}^{K-1} \hat{X}_k^{GDFT}(m) W_K^{(k+k_0)(n+n_0)} \qquad (7.190)$$

where K is the number of frequency channels, M is the decimation ratio of the channel signals $X_k^{GDFT}(m)$, and $h(n)$ and $f(n)$ are the analysis and synthesis filters (windows), respectively. The center frequencies of the channels of this filter bank are located at frequencies

$$\omega_k = \frac{2\pi}{K}(k + k_0), \quad k = 0, 1, ..., K - 1 \qquad (7.191)$$

In a manner similar to that of the DFT filter bank, a complex modulator interpretation of the filter bank structure can be derived by an appropriate interpretation of Eqs. (7.189) and (7.190). This leads to the analyzer and synthesizer models shown in Fig. 7.39 for a single channel k. Alternatively, by defining the complex bandpass filters

$$h_k(n) = h(n) W_k^{(k+k_0)n} \qquad (7.192)$$

and

$$f_k(n) = f(n) W_K^{(k+k_0)n} \qquad (7.193)$$

Eqs. (7.189) and (7.190) can be expressed in the form

$$X_k^{GDFT}(m) = W_K^{-(k+k_0)(mM+n_0)} \sum_{n=-\infty}^{\infty} h_k(n) x(mM - n), \quad k = 0, ..., K - 1 \quad (7.194)$$

and

$$\hat{x}(n) = \frac{1}{K} \sum_{k=0}^{K-1} \sum_{m=-\infty}^{\infty} f_k(n - mM) \hat{X}_k^{GDFT}(m) W_K^{(k+k_0)(mM+n_0)} \qquad (7.195)$$

This suggest the complex bandpass filter structures shown in Fig. 7.40 for channel k.

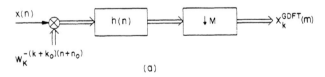

(a)

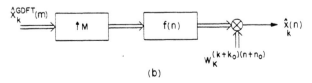

(b)

Figure 7.39 Complex modulator model of the GDFT filter bank for channel k; (a) analyzer; and (b) synthesizer.

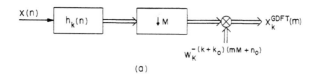

(a)

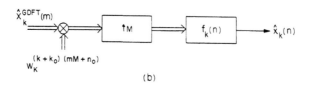

(b)

Figure 7.40 Complex bandpass models for the analyzer and synthesizer of channel k of a GDFT filter bank.

7.5.3 Polyphase Structure for the GDFT Filter Bank

As in the DFT filter bank, the GDFT filter bank can be more efficiently realized by means of polyphase or overlap-add structures, which allow computation to be shared among the channels of the filter bank. These structures take on a similar form to those of the DFT structures discussed in Sections 7.2.3 to 7.2.6. However, due to the time and frequency offsets of the GDFT framework, additional modulation terms appear in these structures. In the next two sections we briefly discuss these differences. Since the derivations of these structures are similar to those of the DFT structures, details of the derivations are omitted.

Recall that for the polyphase structures we define the polyphase branch filters for $h(n)$ and $f(n)$ as

$$\bar{p}_\rho(m) = h(mM - \rho), \quad \rho = 0, 1, ..., M - 1 \tag{7.196}$$

and

$$q_\rho(m) = f(mM + \rho), \quad \rho = 0, 1, ..., M - 1 \tag{7.197}$$

where the clockwise commutator model is used for the analyzer (denoted by the overbar notation) and the counterclockwise commutator model is used for the synthesizer (see Chapter 3 for further detail on these models). According to this framework the polyphase branch signals for the inputs and outputs of the branches in the analyzer and synthesizer, respectively, are

$$x_\rho(m) = x(mM + \rho), \quad \rho = 0, 1, ..., M - 1 \tag{7.198}$$

$$\hat{x}_\rho(m) = \hat{x}(mM + \rho), \quad \rho = 0, 1, ..., M - 1 \tag{7.199}$$

Also recall that the polyphase filter bank structure is most conveniently applied to the case of critically sampled filter banks, so that

$$K = M \tag{7.200}$$

By applying the change of variables

$$n = rM + \rho \tag{7.201}$$

and the definitions in Eqs. (7.196) to (7.200) to Eq. (7.189), we get the form

$$
\begin{aligned}
X_k^{\mathrm{GDFT}}(m) &= \sum_{\rho=0}^{M-1} W_M^{-(k+k_0)(\rho+n_0)} \sum_{r=-\infty}^{\infty} h((m-r)M - \rho) x(rM + \rho) W_M^{-(k+k_0)rM} \\
&= W_M^{-k_0 mM} \sum_{\rho=0}^{M-1} W_M^{-(k+k_0)(\rho+n_0)} \sum_{r=-\infty}^{\infty} \bar{p}_\rho(m-r) W_M^{k_0(m-r)M} x_\rho(r) \\
&= W_M^{-k_0 mM} \sum_{\rho=0}^{M-1} W_M^{-(k+k_0)(\rho+n_0)} [x_\rho(m) * (\bar{p}_\rho(m) W_M^{k_0 mM})],
\end{aligned}
$$
$$k = 0, 1, ..., M - 1 \tag{7.202}$$

In a similar manner, by applying the definitions in Eqs. (7.197) to (7.201) to Eq. (7.190), we get the form for the ρth branch output for the polyphase synthesis structure

$$
\hat{x}_\rho(m) = \left[q_\rho(m) W_M^{k_0 mM} \right] * \left[\frac{1}{M} \sum_{k=0}^{M-1} [\hat{X}_k^{\mathrm{GDFT}}(m) W_M^{k_0 mM}] W_M^{(k+k_0)(\rho+n_0)} \right],
$$
$$\rho = 0, 1, ..., M - 1 \tag{7.203}$$

where $*$ denotes discrete convolution. Figure 7.41 shows the resulting polyphase structures for the GDFT analyzer and synthesizer based on the interpretations of Eqs. (7.202) and (7.203), respectively. In comparison to the polyphase structure for the DFT filter bank in Section 7.2.3, it is seen that the time and frequency offsets add three modifications to the structure. First, each polyphase filter $\bar{p}_\rho(m)$ and $q_\rho(m)$ are modified by the multiplicative term $W_M^{k_0 mM}$. Thus the polyphase filters, in general, become complex filters. Second, each filter bank output from the analyzer is modulated by the factor $W_M^{-k_0 mM}$ and each input to the filter bank synthesizer is modulated by the factor $W_M^{k_0 mM}$. Finally, the DFT is replaced by the GDFT. In an actual implementation the modifying terms above can be rearranged and sometimes combined to form simpler variations of the structure. An example of this is given in Section 7.6. Also when $k_0 = n_0 = 0$ it can be readily verified that the polyphase structure reverts back to the simpler DFT polyphase structure

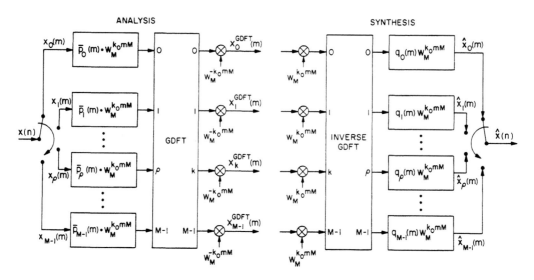

Figure 7.41 Polyphase analysis and synthesis structures for the GDFT filter bank.

discussed in Section 7.2.3. Finally, by letting $K = IM$ and carefully following through the mathematics, a more general form of the GDFT polyphase filter bank structure can be derived, similar to that in Section 7.2.4. The details will not be covered here.

7.5.4 Weighted Overlap-Add Structure for the GDFT Filter Bank

The second class of structures for efficiently realizing filter banks is the weighted overlap-add structure. Again, since the derivation for the GDFT filter bank is similar to that of the DFT filter bank in Section 7.2.5, it is outlined only in brief form.

Recall that in this derivation it is first necessary to convert the time frame from a fixed time frame to a sliding time frame according to the change of variables

$$r = n - mM \qquad (7.204)$$

Applying this change of reference to Eq. (7.189) leads to the expression

$$X_k^{GDFT}(m) = W_K^{-(k+k_0)mM} \underline{X}_k^{GDFT}(m) \qquad (7.205)$$

where $\underline{X}_k^{GDFT}(m)$ refers to the GDFT spectrum in the sliding time frame. We then define

$$\underline{y}_m(r) = h(-r)x(r + mM) \qquad (7.206)$$

to be the short-time windowed data in the sliding time frame r for block time m and $\underline{x}_m(r)$ to be the block-modulated and time-aliased (stacked) version of $\underline{y}_m(r)$ according to

$$\underline{x}_m(r) = \sum_{l=-\infty}^{\infty} W_K^{-k_0 lK} \underline{y}_m(r + lK) \qquad (7.207)$$

Applying Eqs. (7.204) to (7.207) to Eq. (7.189) then leads to the form

$$\underline{X}_k^{\mathrm{GDFT}}(m) = \sum_{r=0}^{K-1} \underline{x}_m(r) W_K^{-(k+k_0)(r+n_0)}, \quad k = 0, 1, ..., K-1 \qquad (7.208)$$

Equations (7.205) to (7.208) form the basis for the weighted overlap-add GDFT analysis structure.

Similarly, the weighted overlap-add structure for the GDFT filter bank synthesizer can be derived from an inverse set of relations. Applying Eq. (7.204) to Eq. (7.190) to convert from the fixed time reference to the sliding time reference for a particular block time $m = m_0$ leads to the expression

$$\hat{x}(r+mM)\Big|_{m=m_0} = f(r)\, \frac{1}{K} \sum_{k=0}^{K-1} \hat{\underline{X}}_k^{\mathrm{GDFT}}(m) W_K^{(k+k_0)(r+n_0)}\Big|_{m=m_0} + \text{(terms for } m \neq m_0) \qquad (7.209)$$

where

$$\hat{\underline{X}}_k^{\mathrm{GDFT}}(m) = W_K^{(k+k_0)mM} \hat{X}_k^{\mathrm{GDFT}}(m) \qquad (7.210)$$

If the filter $f(r)$ spans more than K samples, it is necessary to consider the periodic extension of the inverse GDFT of $X_k^{\mathrm{GDFT}}(m)$. When k_0 is a fractional number, this extension consists of a modulated extension in which each extended block is modulated by a phase factor. For example, if $\hat{x}_m(s)$ is the inverse GDFT of $\hat{X}^{\mathrm{GDFT}}(m)$, that is,

$$\hat{\underline{x}}_m(s) = \sum_{k=0}^{K-1} \hat{\underline{X}}_k^{\mathrm{GDFT}}(m) W_K^{(k+k_0)(s+n_0)}, \quad s = 0, 1, ..., K-1 \qquad (7.211)$$

and $\hat{\underline{y}}_m(r)$ is the modulated periodically extended version of $\hat{x}_m(s)$, then

$$\hat{\underline{y}}_m(r) = \hat{\underline{x}}_m(s + lK)\big|_{r=s+lK}$$

$$= \frac{1}{K} \sum_{k=0}^{K-1} \underline{X}_k^{GDFT}(m) \, W_K^{(k+k_0)(s+lK+n_0)}$$

$$= W_K^{k_0 lK} \, \hat{\underline{x}}_m(s)\big|_{r=s+lK}, \quad s = 0, 1, ..., K - 1; \, l = 0, \pm1, ... \quad (7.212)$$

Equation (7.209) can then be expressed in the form

$$\hat{x}(r + mM)\bigg|_{m=m_0} = f(r) \, \hat{\underline{y}}_m(r)\bigg|_{m=m_0} + \text{(terms for } m \neq m_0) \quad (7.213)$$

Figures 7.42 and 7.43 illustrate the sequence of operations necessary to implement the analysis and synthesis structures, respectively, for the weighted overlap-add realization of the GDFT filter bank, In the analyzer the input data are stored in an N_h-sample shift register where N_h is the size of the analysis window. For each block m, the data in the shift register is copied and weighted by the time-reversed analysis window, $h(-r)$, according to Eq. (7.206). The resulting short-time sequence, $y_m(r)$, is then partitioned into N_h/K blocks of K samples each. Samples in each block l are multiplied (block modulated) by $W_K^{-k_0 lK}$ and

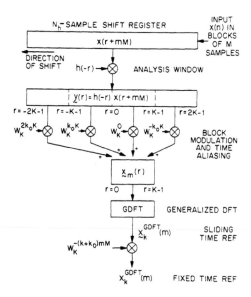

Figure 7.42 Weighted overlap-add analysis structure for the GDFT filter bank.

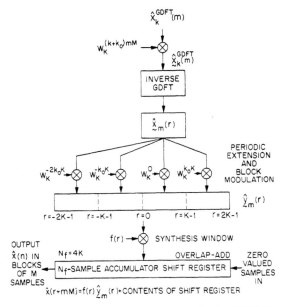

Figure 7.43 Weighted overlap-add synthesis structure for the GDFT filter bank.

time aliased (stacked and added) according to Eq. (7.207) to produce the K-sample sequence $\underline{x}_m(r)$. The GDFT is taken of this sequence and its output is modulated by the factor $W_K^{-(k+k_0)mM}$ according to Eqs. (7.208) and (7.205) to produce the filter bank output $X_k^{GDFT}(m)$ for block m.

In the synthesizer (see Fig. 7.43) essentially the reverse (transpose) sequence of operations takes place. The block modulation and time aliasing becomes a periodic extension of the signal $\hat{\underline{x}}_m(r)$ followed by a block modulation by $W_K^{k_0 l K}$ according to Eq. (7.212), where l corresponds to the block number of the periodic extension. The remaining operations are similar to those discussed for the DFT filter bank. When $n_0 = k_0 = 0$ the structures of Figs. 7.42 and 7.43 become identical to those of Figs. 7.19 and 7.20, respectively, for the conventional DFT filter bank.

7.5.5 Filter Design Criteria for the GDFT Filter Bank

The requirements on the filter design for the GDFT filter bank can be derived in a manner similar to that for the DFT filter bank. To show this similarity we can examine the requirements for the back-to-back analyzer/synthesizer framework. By combining Eqs. (7.189) and (7.190), assuming that

$$\hat{X}_k^{GDFT}(m) = X_k^{GDFT}(m), \quad k = 0, 1, ..., K - 1 \tag{7.214}$$

and following a similar approach to that in Section 7.3.3, we get

$$\hat{x}(n) = \sum_{s=-\infty}^{\infty} \sum_{m=-\infty}^{\infty} x(n - sK) f(n - mM) h(mM - n + sK) W_K^{k_0 sK} \qquad (7.215)$$

If it is desired that the system be a back-to-back identity system, it is required that for the $s = 0$ term

$$\sum_{m=-\infty}^{\infty} f(n - mM) h(mM - n) = 1 \qquad (7.216)$$

and for the $s \neq 0$ terms

$$\sum_{m=-\infty}^{\infty} f(n - mM) h(mM - n + sK) W_K^{k_0 sK} = 0, \; s \neq 0 \qquad (7.217)$$

or, equivalently,

$$\sum_{m=-\infty}^{\infty} f(n - mM) h(mM - n + sK) = 0, \; s \neq 0 \qquad (7.218)$$

Equations (7.216) and (7.218) are seen to be identical to those for the DFT filter bank, as discussed in Section 7.3.3, for elimination of time domain aliasing at the output of the filter bank. Similarly, it can be shown from Fig. 7.40 and the Fourier transforms of Eqs. (7.192) and (7.193) that the criteria for eliminating frequency domain aliasing at the output of a back-to-back GDFT filter bank are identical to those discussed in Section 7.3.1. Therefore, all the filter design techniques and criteria discussed in Section 7.3 can be directly applied to designs of GDFT filter banks. As will be seen in the next section, these design issues become important when applying the GDFT filter bank to applications in single-sideband filter banks.

7.6 UNIFORM SINGLE-SIDEBAND (SSB) FILTER BANKS

In the past four sections we discussed in considerable detail the classes of uniform DFT and GDFT filter banks which are based on the methods of *quadrature modulation*. Another extremely important class of filter banks are those which are based on the principle of *single-sideband (SSB) modulation* (see Fig. 7.2). Unlike the DFT and GDFT filter banks, SSB filter banks have real-valued channel signals instead of complex-valued signals, and therefore they are often preferred in applications involving communications and coding systems [7.8, 7.9, 7.13-7.18]. Many of the basic concepts of uniform SSB filter banks have evolved from efforts to design transmultiplexers for conversion between TDM (time division multiplexed) and FDM (frequency division multiplexed) signal formats for telecommunication systems [7.18, 7.38, 7.47].

As in the case of DFT and GDFT filter banks, it can be shown that uniform SSB filter banks can be realized in an efficient manner by sharing computation among channels, employing efficient methods of decimation and interpolation, and using fast transform algorithms for modulation. These issues are discussed in this section. In Section 7.7 we consider an alternative approach based on cascaded tree structures which are applicable to the design of nonuniform filter banks as well as uniform designs.

7.6.1 Realization of SSB Filter Banks from Quadrature Modulation Designs

As illustrated in Fig. 7.2, the process of single-sideband modulation or demodulation is one in which the frequency locations of two conjugate symmetric frequency bands from a real-valued signal $x(n)$, centered at frequencies $\pm\omega_k$, are translated respectively to the new locations $\omega = \pm\omega_\Delta/2$, where ω_Δ is the width of the bands. Since the resulting signal spectrum remains conjugate symmetric, the SSB modulated channel signal, denoted as $X_k^{SSB}(m)$, is real-valued.[2] Also, since the bandwidth of the SSB channel signal, $X_k^{SSB}(m)$, is less than that of the original signal, it can be reduced in sampling rate by the factor

$$M = \frac{\pi}{\omega_\Delta} \qquad\qquad (7.219)$$

where ω_Δ is defined with respect to the original sampling frequency.

As discussed in Chapter 2, the process of single-sideband modulation can be interpreted in terms of a modification of the quadrature modulation approach. Since the DFT and GDFT filter banks have already been discussed in considerable detail, this approach is particularly convenient and is the one that is used in this section. Figure 7.44 shows a block diagram of this process of modifying a quadrature-modulated signal to form a SSB modulated signal based on the discussion in Chapter 2. For convenience in later discussion the signs on the sine modulation terms have been changed and the scale factors of $\sqrt{2}$ have been dropped, although the basic circuit remains otherwise unchanged.

By carefully expressing Fig. 7.44 in terms of complex signals, the equivalent block diagrams in Fig. 7.45 result. This figure clearly illustrates the relationship between the quadrature modulation of a GDFT (or DFT) filter bank (see Fig. 7.39) and the modification made on the quadrature signals $X_k^{GDFT}(m)$ to form the SSB filter bank structure. Thus a straightforward approach for designing a SSB

---

[2] In this discussion we refer to the channel signal $X_k^{SSB}(m)$ as the single-sideband modulated signal and the input signal $x(n)$ as the original signal. The reader should be cautioned that this is the opposite of that used in the classical sense, where $X_k^{SSB}(m)$ would play the role of the input signal and $x(n)$ would be referred to as the SSB modulated signal.

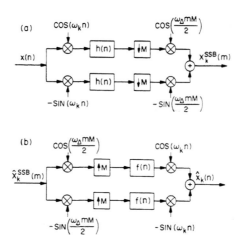

Figure 7.44 Methods of single-sideband modulation and demodulation by means of quadrature modulation.

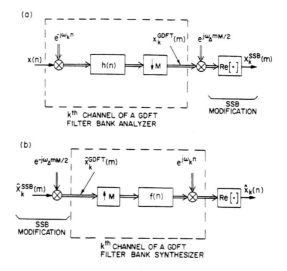

Figure 7.45 Block diagram for realizing a SSB filter bank anlayzer and synthesizer from a GDFT filter bank for the kth channel.

filter bank analyzer or synthesizer is to design an appropriate GDFT filter bank analyzer or synthesizer and then modify the channel signals according to the block diagrams in Fig. 7.44 or 7.45. In this way all the methods of efficiently realizing GDFT filter banks by means of polyphase or weighted overlap-add structures can be directly applied to the efficient realization of uniform SSB filter banks.

Based on Fig. 7.45 and the discussion in Section 7.5, the relationship for expressing the SSB signal, $X_k^{SSB}(m)$, in terms of the GDFT signal, $X_k^{GDFT}(m)$, for the kth channel is

$$X_k^{SSB}(m) = \text{Re } [X_k^{GDFT}(m)] \cos \left(\omega_\Delta \frac{mM}{2} \right) - \text{Im } [X_k^{GDFT}(m)] \sin \left(\omega_\Delta \frac{mM}{2} \right)$$

$$= \text{Re } [X_k^{GDFT}(m)e^{j\omega_\Delta mM/2}] \tag{7.220}$$

where Re [·] denotes the real part of the quantity in brackets and Im [·] denotes the imaginary part of the quantity in brackets. Similarly, it can be shown that for synthesis the input to the GDFT synthesizer can be related to the SSB signal $\hat{X}_k^{SSB}(m)$ according to

$$\hat{X}_k^{GDFT}(m) = \hat{X}_k^{SSB}(m)e^{-j\omega_\Delta mM/2} \tag{7.221}$$

Recall also from Chapter 2 that the synthesis process depends on the property that the synthesis filter $f(n)$ sufficiently removes (or cancels) the signal components outside the range $|\omega| \leqslant \omega_\Delta/2$.

The choice of the GDFT filter bank in Figs. 7.44 or 7.45 directly affects the channel stacking arrangement and the number of channels in the SSB filter bank. The center frequencies ω_k, $k = 1, ...,$ of the SSB filter bank channels, for example, are identical to those of the GDFT filter bank. Since conjugate frequency locations of the GDFT filter bank represent redundant channels [assuming real $x(n)$] not all of the GDFT frequency channels are necessarily applicable. For a K-point GDFT filter bank, as discussed in the preceding section, the center frequencies of the channels are located at

$$\omega_k = \frac{2\pi}{K}(k + k_0), \quad k = 0, 1, ..., K - 1 \tag{7.222}$$

These frequencies represent K equispaced points on the unit circle of the z-transform domain.

For example, Fig. 7.46 shows three possible choices of SSB filter bank designs for a $K = 8$ point GDFT with respective choices of $k_0 = 0$, 1/2, and 1/4. Figure 7.46(a) shows the choice of center frequencies for $k = 0$ (an eight-point DFT) and the corresponding choices of SSB filter bank channels. As seen in the figure, there are $(K/2-1) = 3$ SSB channels of width $\omega_\Delta = 2\pi/K$ plus two half-width channels, at $\omega = 0$ and $\omega = \pi$.

By choosing $k_0 = 1/2$ the design in Fig. 7.46(b) results. In this case there are $K/2$ uniform SSB channels spanning the range of $0 \leqslant \omega \leqslant \pi$ with potential bandwidths of $\omega_\Delta = 2\pi/K$.

Finally, Fig. 7.46(c) shows an example for which $k_0 = 1/4$ where solid dots represent center-frequencies and open dots represent their conjugate locations. In this case none of the center frequencies in Eq. (7.222) occur in conjugate pairs.

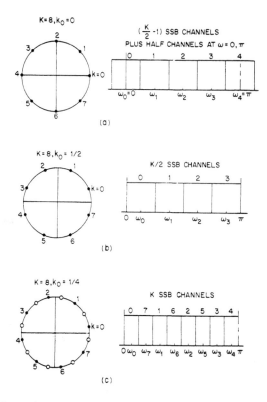

Figure 7.46 Center frequencies and band locations of SSB filter bank designs for $k_0 = 0$, 1/2, and 1/4.

Therefore, all the channels of the GDFT can be used to produce a $K = 8$-channel SSB filter bank with equally spaced channels in the range $0 \leqslant \omega \leqslant \pi$. Channels 0 through 3 are obtained from GDFT filter bank channels whose center frequencies are located in the upper half of the unit circle and channels 4 through 7 are obtained from GDFT filter bank channels with center frequencies in the lower half of the unit circle. The resulting SSB channels represent an interleaved set of channels in which alternate channels are inverted in frequency. This can be easily corrected by a simple modulation of the channel signals for channels 4 through 7 by the factor $(-1)^m$ [or alternatively by using the conjugate values of $X_k^{GDFT}(m)$ for generating these channels]. Since there are $K = 8$ uniformly spaced SSB channels in this approach, each channel has a potential bandwidth of $\omega_\Delta = \pi/K$.

From the examples above it can be seen that the choice of GDFT with $k_0 = 1/4$ offers the largest number of SSB channels for a given transform size. In the next two sections we examine in more detail two specific examples of SSB filter bank designs for $k_0 = 1/4$ and $k_0 = 1/2$.

7.6.2 Critically Sampled SSB Filter Banks with $k_0 = 1/4$

A particularly important class of single-sideband filter banks are those for which the channel signals are *critically sampled* and uniformly spaced in the frequency range $0 \leqslant \omega \leqslant \pi$. As seen from the discussion above, it is convenient to develop this class of filter banks around the choice of GDFT for which $k_0 = 1/4$, so that the number of SSB channels is equal to the transform size K. In this case the decimation ratio M is identical to the transform size, that is,

$$M = K \tag{7.223}$$

and the widths of the frequency channels [see Fig. 7.46(c)] become

$$\omega_\Delta = \frac{\pi}{M} = \frac{\pi}{K} \tag{7.224}$$

By applying Eqs. (7.223) and (7.224) to the definitions of $X_k^{SSB}(m)$ and $X_k^{GDFT}(m)$ in Eqs. (7.220) and (7.189), respectively, we get

$$X_k^{SSB}(m) = \text{Re}\left[X_k^{GDFT}(m)e^{j(\pi m/2)}\right]$$

$$= \text{Re}\left[\sum_{n=-\infty}^{\infty} h(mK - n)x(n)e^{-j(2\pi/K)(k+1/4)(n+n_0)}e^{j(\pi m/2)}\right] \tag{7.225}$$

By further manipulation of terms this can be expressed in the form

$$X_k^{SSB}(m) = \text{Re}\left[\sum_{n=-\infty}^{\infty} h(mK - n)x(n)e^{j(2\pi/K)(k+1/4)(mK-n-n_0)}\right]$$

$$= \sum_{n=-\infty}^{\infty} h_k(mK - n)x(n) \tag{7.226}$$

where

$$h_k(n) = h(n)\cos\left[\frac{2\pi}{K}\left(k + \frac{1}{4}\right)(n - n_0)\right] \tag{7.227}$$

Similarly, for the filter bank synthesis we can apply Eqs. (7.223) and (7.224) to the relationships expressed in Fig. 7.45 to obtain

$$\hat{x}(n) = \frac{1}{K}\sum_{k=0}^{K-1} \text{Re}\left[\hat{x}_k(n)\right] \tag{7.228}$$

$$= \frac{1}{K} \sum_{k=0}^{K-1} \sum_{m=-\infty}^{\infty} \mathrm{Re} \left[f(n - mK) \hat{X}_k^{\mathrm{SSB}}(m) e^{-j\pi m/2} e^{j(2\pi/K)(k+1/4)(n+n_0)} \right]$$

After further manipulation, this can be expressed as

$$\hat{x}(n) = \frac{1}{K} \sum_{k=0}^{K-1} \sum_{m=-\infty}^{\infty} f_k(n - mK) \hat{X}_k^{\mathrm{SSB}}(m) \tag{7.229}$$

where

$$f_k(n) = f(n) \cos \left[\frac{2\pi}{K} \left(k + \frac{1}{4} \right)(n + n_0) \right] \tag{7.230}$$

The form of Eqs. (7.226) and (7.230) suggests the simplified SSB channel model shown in Fig. 7.47 for the kth channel, where $h_k(n)$ and $f_k(n)$ correspond to real bandpass filters with center frequencies at $\omega_k = (2\pi/K)(k + 1/4)$. This model is in fact identical to the integer-band sampling model discussed in Chapter 2.

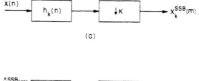

(a)

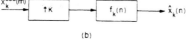

(b)

Figure 7.47 Integer-band sampling model for the kth channel of a critically sampled SSB filter bank with $k_0 = 1/4$.

Although the integer-band form of this model suggests a simple interpretation of the critically sampled SSB filter bank, it does not lead to an efficient realization for sharing computation among multiple channels in a filter bank. For this we must go back to the implementation suggested by Fig. 7.45. As an example of this approach we consider, for simplicity, the case where $n_0 = 0$ and where the GDFT quadrature filter bank (the dashed box in Fig. 7.45) is realized by the polyphase structure in Fig. 7.41. Because of the conditions given by Eqs. (7.223) and (7.224), a number of simplifications occur in the overall structure.

First we can observe in the polyphase filters (see Fig. 7.41) that the modulation terms $W_M^{k_0 mM}$ simplify to

$$W_M^{k_0 mM} = e^{j\frac{2\pi}{K}\frac{mM}{4}} \Big|_{M=K} = (j)^m \tag{7.231}$$

Next the modulation term $W_M^{-k_0 mM}$ at the outputs of the GDFT in Fig. 7.41 can be seen to cancel with the SSB modification $e^{j\omega_\Delta mM/2}$ in Fig. 7.45. A similar cancellation property occurs at the input to the inverse GDFT of the synthesizers. By applying these conditions and combining the results of Figs. 7.38, 7.41, and 7.45, we get the structure shown in Fig. 7.48 for the critically sampled SSB filter bank with $k_0 = 1/4$ and $n_0 = 0$.

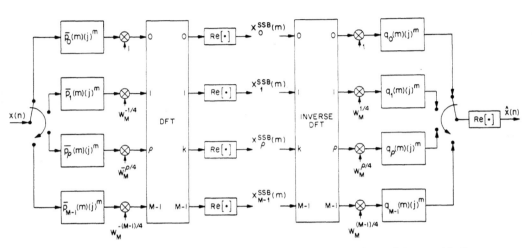

Figure 7.48 Polyphase analysis and synthesis structures for realizing a critically sampled SSB filter bank with $k_0 = 1/4$ and $n_0 = 0$.

Further efficiency in the realization of the structure in Fig. 7.48 can be obtained by recognizing that the coefficients of the polyphase filters are either pure real or pure imaginary. Thus by carefully separating real components from imaginary components the total number of multiplications and additions necessary in the polyphase filters is no greater than that for filters with only real-valued coefficients. Also by recognizing that only the real outputs of the DFT are needed and because it is known that the inputs of the inverse DFT are real, greater efficiency in computing the DFT and inverse DFT can be obtained by using this information [7.29].

7.6.3 SSB Filter Banks Based on $k_0 = 1/2$ Designs

Another important class of SSB filter banks are those for which $k_0 = 1/2$, as illustrated in Fig. 7.46(b). In this case there are $K/2$ independent filter bank channels [assuming a real input, $x(n)$] with center frequencies at

$$\omega_k = \frac{2\pi}{K}\left(k + \frac{1}{2}\right), \quad k = 0, 1, ..., \frac{K}{2} - 1 \qquad (7.232)$$

and potential bandwidths of

$$\omega_\Delta = \frac{2\pi}{K} \tag{7.233}$$

Again, a case of particular interest for this class of filter banks occurs when the filter bank is critically sampled. Since there are only $K/2$ unique bands in this framework with bandwidths given by Eq. (7.233), critical sampling in this case implies the relationship

$$M = \frac{K}{2} \tag{7.234}$$

By applying Eqs. (7.232) to (7.234) to the definitions in Eqs. (7.220) and (7.189), respectively, we get for the SSB analyzer

$$X_k^{SSB}(m) = \text{Re}\left[X_k^{GDFT}(m)e^{j\pi m/2} \right] \tag{7.235}$$

$$= \text{Re}\left[\sum_{n=-\infty}^{\infty} h(mM-n)x(n)e^{-j(2\pi/K)(k+1/2)(n+n_0)}e^{j\pi m/2} \right],$$

$$k = 0, 1, ..., \frac{K}{2} - 1$$

Similarly, for the SSB synthesizer we can apply Eqs. (7.232) to (7.234) to Eqs. (7.190) and (7.221) to get the form

$$\hat{x}(n) = \text{Re}\left[\sum_{m=-\infty}^{\infty} f(n-mM)\frac{1}{K}\sum_{k=0}^{(K/2)-1} \hat{X}_k^{SSB}(m)e^{-j(\pi m/2)}e^{j(2\pi/K)(k+1/2)(n+n_0)} \right]$$

$$\tag{7.236}$$

By careful manipulation of Eqs. (7.235) and (7.236), it can be shown that they can be expressed in the form

$$X_k^{SSB}(m) = (-1)^{km}\sum_{n=-\infty}^{\infty} h_k(mM-n)x(n), \quad k = 0, 1, ..., \frac{K}{2} - 1 \tag{7.237}$$

and

$$\hat{x}(n) = \sum_{m=-\infty}^{\infty}\sum_{k=0}^{\frac{K}{2}-1} (-1)^{km}f_k(n-mM)\hat{X}_k^{SSB}(m) \tag{7.238}$$

where

$$h_k(n) = h(n) \cos\left[\frac{2\pi}{K}\left(k + \frac{1}{2}\right)(n - n_0)\right] \qquad (7.239)$$

and

$$f_k(n) = f(n) \cos\left[\frac{2\pi}{K}\left(k + \frac{1}{2}\right)(n + n_0)\right] \qquad (7.240)$$

This leads to the integer-band sampling model shown in Fig. 7.49 for the kth channel of the analyzer and synthesizer. Again, this model is useful for insight and for realization where only a small number of filter bank channels are needed.

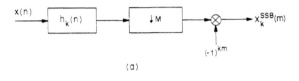

(a)

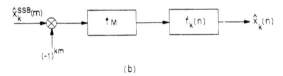

(b)

Figure 7.49 Integer-band sampling model for the kth band of a critically sampled SSB filter bank with $k_0 = 1/2$.

For filter banks with a large number of channels, it is more efficient to use a structure in which computation can be shared among bands and where a fast transform algorithm can aid in the computation of the modulation terms. A number of related structures for accomplishing this have been proposed [7.16-7.18, 7.38]. For example, an approach starting from the form of Eqs. (7.236) to (7.240), with $n_0 = 0$, leads to a structure based on the discrete cosine transform [7.16]. This transform has the form

$$Y_k = \sum_{n=0}^{K'-1} y(n) \cos\left[\pi(2k + 1)\frac{n}{2K'}\right] \qquad (7.241)$$

where K' is related to $K/2$ in the development of this formulation.

Another approach which leads to a similar structure can be developed by using the modified GDFT method discussed in Section 7.6.1. As an example of this approach we consider the case where $n_0 = 0$, where the GDFT quadrature filter bank is realized with the weighted overlap-add structure, and where the length of the filters $h(n)$ and $f(n)$ are $N_h = N_f = 4K$. By applying Eqs. (7.232) to (7.235) to the structures in Figs. 7.38 and 7.43, we get the resulting structures in Figs. 7.50 and 7.51 for the SSB analysis and synthesis, respectively. In these structures the realizations of the GDFT and the inverse GDFT are replaced by their equivalent DFT and inverse DFT realizations, respectively.

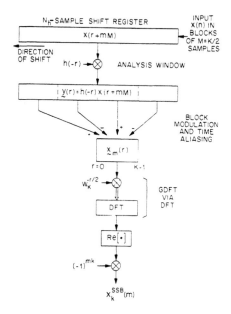

Figure 7.50 Weighted overlap-add analysis structure for the SSB filter bank with $k_0 = 1/2$.

Numerous other forms of these structures can also be generated, for example, by applying a polyphase structure instead of the weighted overlapp-add structure to the GDFT and by using more efficient real or symmetric forms of the DFT realization. Details of these variations are beyond the scope of this book.

7.7 FILTER BANKS BASED ON CASCADED REALIZATIONS AND TREE STRUCTURES

Thus far we have considered designs of *uniform* filter banks with quadrature or single-sideband modulation, where each channel in the filter bank is a spectral translation of a common baseband filter. It was shown that this property allows the filter bank structure to be factored so that the processing for all the channels can be shared and the modulation or frequency translation of the bands is efficiently achieved with the aid of a fast transform algorithm. Additional efficiency in

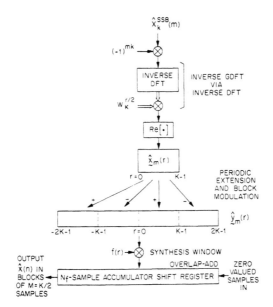

Figure 7.51   Weighted overlap-add synthesis structure for the SSB filter bank with $k_0 = 1/2$.

computation is achieved in these designs with the aid of *single-stage* decimation or interpolation techniques.

In this section we consider a different approach to filter bank designs based on *tree structures* in which the input signal is successively divided into smaller frequency bands at each stage of the tree [7.14, 7.17, 7.18, 2.23, 2.24, 7.38]. Efficiency in these designs is achieved through the use of the *multistage* decimation and interpolation methods discussed in Chapters 5 and 6. In fact it has been shown [7.23] that the computational efficiency obtained with this tree structure approach is highly competitive with the DFT type structures discussed above; especially for low order designs. The resulting filter bank designs are not restricted to the case where all the channels are frequency translations of a common lowpass design. They also need not be restricted to the case of *uniform* filter banks. Therefore, they offer greater flexibility in the choice of filter bank designs.

Figure 7.52(a) illustrates a simple example of a two-stage tree structure for a four-band, critically sampled, uniform integer-band SSB filter bank design. The input signal $x(n)$ is filtered in the first stage of the tree by complementary highpass and lowpass filters HP_1 and LP_1, respectively, to divide it into two equal parts as illustrated in Fig. 7.52(b). Each of these signals is reduced in sampling rate by a factor of 2 and filtered again in the second stage of the tree by complementary filters HP_2 and LP_2. The resulting signals are again reduced in sampling rate by a factor of 2 to produce the four-band uniform filter bank arrangement shown in Fig. 7.52(b). Since the process of highpass filtering and reducing the sampling rate by two (integer-band sampling) produces an inverted spectrum, the roles of the

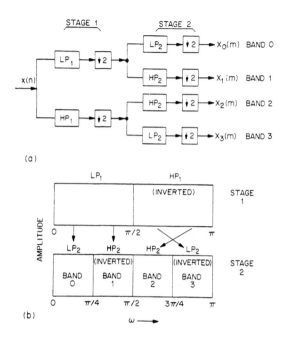

Figure 7.52 Example of a two-stage integer-band tree structure for a four-band uniform SSB filter bank.

lowpass and highpass filters LP_2 and HP_2 in the upper pair of filters (bands 2 and 3) in stage 2 are interchanged to compensate for this inversion. Also, since the outputs of the filter bank, $X_0(m)$ to $X_3(m)$, are reduced in sampling rate by a net factor of 4, the resulting design is a critically sampled filter bank.

Figure 7.53(a) shows a similar example of a four-band, critically sampled, octave-spaced SSB filter bank based on a three-stage tree structure. In this structure only the upper branch (i.e., the lower band) of each stage is divided equally to produce the nonuniform band arrangement shown in Fig. 7.53(b). Note that in this approach each band (except the bottom two bands) are sampled at different sampling rates comensurate with the width of the bands.

In this section we consider a number of aspects of these tree structure filter bank designs. In particular, we focus on the design of quadrature mirror filters (QMF), which play a key role in designs where critical sampling and analysis/synthesis reconstruction is desired [7.14, 7.39]. We conclude by showing how these tree structure designs relate to equivalent, parallel, integer-band sampling designs.

7.7.1 Quadrature Mirror Filter (QMF) Bank Design

The technique of quadrature mirror filter bank (QMF) design can be developed from the two-band integer-band sampling analysis/synthesis structure shown in Fig.

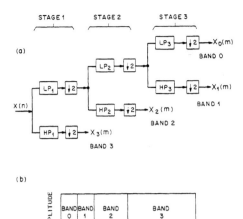

Figure 7.53 Example of a three-stage tree structure for an octave spaced SSB filter bank.

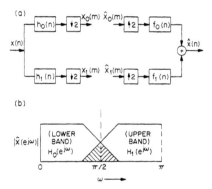

Figure 7.54 Two-band integer-band analysis/synthesis framework for quadrature mirror filter (QMF) design.

7.54(a) [7.14, 7.39]. In this process the filters $h_0(n)$ and $h_1(n)$ represent the lowpass and highpass analysis filters, respectively, for the two integer bands. Similarly, the filters $f_0(n)$ and $f_1(n)$ represent the respective lowpass and highpass synthesis filters and the signals $X_0(m)$ and $X_1(m)$ represent the respective lowpass and highpass integer-band signals.

By letting $X(e^{j\omega})$, $X_0(e^{j\omega})$, $X_1(e^{j\omega})$, $H_0(e^{j\omega})$, and $H_1(e^{j\omega})$ be the Fourier transforms of $x(n)$, $X_0(m)$, $X_1(m)$, $h_0(n)$, and $h_1(n)$, respectively, we get (from Chapter 2)

$$X_0(e^{j\omega}) = \frac{1}{2}\left[X(e^{j\omega/2})H_0(e^{j\omega/2}) + X(e^{j(\omega+2\pi)/2})H_0(e^{j(\omega+2\pi)/2})\right] \quad (7.242a)$$

and

$$X_1(e^{j\omega}) = \frac{1}{2}\left[X(e^{j\omega/2})H_1(e^{j\omega/2}) + X(e^{j(\omega+2\pi)/2})H_1(e^{j(\omega+2\pi)/2})\right] \quad (7.242b)$$

Similarly, letting $\hat{X}_0(e^{j\omega})$, $\hat{X}_1(e^{j\omega})$, $F_0(e^{j\omega})$, $F_1(e^{j\omega})$, and $\hat{X}(e^{j\omega})$ be the transforms of $\hat{X}_0(m)$, $\hat{X}_1(m)$, $f_0(n)$, $f_1(n)$, and $\hat{x}(n)$, respectively, leads to the synthesis relation

$$\hat{X}(e^{j\omega}) = X_0(e^{j2\omega})F_0(e^{j\omega}) + X_1(e^{j2\omega})F_1(e^{j\omega}) \quad (7.243)$$

Finally, by combining Eqs. (7.242) and (7.243) and assuming a back-to-back filter bank arrangement such that $\hat{X}_0(e^{j\omega}) = X_0(e^{j\omega})$ and $\hat{X}_1(e^{j\omega}) = X_1(e^{j\omega})$, we get the input-to-output frequency domain relationship of the filter bank in the form

$$\hat{X}(e^{j\omega}) = \frac{1}{2}[H_0(e^{j\omega})F_0(e^{j\omega}) + H_1(e^{j\omega})F_1(e^{j\omega})]X(e^{j\omega})$$

$$+ \frac{1}{2}[H_0(e^{j(\omega+\pi)})F_0(e^{j\omega}) + H_1(e^{j(\omega+\pi)})F_1(e^{j\omega})]X(e^{j(\omega+\pi)}) \quad (7.244)$$

The first term in this equation represents the desired signal translation from $X(e^{j\omega})$ to $\hat{X}(e^{j\omega})$, whereas the second term represents the undesired frequency domain aliasing components.

To eliminate these aliasing components, the second term must cancel; that is, it is required that

$$H_0(e^{j(\omega+\pi)})F_0(e^{j\omega}) + H_1(e^{j(\omega+\pi)})F_1(e^{j\omega}) = 0 \quad (7.245)$$

This condition can be achieved by carefully choosing the filter designs so that the two terms in this relation cancel.

The first requirement that is made [7.14] is that the lowpass and highpass filters $h_0(n)$ and $h_1(n)$ be frequency translations of a common lowpass filter design denoted as $h(n)$, that is,

$$h_0(n) = h(n) \qquad \text{for all } n \qquad (7.246)$$

and

$$h_1(n) = (-1)^n h(n) \quad \text{for all } n \qquad (7.247)$$

Equivalently, in the frequency domain, this implies that

$$H_0(e^{j\omega}) = H(e^{j\omega}) \qquad (7.248)$$

and

$$H_1(e^{j\omega}) = H(e^{j(\omega+\pi)}) \tag{7.249}$$

where $H(e^{j\omega})$ is the Fourier transform of $h(n)$. The condition above implies that the filters $H_0(e^{j\omega})$ and $H_1(e^{j\omega})$ have *mirror-image* symmetry about the frequency $\omega = \pi/2$. This property is illustrated in Fig. 7.54(b) where the shaded regions illustrate the regions where aliasing and imaging between the two bands occur.

Applying Eqs. (7.248) and (7.249) to Eq. (7.245) leads to

$$H(e^{j(\omega+\pi)})F_0(e^{j\omega}) + H(e^{j\omega})F_1(e^{j\omega}) = 0 \tag{7.250}$$

The requirements on $F_0(e^{j\omega})$ and $F_1(e^{j\omega})$ can now be stated. Since $F_0(e^{j\omega})$ must be a lowpass filter, we can assign

$$F_0(e^{j\omega}) = 2H(e^{j\omega}) \tag{7.251a}$$

or equivalently

$$f_0(n) = 2h(n) \quad \text{for all } n \tag{7.251b}$$

where the factor of 2 applies to the gain factor associated with the interpolation filter as discussed in Chapter 2. Then, by applying this to Eq. (7.250), it is clear that $F_1(e^{j\omega})$ must have the highpass form

$$F_1(e^{j\omega}) = -2H(e^{j(\omega+\pi)}) \tag{7.252a}$$

or equivalently

$$f_1(n) = -2(-1)^n h(n) \quad \text{for all } n \tag{7.252b}$$

where the minus sign is necessary to obtain proper cancellation.

The design of the analysis and synthesis filters in Fig. 7.54 is now translated to a common lowpass filter design $H(e^{j\omega})$ through the conditions in Eqs. (7.248), (7.249), (7.251), and (7.252). By applying these definitions to Eq. (7.244), we get the resulting input-to-output relation of the filter bank in the form

$$\hat{X}(e^{j\omega}) = [H^2(e^{j\omega}) - H^2(e^{j(\omega+\pi)})]X(e^{j\omega}) \tag{7.253}$$

where the second term, due to aliasing, is now canceled. This implies that the aliasing due to the decimation in the analysis structure is exactly canceled by the imaging due to the interpolation in the synthesis structure. From Eq. (7.253) it can be seen that the back-to-back filter bank becomes a unity gain system if we impose the further condition that the prototype filter $H(e^{j\omega})$ satisfy the relation

$$|H^2(e^{j\omega}) - H^2(e^{j(\omega+\pi)})| = 1 \text{ for all } \omega \qquad (7.254)$$

Additionally, it is desired that $H(e^{j\omega})$ approximate the ideal lowpass condition

$$|H(e^{j\omega})| = \begin{cases} 1, & 0 \leqslant \omega \leqslant \dfrac{\pi}{2} \\ 0, & \dfrac{\pi}{2} \leqslant \omega \leqslant \pi \end{cases} \qquad (7.255)$$

The conditions in Eqs. (7.254) and (7.255) cannot be met exactly; however, they can be closely approximated for modest-size filters, as will be seen in the next section. For the special case where $H(e^{j\omega})$ is a two-tap finite impulse response filter, Eq. (7.254) is exactly satisfied, although the resulting filter response approximates Eq. (7.255) very poorly.

7.7.2 Finite Impulse Response (FIR) Filter Designs for QMF Filter Banks

In the preceding section it was shown that the design of the quadrature mirror filters $h_0(n)$, $h_1(n)$, $f_0(n)$, and $f_1(n)$ in Fig. 7.54 can be translated to that of a single lowpass filter $h(n)$ which must approximate the constraints in Eqs. (7.254) and (7.255). Several approaches have been suggested for this design, including symmetric finite impulse response (FIR) filters [7.14, 7.39], infinite impulse response (IIR) filters [7.40, 7.41], and variations of half-band filters [7.42]. Also, slightly more general variations of the translations in Eqs. (7.246), (7.247), (7.251), and (7.252) are possible [7.43].

Although all the categories of filter design above lead to appropriate methods of quadrature mirror filter bank design, a particularly useful class of designs are those based on symmetric FIR filters. They have the advantage of linear phase response (i.e., flat group delay), which allows the two-band design to be conveniently cascaded in the tree structures (such as those in Fig. 7.52 and 7.53) without the necessity for phase compensation. In this section we consider a number of issues involved in these designs and show practical examples of such designs.

If we assume that $h(n)$ is a symmetric FIR filter with N taps in the range $0 \leqslant n \leqslant N - 1$, the symmetry property implies that

$$h(n) = h(N - 1 - n), \quad n = 0, 1, ..., N - 1 \qquad (7.256)$$

The Fourier transform of $h(n)$ then has the form [7.26, 7.27]

$$H(e^{j\omega}) = H_r(e^{j\omega})e^{-j\omega(N-1)/2} \qquad (7.257a)$$

where $H_r(e^{j\omega})$ is a *real* function such that

$$H_r^2(e^{j\omega}) = |H(e^{j\omega})|^2 \qquad (7.257\text{b})$$

and where the term $e^{-j\omega(N-1)/2}$ is the linear phase term due to the flat delay of $(N-1)/2$ samples in the response.

Applying Eq. (7.257) to Eq. (7.253) gives the input-to-output frequency response of the back-to-back filter bank in the form

$$\hat{X}(e^{j\omega}) = [|H(e^{j\omega})|^2 e^{-j\omega(N-1)} - |H(e^{j(\omega+\pi)(N-1)})|^2 e^{-j(\omega+\pi)(N-1)}]X(e^{j\omega})$$

$$= [|H(e^{j\omega})|^2 - (-1)^{(N-1)}|H(e^{j(\omega+\pi)})|^2]e^{-j\omega(N-1)}X(e^{j\omega}) \qquad (7.258)$$

This form shows that the overall filter bank has a linear phase delay due to the term $e^{-j\omega(N-1)}$ (i.e., it has a flat delay of $N-1$ samples). The term in brackets defines the magnitude response of the filter bank, which should approximate a unity gain. In practice this magnitude response can only be approximated as being a constant as a function of ω and it is strongly affected by the choice of whether N is even or odd. If N is even, Eq. (7.258) has the form

$$\hat{X}(e^{j\omega}) = [|H(e^{j\omega})|^2 + |H(e^{j(\omega+\pi)})|^2]e^{j\omega(N-1)}X(e^{j\omega}) \qquad (7.259)$$

and if N is odd it has the form

$$\hat{X}(e^{j\omega}) = [|H(e^{j\omega})|^2 - |H(e^{j(\omega+\pi)})|^2]e^{j\omega(N-1)}X(e^{j\omega}) \qquad (7.260)$$

In the latter case it is seen that at $\omega = \pi/2$ the overall magnitude response of the filter is zero due to the fact that $|H(e^{j\pi/2})| = |H(e^{j3\pi/2})|$. Therefore, only designs for N even are useful[3] and we assume that N is even in the remainder of this section.

The process of designing the filter $h(n)$ is a nontrivial one in that it must simultaneously approximate the ideal lowpass characteristic in Eq. (7.255) as well as approximating the flat reconstruction constraint above, that is,

$$|H(e^{j\omega})|^2 + |H(e^{j(\omega+\pi)})|^2 = 1 \quad \text{for all } \omega \qquad (7.261)$$

This dual requirement imposes constraints not only on the behavior of $H(e^{j\omega})$ in the passband and stopband regions but also on its behavior in the transition band. Thus the lowpass filter designs discussed in Chapter 4 are generally not directly applicable to this design problem.

One approach to the design of $h(n)$, suggested in Ref. 7.44, applies a conventional Hanning window design and it is one of the few methods from Chapter 4 that leads directly to a suitable (but suboptimal) design. Its advantage

---

[3] An alternative set of conditions for quadrature mirror filter designs can be derived for the case where N is odd [7.43]. The details are beyond the scope of this book.

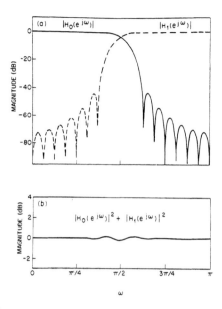

Figure 7.55 (a) Frequency response of a 32-tap QMF Hanning window design and
(b) the analysis/synthesis reconstruction error due to this design.

is that it is relatively easy to apply. Figure 7.55(a) shows an example of an
$N = 32$-tap Hanning window design. The solid curve shows the lowpass frequency
response of $H_0(e^{j\omega}) = H(e^{j\omega})$, and the dashed curve shows its mirror image
$H_1(e^{j\omega}) = H(e^{j(\omega+\pi)})$. Figure 7.55(b) shows the overall frequency response,
$|H_0(e^{j\omega})|^2 + |H_1(e^{j\omega})|^2$, of the back-to-back filter bank based on this design. As
seen from this figure, the reconstruction error is less than ± 0.2 dB, which is quite
acceptable for many applications. Because of the nearly flat frequency response of
the Hanning window design in the passband, the largest error in reconstruction
occurs in the transition region of the filters.

Quadrature mirror filter designs with better lowpass characteristics and
smaller reconstruction errors for a given filter length N can be achieved with the
use of computer-aided optimization techniques [7.45, 7.48]. The problem can be
formulated in the framework of a nonlinear optimization problem by defining a
single-valued error function which encompases the desired criteria in Eqs. (7.255)
and (7.261). An optimization program can then be applied which systematically
searches the coefficient space $h(n)$ for the choice of coefficients that minimize this
error function.

The error function can be defined as a weighted sum of two terms which
express the error in approximating the conditions in Eqs. (7.255) and (7.261),
respectively. The first term can be defined as

$$E_s(\omega_s) = \int_{\omega-\omega_s}^{\pi} |H(e^{j\omega})|^2 \, d\omega \qquad (7.262)$$

and it describes the energy in the stopband region of the frequency response of $H(e^{j\omega})$. The cutoff frequency ω_s in this definition defines the stopband edge where $\omega_s > \pi/2$. The second term can be defined as

$$E_r = 2 \int_{\omega-0}^{\pi} [|H(e^{j\omega})|^2 + |H(e^{j(\omega+\pi)})|^2 - 1] \, d\omega \qquad (7.263)$$

and it represents the error in approximating Eq. (7.261). In both cases the evaluation of the errors in Eqs. (7.262) and (7.263) can be obtained in computer simulations by replacing the integrals with summations over a sufficiently dense set of frequency samples (obtained with the aid of a large-order FFT).

The error functions in Eq. (7.262) and (7.263) are combined into a single error function of the form

$$E = E_r + \alpha E_s(\omega_s) \qquad (7.264)$$

where the weighting factor α determines the relative importance of the two terms. Thus the choice of α, ω_s, and the filter size N determine the trade-offs in the design of $h(n)$.

In principle, the design procedure described above is relatively straightforward. An initial filter is chosen as a starting point and the optimization routine iteratively evaluates the error E and searches the coefficient space in a systematic way until it finds the coefficients $h(n)$ that minimize this error. In practice, the error surface E, as a function of the coefficients, has many local minima and the optimization routine can get trapped in these local minima rather than in the global minimum (the true minimum). This difficulty is mitigated to some extent in the method of Ref. 7.45 with the use of the Hooke and Jeeves optimization routine, which is somewhat robust to the presence of local minima. It is also overcome by a careful choice of several initial starting points (the Hanning window design is a good starting point) and by manual intervention in terms of selecting proper parameters for the search algorithm and monitoring of intermediate results of the algorithm.

In an alternative design method [7.48] another form of an iterative algorithm is described which converges more rapidly and is numerically stable (i.e., it does not require manual intervention or multiple starting points). The technique is based on the following concept: (a) for the analysis filter bank use the filter tap weights from the previous iteration and optimize the filter tap weights for the synthesis filter bank (with respect to the criterion E in Eq. (7.264); and then (b) reverse the strategy, that is, use the known tap weights for the synthesis filter bank and find the optimum tap weights for the analysis filter bank. This process can be formulated in the form of a constrained optimization of a quadratic function which has a well-known solution for each iteration. The optimum coefficients in each iteration are directly related to the eigenvector of a real symmetric matrix corresponding to its minimum eigenvalue. In principle this approach can be applied to a large class of filter bank analysis/synthesis designs. For the specific case of the QMF design problem where the filters for analysis and synthesis are derived from a common lowpass prototype, a simplification accrues such that only step (a) need be performed in each iteration.

A set of quadrature mirror filter designs has been tabulated using the design procedure in Ref. 7.45 for values of N of 8, 12, 16, 24, 32, 48, and 64 and for various choices of the parameters α and ω_s. The properties and coefficients for these designs are presented in Appendix 7.1. Figure 7.56 shows an example of the frequency response (and its mirror image) and the reconstruction error for a 32-tap filter design (filter 32D) from this set. The design was obtained for a value of $\omega_s = 0.5\pi + (0.043)2\pi = 0.586\pi$ and a value of $\alpha = 2$. As seen from the figure, the filter has a stopband attenuation peak of -38 dB and a reconstruction error of less than ± 0.025 dB. Various alternatives to this design can be found in the tables in Appendix 7.1.

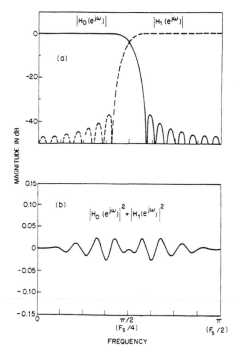

Figure 7.56   Frequency response and reconstruction error for the 32D quadrature mirror filter.

7.7.3 Polyphase Realization of Quadrature Mirror Filter Banks

While the two-band integer-band structure in Fig. 7.54 suggests a straightforward way of realizing the quadrature mirror filter bank, a more efficient realization can be achieved by using a polyphase structure [7.14]. This is a consequence of the fact that the filters $h_0(n)$, $h_1(n)$, $f_0(n)$, and $f_1(n)$ are derived from a common lowpass design $h(n)$ as discussed above. The development of this structure is similar to that for a two-band polyphase DFT structure (see Section 7.2) with a number of subtle modifications to accommodate several differences in the QMF formulation. For example, in the analysis structure it is more convenient to start with a counterclockwise formulation of the polyphase structure because the filter $h(n)$ is defined over the range $0 \leqslant n < N - 1$. In the synthesis structure an additional minus sign occurs due to the definition of $F_1(e^{j\omega})$ in Eq. (7.252). In this section we consider these issues in more detail and derive the polyphase structure for the QMF filter bank. We then show examples of how this structure is applied to the tree structure to obtain an efficient computational approach to QMF filter bank designs with more than two bands.

From the structure in Fig. 7.54 and the discussion in Chapter 2 it can be seen that the filter bank signals $X_0(m)$ and $X_1(m)$ can be expressed as

$$X_k(m) = \sum_{n=-\infty}^{\infty} h_k(n) x(2m - n), \quad k = 0, 1 \tag{7.265}$$

By applying Eqs. (7.246) and (7.247), this can be expressed in the form

$$X_k(m) = \sum_{n=-\infty}^{\infty} (-1)^{kn} h(n) x(2m - n) \tag{7.266}$$

and by the change of variables

$$n = 2r + \rho \tag{7.267}$$

it becomes

$$X_k(m) = \sum_{\rho=0}^{1} \sum_{r=-\infty}^{\infty} (-1)^{k(2r+\rho)} h(2r + \rho) x(2(m - r) - \rho) \tag{7.268}$$

We can now define the (counterclockwise) polyphase filters and signals for an $M = 2$ decimator as (see Chapter 3)

$$p_\rho(m) = h(2m + \rho), \quad \rho = 0, 1 \tag{7.269}$$

and

$$x_\rho(m) = x(2m - \rho), \quad \rho = 0, 1 \tag{7.270}$$

where $\rho = 0$ is associated with the even samples of $h(n)$ and $x(n)$ and $\rho = 1$ is associated with the odd samples. For $h(n)$ defined as an N-tap filter (N even) with coefficients in the range $0 \leqslant n \leqslant N - 1$, the polyphase filters $p_\rho(m)$ each have $N/2$ coefficients in the range $0 \leqslant m \leqslant N/2 - 1$.

By applying these polyphase definitions to Eq. (7.268), we get the form

$$X_k(m) = \sum_{\rho=0}^{1} \sum_{r=0}^{(N/2)-1} (-1)^{k\rho} p_\rho(r) x_\rho(m - r)$$

$$= p_0(m) * x_0(m) + (-1)^k p_1(m) * x_1(m), \quad k = 0, 1 \tag{7.271}$$

where $*$ denotes discrete convolution. This form suggests the polyphase analysis structure shown on the left side of Fig. 7.57. Note that this structure differs from that in the two-band DFT filter bank in that it is based on a counterclockwise commutator model of the polyphase structure. This choice is more convenient in the QMF filter bank. If the clockwise model were used, the $p_1(m)$ filter would have a zero-valued coefficient for $p_1(0)$ and nonzero coefficient values for the range $1 \leqslant m \leqslant N/2$. This definition is feasible but not as convenient.

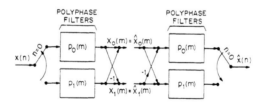

Figure 7.57 Polyphase realization of the quadrature mirror filter bank.

The derivation of the polyphase synthesis structure for the QMF filter bank can be derived by noting that (see Fig. 7.54 and Chapter 2)

$$\hat{x}(n) = \sum_{k=0}^{1} \sum_{m=-\infty}^{\infty} \hat{X}_k(m) f_k(n - 2m) \tag{7.272}$$

By applying the definitions of $f_0(n)$ and $f_1(n)$ in Section 7.7.1, that is,

$$f_k(n) = (-1)^k (-1)^{kn} h(n) \tag{7.273}$$

we get the form

$$\hat{x}(n) = \sum_{k=0}^{1} \sum_{m=-\infty}^{\infty} \hat{X}_k(m)(-1)^k(-1)^{kn}h(n-2m) \qquad (7.274)$$

The polyphase signals for even ($\rho = 0$) and odd ($\rho = 1$) outputs of the filter bank can now be defined as

$$\hat{x}_\rho(r) = \hat{x}(2r+\rho), \quad \rho = 0, 1$$

$$= \sum_{k=0}^{1} \sum_{m=-\infty}^{\infty} \hat{X}_k(m)(-1)^k(-1)^{k(2r+\rho)}h(2(r-m)+\rho) \qquad (7.275)$$

By applying the definition of the polyphase filters $p_\rho(m)$ in Eq. (7.269), Eq. (7.275) becomes

$$\hat{x}_\rho(r) = \sum_{k=0}^{1} \sum_{m=-\infty}^{\infty} \hat{X}_k(m)(-1)^k(-1)^{k\rho}p_\rho(r-m)$$

$$= \sum_{m=0}^{\frac{N}{2}-1} p_\rho(r-m) \sum_{k=0}^{1} (-1)^k(-1)^{k\rho}\hat{X}_k(m)$$

$$= p_\rho(m) * [\hat{X}_0(m) - (-1)^\rho\hat{X}_1(m)], \quad \rho = 0, 1 \qquad (7.276)$$

Equation (7.276) suggests the polyphase synthesis structure on the right side of Fig. 7.57. This structure also differs slightly from that of the two-band DFT polyphase structure due to the additional (-1) term that appears in the QMF structure.

By rearranging the structure in Fig. 7.57, the equivalent structure in Fig. 7.58(a) can be obtained, and if the sum and difference terms are combined, the merged structure of Fig. 7.58(b) results. This form more clearly illustrates what is happening in the QMF filter bank. For example, it is seen that in order for the filter bank to have a flat response it is required that the polyphase filters satisfy the condition

$$p_0(m) * p_1(m) = u_0(n) = \begin{cases} 1, & n = 0 \\ 0, & \text{otherwise} \end{cases} \qquad (7.277)$$

(excluding any scale factors). This is equivalent to the flat spectrum requirement in Eq. (7.261). For even-length, symmetric filters it can be readily shown that $p_0(m)$ and $p_1(m)$ are time reversed versions of each other [i.e., $p_0(0) = p_1(N/2-1)$, $p_0(1) = p_1(N/2-2)$, etc.]. Therefore, they have exactly inverse phase relations and equal magnitude responses. Thus if the filter bank is to

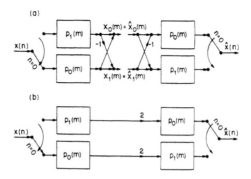

Figure 7.58 Rearranged and merged forms of the QMF polyphase filter bank structure to show its properties.

have a flat response, these polyphase filters must approximate all-pass functions. For the special case where $N = 2$, it is readily seen that

$$p_0(m) = p_1(m) = u_0(n) \qquad (7.278)$$

and Eq. (7.277) is satisfied exactly. It is also seen from the structure in Fig. 7.58(b) that input samples at even sample times occur at odd sample times at the output [i.e., $\hat{x}(0) = x(-1)$, $\hat{x}(1) = x(0)$, etc.]. Thus there is an inherent delay of one sample in this framework in addition to the delay of the filters. In contrast, the DFT polyphase structure, defined for N odd, has no such inherent delay.

The polyphase structure shown in Fig. 7.57 or 7.58(a) is more efficient than a direct application of the structure in Fig. 7.54 by approximately a factor of 4. A factor of 2 is obtained by the use of the polyphase decimation/interpolation structure and a second factor of 2 is obtained by sharing the same computation between the upper and lower bands of the filter bank.

This polyphase structure can be readily extended to the case of the cascaded tree structures to achieve higher-order filter banks as discussed earlier. For example, Fig. 7.59 shows block diagrams of the analysis and synthesis structures for a three-stage, five-band nonuniform QMF filter bank based on a cascade of polyphase structures [7.46]. The input signal $x(n)$ is divided into two bands by stage 1 with polyphase filters $p_0^{(1)}(m)$ and $p_1^{(1)}(m)$. These bands are each divided into two more bands in stage 2 with polyphase filter designs $p_0^{(2)}(m)$ and $p_1^{(2)}(m)$. In stage 3 only the bottom band is divided with polyphase filters $p_0^{(3)}(m)$ and $p_1^{(3)}(m)$ giving the choice of five bands with frequency ranges $0 - \pi/8$, $\pi/8 - \pi/4$, $\pi/4 - \pi/2$, $\pi/2 - 3\pi/4$, and $3\pi/4 - \pi$. A similar inverse process takes place in the synthesizer to reconstruct the output $\hat{x}(n)$ from the filter bank signals. The bottom two bands are each one-eighth of the total bandwidth and have one-eighth of the original sampling rate. The top three bands each have one-fourth of the original bandwidth and one-fourth of the original sampling rate. Delays are included in these bands to compensate for the delay of the filters $p_0^{(3)}(m)$ and $p_1^{(3)}(m)$ in the third stage.

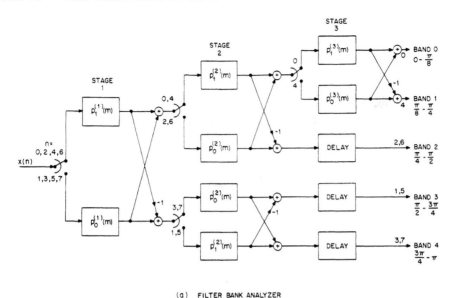

(a) FILTER BANK ANALYZER

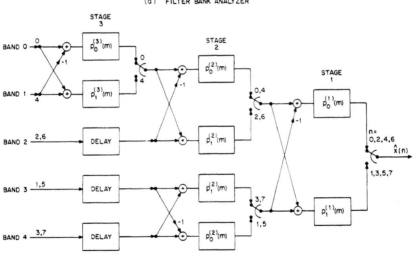

(b) FILTER BANK SYNTHESIZER

Figure 7.59 Example of a five-band nonuniform QMF filter bank based on a tree structure design; (a) analyzer, and (b) synthesizer.

The computation in this filter bank structure can be accomplished in an eight-cycle process since the maximum decimation ratio in the structure is 8. In each cycle of the process, one input sample, $x(n)$, is obtained, one branch of the tree is computed, and one output band is computed. The numbers on the switches in Fig. 7.59 show the positions of the switches for each cycle of this process. Paths through the tree with higher sampling rates are computed more often and paths with lower sampling rates are computed less often. The outputs of the filter analyzer (and inputs to the filter bank synthesizer) are computed in the order: bands 0, 3, 2, 4, 1, 3, 2, 4 for cycles 0 to 7, respectively. As seen, bands 0 and 1 are computed only once in this process, whereas bands 2, 3, and 4 are computed twice in each eight-cycle process since they have twice the sampling rate of bands 0 and 1. After one complete eight-cycle iteration, the complete process is repeated in the next eight-cycle iteration. In this way approximately one-eighth of the filter bank is computed in each cycle and the computing is distributed in time. This approach is particularly effective for real-time computing requirements [7.46].

Figure 7.60(a) shows an example of the frequency response of this five-band filter bank for the choice of QMF filter designs 32X, 16C, and 16C (see Appendix 7.1) for stages 1, 2, and 3, respectively. Figure 7.60(b) shows the resulting reconstruction error of the back-to-back filter bank and it can be seen that this error is less than ±0.125 dB across the entire frequency band. This design has been used in a speech coding application [7.46].

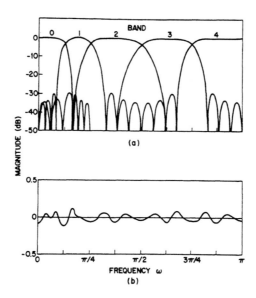

Figure 7.60 Frequency response of a five-band nonuniform QMF filter bank.

7.7.4 Equivalent Parallel Realizations of Cascaded Tree Structures

In the above discussion we have considered filter banks based on designs obtained from cascaded tree structures. While this design approach intuitively suggests a tree structure realization, it can be shown that such designs can also be readily

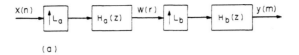

(a)

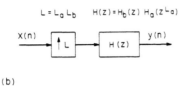

(b)

Figure 7.61   (a) A two-stage interpolator; and (b) its single-stage identity system.

implemented in terms of parallel integer-band structures. For low order designs this approach is sometimes preferred because of its structural simplicity [7.43].

To show this parallel structure relationship, consider the example of a two-stage interpolator in Fig. 7.61(a) where $x(n)$ is the input, $y(m)$ is the output, and $w(r)$ is the intermediate signal. Let $X(z)$, $Y(z)$, and $W(z)$ be their z-transforms respectively. Also let $H_a(z)$ and $H_b(z)$ be the interpolation filters and L_a and L_b be the interpolation ratios for these two stages as illustrated in Fig. 7.61(a). Then from Chapter 2 it can be shown that

$$W(z) = H_a(z)X(z^{L_a}) \qquad (7.279)$$

and

$$Y(z) = H_b(z)W(z^{L_b}) \qquad (7.280)$$

Combining Eqs. (7.279) and (7.280) leads to the form

$$Y(z) = H(z)X(z^L) \qquad (7.281)$$

where

$$H(z) = H_b(z)H_a(z^{L_a}) \qquad (7.282)$$

and

$$L = L_aL_b \qquad (7.283)$$

This form suggests the single-stage interpolator model shown in Fig. 7.61(b) and it is therefore an identity to the two-stage system in Fig. 7.61(a). A similar identity can be derived for the case of cascaded decimators by applying the principle of

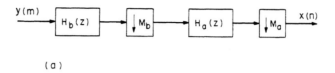

(a)

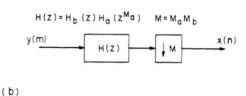

$$H(z) = H_b(z) H_a(z^{M_a}) \qquad M = M_a M_b$$

(b)

Figure 7.62 (a) A two-stage decimator; and (b) its single-stage identity system.

network transposition (see Chapter 3). This identity is illustrated in Fig. 7.62. Also, by applying the above identities repeatedly, it can be shown that multiple cascades of integer decimators (or integer interpolators) can be collapsed to single-stage integer decimator (or interpolator) designs for any number of stages.

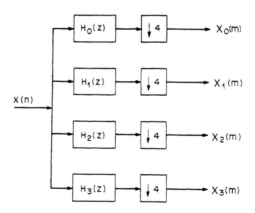

Figure 7.63 Parallel equivalent of the four-band filter-bank structure.

The conversion of a cascaded tree structure to an equivalent parallel structure can now be readily accomplished by applying the above identities to each branch of the tree structure. For example, the two-stage, four-band structure in Fig. 7.52(a) can be translated into the equivalent parallel structure shown in Fig. 7.63 by this process. Let $H_{LP1}(z)$, $H_{HP1}(z)$, $H_{LP2}(z)$ and $H_{HP2}(z)$ be the respective z-transforms of the filters denoted as LP1, HP1, LP2, and HP2 in Fig. 7.52. Also note that if these are QMF designs they satisfy the relations

$$H_{\text{HP1}}(z) = H_{\text{LP1}}(-z) \qquad\qquad (7.284)$$

and

$$H_{\text{HP2}}(z) = H_{\text{LP2}}(-z) \qquad\qquad (7.285)$$

Then the filters $H_0(z)$ to $H_3(z)$ for the equivalent parallel structure in Fig. 7.63 have the form

$$H_0(z) = H_{\text{LP1}}(z)H_{\text{LP2}}(z^2) \qquad\qquad (7.286)$$

$$H_1(z) = H_{\text{LP1}}(z)H_{\text{HP2}}(z^2)$$

$$= H_{\text{LP1}}(z)H_{\text{LP2}}(-z^2) \qquad\qquad (7.287)$$

$$H_2(z) = H_{\text{HP1}}(z)H_{\text{HP2}}(z^2)$$

$$= H_{\text{LP1}}(-z)H_{\text{LP2}}(-z^2) \qquad\qquad (7.288)$$

and

$$H_3(z) = H_{\text{HP1}}(z)H_{\text{LP2}}(z^2)$$

$$= H_{\text{LP1}}(-z)H_{\text{LP2}}(z^2) \qquad\qquad (7.289)$$

In practice these designs often have impulse responses that are characterized by very long but low amplitude "tails." As such these tails are sometimes carefully truncated about their center of symmetry to obtain shorter length modified filter designs for greater efficiency in implementation [7.43].

7.8 SUMMARY

In this chapter we have focused on the application of multirate techniques to the design of filter banks and short-time spectral analyzers and synthesizers. Two efficient multirate structures were derived for realization of these systems, namely the polyphase and weighted overlap-add filter bank structures. We then considered issues in the design of filters (or windows) for these systems. For nonoverlapping filter banks it was shown that these filters can be directly designed using the techniques discussed in Chapter 4.

For overlapping designs, and particularly for designs in which a back-to-back identity system is required, additional constraints are necessary. We identified these constraints in terms of requirements on the time and/or frequency domain aliasing effects in the filter bank. It was shown that filter designs based on the removal of frequency domain aliasing in the filter bank signals (with cancelation of time domain aliasing effect) leads to a generalized approach to the filter-bank-sum method. Alternatively designs based on removal of time domain aliasing in the filter bank signals (with cancelation of frequency domain aliasing effects) leads to a generalized approach to the overlap-add method. For designs in which the effects of both frequency domain and time domain aliasing are removed, it was shown that the filter bank signals must be oversampled by a factor of approximately four or more.

We next considered effects of modifications of the filter bank signals in this DFT analysis/synthesis framework. In particular, multiplicative modifications were considered and modeled. In the limit of appropriate rectangular analysis and synthesis windows, it was shown that this model leads to a generalized approach to the well-known methods of overlap-add and overlap-save for fast convolution.

A generalized definition of the DFT filter bank was then developed which accounts for offsets in time and frequency in the DFT model. Modifications in the polyphase and weighted overlap-add structures were derived for this modified DFT. It was applied to designs and structures for uniform single-sideband (SSB) filter banks. Two variations of this approach were considered for critically sampled SSB filter banks. The first is based on an offset of a quarter sample in the frequency domain and the second is based on a half-sample frequency domain offset.

Finally, filter bank designs based on tree structures instead of the polyphase or weighted overlap-add structures have been considered. It was shown that this approach applies to nonuniform filter bank designs as well as uniform designs. A particular emphasis was placed on designs based on two-band quadrature mirror filters (QMF) which apply to critically sampled analysis/synthesis systems. Polyphase structures were defined for these QMF designs and applied to the tree structure approach as illustrated by example.

REFERENCES

7.1   D. Gabor, "Theory of Communication," *J. IEE*, No. 93, pp. 429-457, November 1946.

7.2   J. L. Flanagan and R. Golden, "Phase Vocoder," *Bell Syst. Tech. J.*, Vol. 45, pp. 1493-1509, November 1966.

7.3   R. W. Schafer and L. R. Rabiner, "Design of Digital Filter Banks for Speech Analysis," *Bell Syst. Tech. J.*, Vol. 50, No. 10, pp. 3097-3115, December 1971.

7.4   S. L. Freeny, R. B. Kieburtz, K. V. Mina, and S. K. Tewksbury, "Design of Digital Filters for an All Digital Frequency Division Multiplex-Time

Division Multiplex Translator," *IEEE Trans. Circuit Theory*, Vol. CT-18, pp. 702-711, November 1971.

7.5   S. L. Freeny, R. B. Kieburtz, K. V. Mina, and S. K. Tewksbury, "Systems Analysis of a TDM-FDM Translator/Digital A-Type Channel Bank," *IEEE Trans. Commun.*, Vol. COM-19, pp. 1050-1059, December 1971.

7.6   R. W. Schafer and L. R. Rabiner, "Design and Simulation of a Speech Analysis-Synthesis System Based on Short-Time Fourier Analysis," *IEEE Trans. Audio Electroacoust.*, Vol. AU-21, No. 3, pp. 165-174, June 1973.

7.7   R. W. Schafer, L. R. Rabiner, and O. Herrmann, "FIR Digital Filter Banks for Speech Analysis," *Bell Syst. Tech. J.*, Vol. 54, No. 3, pp. 531-544, March 1975.

7.8   M. Bellanger and J. L. Daguet, "TDM-FDM Transmultiplexer: Digital Polyphase and FFT," *IEEE Trans. Commun.*, Vol. COM-22, No. 9, pp. 1199-1205, September 1974.

7.9   M. G. Bellanger, G. Bonnerot, and M. Coudreuse, "Digital Filtering by Polyphase Network: Application to Sample Rate Alternation and Filter Banks," *IEEE Trans. Acoust. Speech Signal Process.*, Vol. ASSP-24, No. 2, pp. 109-114, April 1976.

7.10   M. R. Portnoff, "Implementation of the Digital Phase Vocoder Using the Fast Fourier Transform," *IEEE Trans. Acoust. Speech Signal Process.*, Vol. ASSP-24, No. 3, pp. 243-248, June 1976.

7.11   J. B. Allen, "Short-Time Spectral Analysis, Synthesis and Modification by Discrete Fourier Transform," *IEEE Trans. Acoust. Speech Signal Process.*, Vol. ASSP-25, pp. 235-238, June 1977.

7.12   J. B. Allen and L. R. Rabiner, "A Unified Approach to Short-Time Fourier Analysis and Synthesis," *Proc. IEEE*, Vol. 65, pp. 1558-1564, November 1977.

7.13   R. E. Crochiere, S. A. Webber, and J. L. Flanagan, "Digital Coding of Speech in Sub-bands," *Bell Syst. Tech. J.*, Vol. 55, pp. 1069-1085, October 1976.

7.14   D. Esteban and C. Galand, "Application of Quadrature Mirror Filters to Split Band Voice Coding Schemes," *Proc. 1977 IEEE Int. Conf. Acoust. Speech Signal Process.*, Hartford, Conn., pp. 191-195, May 1977.

7.15   J. M. Tribolet and R. E. Crochiere, "Frequency Domain Coding of Speech," *IEEE Trans. Acoust. Speech Signal Process.*, Vol. ASSP-27, No. 5, pp. 512-530, October 1979.

7.16   M. J. Narasimha and A. M. Peterson, "Design and Application of Uniform

Digital Bandpass Filter-Banks," *Proc. 1978 IEEE Int. Conf. Acoust. Speech Signal Process.*, Tulsa, Okla., pp. 499-503, April 1978.

7.17   T. A. Claasen and W. F. Mecklenbrauker, "A Generalized Scheme for an All Digital Time-Division Multiplex to Frequency-Division Multiplex Translator," *IEEE Trans. Circuits Sys.*, Vol. CAS-25, No. 5, pp. 252-259, May 1978.

7.18   Special Issue on TDM-FDM Conversion, *IEEE Trans. Commun.*, Vol. COM-26, No. 5, May 1978.

7.19   M. R. Portnoff, "Time-Frequency Representation of Digital Signals and Systems Based on Short-Time Fourier Analysis," *IEEE Trans. Acoust. Speech Signal Process.*, Vol. ASSP-28, No. 1, pp. 55-69, February 1980.

7.20   R. E. Crochiere, "A Weighted Overlap-Add Method of Fourier Analysis/Synthesis," *IEEE Trans. Acoust. Speech Signal Process.*, Vol. ASSP-28, No. 1, pp. 99-102, February 1980.

7.21   P. Vary, "On the Design of Digital Filter Banks Based on a Modified Principle of Polyphase," *Arch. Elekt. Ubertrag. (AEU)*, Vol. 33, pp. 293-300, 1979.

7.22   P. Vary and U. Heute, "A Short-Time Spectrum Analyzer with Polyphase-Network and DFT," *Signal Process.*, Vol. 2, No. 1, pp. 55-65, January 1980.

7.23   G. A. Nelson, L. L. Pfeifer, and R. C. Wood, "High Speed Octave Band Digital Filtering," *IEEE Trans. Audio Electroacoust.*, Vol. AU-20, No. 1, pp. 8-65, March 1972.

7.24   A. J. Barabell and R. E. Crochiere, "Sub-band Coder Design Incorporating Quadrature Filters and Pitch Prediction," *Proc. 1979 IEEE Int. Conf. Acoust. Speech Signal Process.*, Washington, DC, pp. 530-533, April 1979.

7.25   L. R. Rabiner and R. W. Schafer, *Digital Processing of Speech Signals*, Englewood Cliffs, N.J.: Prentice-Hall, 1978.

7.26   L. R. Rabiner and B. Gold, *Theory and Application of Digital Signal Processing*, Englewood Cliffs, N.J.: Prentice-Hall, 1975.

7.27   A. V. Oppenheim and R. W. Schafer, *Digital Signal Processing*, Englewood Cliffs, N.J.: Prentice-Hall, 1975.

7.28   J. H. McClellan and C. M. Rader, *Number Theory in Digital Signal Processing*, Englewood Cliffs, N.J.: Prentice-Hall, 1979.

7.29   L. R. Rabiner, "On the Use of Symmetry in FFT Computation," *IEEE Trans. Acoust. Speech Signal Process.*, Vol. ASSP-27, No. 3, pp. 233-239, June 1979.

7.30   R. Yarlagadda and J. B. Allen, "Digital Poisson Summation Formula and an Application," *Signal Process.*, Vol. 4, No. 1, pp. 79-84, January 1982.

7.31   A. Papoulis, "Generalized Sampling Expansion," *IEEE Trans. Circuits Syst.*, Vol. CAS-24, pp. 652-654, November 1977.

7.32   J. L. Brown, "Multi-channel Sampling of Lowpass Signals," *IEEE Trans. Circuits Syst.*, Vol. CAS-28, No. 2, pp. 101-106, February 1981.

7.33   D. Malah and J. L. Flanagan, "Frequency Scaling of Speech Signals by Transform Techniques," *Bell Syst. Tech. J.*, Vol. 60, No. 9, pp. 2107-2156, November 1981.

7.34   D. Malah, "Time Domain Algorithms for Harmonic Bandwidth Reduction and Time Scaling of Speech Signals," *IEEE Trans. Acoust. Speech Signal Process.*, Vol. ASSP-27, No. 2, pp. 121-133, April 1979.

7.35   J. B. Allen, "Applications of the Short-Time Fourier Transform to Speech Processing and Spectral Analysis," *Proc of the First ASSP Workshop on Spectral Estimation*, pp. 6.3.1-6.3.5, August 1981.

7.36   J. L. Vernet, "Real Signals Fast Fourier Transform: Storage Capacity and Step Number Reduction by Means of an Odd Discrete Fourier Transform," *Proc. IEEE*, Vol. 59, pp. 1531-1532, October 1971.

7.37   G. Bonnerot and M. Bellanger, "Odd-Time Odd-Frequency Discrete Fourier Transform for Symmetric Real-Valued Series," *Proc. IEEE*, Vol. 64, pp. 392-393, March 1976.

7.38   H. Scheuermann and H. Gockler, "A Comprehensive Survey of Digital Transmultiplexing Methods," *Proc. IEEE*, Vol. 69, No. 11, pp. 1419-1450, November 1981.

7.39   A. Croisier, D. Esteban, and C. Galand, "Perfect Channel Splitting by Use of Interpolation/Decimation/Tree Decomposition Techniques," *Int. Conf. Inf. Sci. Sys.*, Patras, August 1976.

7.40   T. A. Ramstad and O. Foss, "Sub-Band Coder Design Using Recursive Quadrature Mirror Filters," *EUSPICO-80*, 1980.

7.41   T. P. Barnwell, "An Experimental Study of Sub-band Coder Design Incorporating Recursive Quadrature Filters and Optimum APDCM," *Proc. 1981 IEEE Int. Conf. Acoust. Speech Signal Process.*, pp. 808-811, April 1981.

7.42   D. Esteban and C. Galand, "HQMF: Halfband Quadrature Mirror Filters," *Proc. 1981 Int. Conf. Acoust. Speech Signal Process.*, pp. 220-223, April 1981.

7.43   D. Esteban and C. Galand, "Direct Approach to Quasi Perfect

Decomposition of Speech in Sub-bands," *9th Int. Cong. Acoust.*, Madrid, 1977.

7.44   J. D. Johnston and R. E. Crochiere, "An All Digital Commentary Grade Sub-band Coder," *J. Audio Eng. Soc.*, Vol. 27, No. 11, pp. 855-865, November 1979.

7.45   J. D. Johnston, "A Filter Family Designed for Use in Quadrature Mirror Filter Banks," *Proc. 1980 IEEE Int. Conf. Acoust. Speech Signal Process.*, pp. 291-294, April 1980.

7.46   R. E. Crochiere, R. V. Cox, and J. D. Johnston, "Real-Time Speech Coding," *IEEE Trans. Commun.*, pp. 621-634, April 1982.

7.47   Special Issue on Transmultiplexers, *IEEE Trans. Commun.*, Vol. COM-30, No. 7, July 1982.

7.48   V. K. Jain, and R. E. Crochiere, "A Novel Approach to the Design of Analysis/Synthesis Filter Banks," *Proc. 1983, IEEE Int. Conf. Acoust. Speech Signal Process.*, April 1983.

APPENDIX 7.1

This appendix tabulates the properties and coefficient values for a set of quadrature mirror filter designs based on material in Ref. 7.45. The designs have been obtained by the computer-aided optimization procedure discussed in Section 7.7.2. Table 7.1, column 1, lists the code number of the filter. The second column gives the number of taps, N, in the filter, and the third column gives the width of the transition band (normalized to 2π); that is, it is the difference between the 3-dB point ($\pi/2$) and the stopband cutoff ω_s, normalized to 2π. The fourth column gives the value of the weighting factor α used in the error function [see Eq. (7.264)]. The fifth column tabulates the resulting reconstruction error in the back-to-back filter bank [e.g., the filter 32D, seen in Fig. 7.56(b), has a reconstruction error of ± 0.025 dB]. Finally, the last two columns give the values of stopband attenuation (in dB) for the first and last ripples in the stopband [e.g., the filter 32D in Fig. 7.56(a) has a -38 dB ripple near the edge of the stopband and a -48 dB ripple near $\omega = \pi$).

TABLE 7.1   PROPERTIES OF QMF FILTER DESIGNS

Filter Number	Number of Taps N	Transition Bandwidth $\dfrac{\omega_s-(\pi/2)}{2\pi}$	Weighting Factor α	Reconstruction Error (dB)	Stopband Attenuation (First Peak) (dB)	Stopband Attenuation (Last Peak) (dB)
8A	8	0.14	1	0.06	31	31
12A	12	0.14	1	0.025	48	50
16A	16	0.14	1	0.008	60	75
12B	12	0.10	1	0.04	33	36
16B	16	0.10	1	0.02	44	48
24B	24	0.10	1	0.008	60	78
16C	16	0.0625	1	0.07	30	36
24C	24	0.0625	1	0.02	44	49
32C	32	0.0625	2	0.009	51	60
32X	32	0.0625	2	0.025	52	60
48C	48	0.0625	2	0.002	63	80
24D	24	0.043	1	0.1	30	38
32D	32	0.043	2	0.025	38	48
48D	48	0.043	2	0.006	50	66
64D	64	0.043	5	0.002	65	80
32E	32	0.023	2	0.25	27	36
48E	48	0.023	2	0.07	32	46
64E	64	0.023	5	0.025	40	51

Table 7.2 gives coefficient values for the various designs. Only the symmetrical half of the coefficients are given, starting from the center (i.e., the largest) coefficient. For example, in the 8A design, coefficients are given in the order $h(3) = h(4)$, $h(2) = h(5)$, $h(1) = h(6)$, and $h(0) = h(7)$ for a filter $h(n)$, $n = 0, 1, 2, ..., 7$.

TABLE 7.2 COEFFICIENTS FOR QMF FILTER DESIGNS

8A	12A	12B
.48998080E 00	.48438940E 00	.48079620E 00
.69428270E-01	.88469920E-01	.98085220E-01
−.70651830E-01	−.84695940E-01	−.91382500E-01
.93871500E-02	−.27103260E-02	−.75816400E-02
	.18856590E-01	.27455390E-01
	−.38096990E-02	−.64439770E-02

16A	16B	16C
.48102840E 00	.47734690E 00	.47211220E 00
.97798170E-01	.10679870E 00	.11786660E 00
−.90392230E-01	−.95302340E-01	−.99295500E-01
−.96663760E-02	−.16118690E-01	−.26275600E-01
.27641400E-01	.35968530E-01	.46476840E-01
−.25897560E-02	−.19209360E-02	.19911500E-02
−.50545260E-02	−.99722520E-02	−.20487510E-01
.10501670E-02	.28981630E-02	.65256660E-02

24B	24C	24D
.47312890E 00	.46864790E 00	4.6542880E-01
.11603550E 00	.12464520E 00	1.3011210E-01
−.98297830E-01	−.99878850E-01	−9.9844220E-02
−.25615330E-01	−.34641430E-01	−4.0892220E-02
.44239760E-01	.50881620E-01	5.4029850E-02
.38915220E-02	.10046210E-01	1.5473930E-02
−.19019930E-01	−.27551950E-01	−3.2958390E-02
.14464610E-02	−.65046690E-03	−4.0137810E-03
.64858790E-02	.13540120E-01	1.9763800E-02
−.13738610E-02	−.22731450E-02	−1.5714180E-03
−.13929110E-02	−.51829780E-02	−1.0614000E-02
.38330960E-03	.23292660E-02	4.6984260E-03

TABLE 7.2 (Continued)

32C	32X	32D
.46640530E 00	4.6645830E-01	4.6367410E-01
.12855790E 00	1.2846510E-01	1.3297250E-01
−.99802430E-01	−9.9800110E-02	−9.9338590E-02
−.39348780E-01	−3.9244910E-02	−4.4524230E-02
.52947450E-01	5.2909300E-02	5.4812130E-02
.14568440E-01	1.4468810E-02	1.9472180E-02
−.31238620E-01	−3.1155320E-02	−3.4964400E-02
−.41874830E-02	−4.1094160E-03	−7.9617310E-03
.17981450E-01	1.7881950E-02	2.2704150E-02
−.13038590E-03	−1.7219030E-04	2.0694700E-03
−.94583180E-02	−9.3636330E-03	−1.4228990E-02
.14142460E-02	1.4272050E-03	8.4268330E-04
.42341950E-02	4.1581240E-03	8.1819410E-03
−.12683030E-02	−1.2601150E-03	−1.9696720E-03
−.14037930E-02	−1.3508480E-03	−3.9711520E-03
.69105790E-03	6.5064660E-04	2.2451390E-03

32E	48C	48D
.45964550E 00	.46424160E 00	.46139480E 00
.13876420E 00	.13207910E 00	.13639810E 00
−.97683790E-01	−.99384370E-01	−.98437790E-01
−.51382570E-01	−.43596380E-01	−.48731140E-01
.55707210E-01	.54326010E-01	.55379000E-01
.26624310E-01	.18809490E-01	.24020070E-01
−.38306130E-01	−.34090220E-01	−.36906340E-01
−.14569000E-01	−.78016710E-02	−.12422540E-01
.28122590E-01	.21736090E-01	.25813150E-01
.73798860E-02	.24626820E-02	.60226430E-02
−.21038230E-01	−.13441620E-01	−.18121920E-01
−.26120410E-02	−.61169920E-04	−.23574670E-02
.15680820E-01	.78402940E-02	.12465680E-01
−.96245920E-03	−.75614990E-03	.33292710E-03
−.11275650E-01	−.42153860E-02	−.82474350E-02
.51232280E-02	.78333890E-03	.63647700E-03
	.20340170E-02	.51489700E-02
	−.52055740E-03	−.95592250E-03
	−.85293900E-03	−.29611340E-02
	.24225190E-03	.89979030E-03
	.30117270E-03	.15016570E-02
	−.56157570E-04	−.66471290E-03
	−.92054790E-04	−.61083240E-03
	−.14619070E-04	.40829340E-03

TABLE 7.2 (Continued)

48E	64D	64E
.45817950E 00	.46009810E 00	.45725790E 00
.14082370E 00	.13823630E 00	.14202200E 00
−.96727910E-01	−.97790960E-01	−.96089540E-01
−.53990540E-01	−.50954870E-01	−.55443150E-01
.55307280E-01	.55432450E-01	.54974440E-01
.29675180E-01	.26447000E-01	.31331460E-01
−.38442230E-01	−.37649730E-01	−.38449870E-01
−.18039380E-01	−.14853970E-01	−.19830160E-01
.28813790E-01	.27160550E-01	.29128720E-01
.11160250E-01	.82875600E-02	.13062040E-01
−.22285690E-01	−.19943650E-01	−.22925640E-01
−.66961680E-02	−.43136740E-02	−.86300040E-02
.17437190E-01	.14593960E-01	.18350770E-01
.36242470E-02	.18947140E-02	.55598240E-02
−.13593290E-01	−.10506890E-01	−.14772120E-01
−.15049380E-02	−.48579350E-03	−.33824410E-02
.10453240E-01	.73671710E-02	.11865500E-01
.25450090E-04	−.25847670E-03	.18165760E-02
−.78527550E-02	−.49891470E-02	−.94456860E-02
.99124640E-03	.57431590E-03	−.70816120E-03
.56447870E-02	.32358770E-02	.74161550E-02
−.16909870E-02	−.62437240E-03	−.57608340E-04
−.37667050E-02	−.19861770E-02	−.57047680E-02
.25404290E-02	.53085390E-03	.55112530E-03
	.11382600E-02	.42629060E-02
	−.38236310E-03	−.84474820E-03
	−.59535630E-03	−.30644720E-02
	.22984380E-03	.98342750E-03
	.27902770E-03	.20745920E-02
	−.11045870E-03	−.10196710E-02
	−.11235150E-03	−.12577780E-02
	.35961890E-04	.10798060E-02

Index